# CSR, Sustainability, Ethics & Governance

In recent years the discussion concerning the relation between business and society has made immense strides. This has in turn led to a broad academic and practical discussion on innovative management concepts, such as Corporate Social Responsibility, Corporate Governance and Sustainability Management. This series offers a comprehensive overview of the latest theoretical and empirical research and provides sound concepts for sustainable business strategies. In order to do so, it combines the insights of leading researchers and thinkers in the fields of management theory and the social sciences – and from all over the world, thus contributing to the interdisciplinary and intercultural discussion on the role of business in society. The underlying intention of this series is to help solve the world's most challenging problems by developing new management concepts that create value for business and society alike. In order to support those managers, researchers and students who are pursuing sustainable business approaches for our common future, the series offers them access to cutting-edge management approaches.

*CSR, Sustainability, Ethics & Governance* is accepted by the Norwegian Register for Scientific Journals, Series and Publishers, maintained and operated by the Norwegian Social Science Data Services (NSD)

Irene Guia Arraiano • Belén Díaz •
Mara Del Baldo • René Schmidpeter •
Samuel O. Idowu
Editors

# Corporate Social Responsibility in a Dynamic Global Environment

## Sustainable Management in Challenging Times

Springer

*Editors*
Irene Guia Arraiano
Lisbon Accounting and Business School
Lisbon, Portugal

Belén Díaz
Department of Business Administration and Santander Financial Institute (SANFI)
University of Cantabria
Santander, Spain

Mara Del Baldo
Department of Economics, Society and Politics
University of Urbino Carlo Bo
Urbino, Italy

René Schmidpeter
BFH
Bern, Switzerland

Samuel O. Idowu
Guildhall School of Business and Law
London Metropolitan University
London, UK

ISSN 2196-7075 ISSN 2196-7083 (electronic)
CSR, Sustainability, Ethics & Governance
ISBN 978-3-031-24646-3 ISBN 978-3-031-24647-0 (eBook)
https://doi.org/10.1007/978-3-031-24647-0



This Springer imprint is published by the registered company Springer Nature Switzerland AG
The registered company address is: Gewerbestrasse 11, 6330 Cham, Switzerland

*Dedicated to All Members of the Global Corporate Governance Institute (GCGI) Worldwide*

# Foreword

In the last few years, the world has watched, with bewilderment, the uprise of the dark side of globalization. A global pandemic with devastating human and economic consequences was followed by a cruel and brutal armed conflict in Eastern Europe, with widespread implications that are yet to be fully uncovered. This is also the age of misinformation, an age where democratic political systems face unprecedented risks, and an era in which the wonders of technology have been transfigured, in many cases, into serious threats to our well-being and our way of life. Basic human rights, which we once took for granted, are being increasingly contested in many geographies. The depletion of natural resources and climate change are additional, and perhaps the most emblematic, manifestations of this not so bright picture.

It is against this backdrop that, more than ever, responsible business practices and ethical corporate behaviour acquire an unquestionable importance. Whether big or small, every corporation must play its part in helping to turn this world—the only one we have—into a good place to live. Regulation and supervision by public authorities are welcome, and in many cases mandatory, but they cannot fully replace the individual responsibility of market players in promoting transparency of procedures and good practices at the economic, environmental, and political levels. We are all agents for good governance, social responsibility, and ethical behaviour. The good practices of many are the only way forward to overturn the bad and damaging conduct of a few.

Sensible and effective action requires knowledge. This is an undeniable truth in any aspect of life, and social responsibility is no exception. In particular, academic and scientific knowledge is fundamental to set the basis for an overarching and rigorous understanding of the challenges faced by practitioners. Quality research on corporate social responsibility and socially responsible investment is of extreme relevance for society and must be stimulated. Fortunately, the last few years have witnessed a proliferation of academic work in this field, what is a clear symptom of how relevant it has become.

This book, *Corporate Social Responsibility in a Dynamic Global Environment: Sustainable Management in Challenging Times*, is an extraordinary example of the current vitality of research on social responsibility. Edited by Irene Guia Arraiano,

Belén Díaz Díaz, Mara Del Baldo, René Schmidpeter, and Samuel O. Idowu, the volume collects contributions of eminent scholars in the field, contributions that have been presented in the Seventh International Conference on Corporate Social Responsibility, Sustainability, Ethics and Governance, which took place in 2021 in Lisbon, Portugal. The conference has been organized by the Lisbon Accounting and Business School in partnership with the Global Corporate Governance Institute, a prestigious international association engaged in the study and promotion of responsible business.

These are, no doubt, challenging times, and I believe that the best way to face them is through the empowerment that science brings. The collection of studies in this book provides the reader with an exceptional opportunity to approach the knowledge frontier regarding corporate social responsibility, an opportunity that no one engaged in the field will certainly want to miss.

Lisbon Accounting and Business School, Lisbon, Portugal

Orlando Gomes

# Preface

The COVID-19 pandemic that besieged our world in 2019 has affected our world in an unprecedented way. It has changed the way we live and function; many things will never be the same again. However, along with many critical and extraordinary changes, a new normal has been triggered, thus eliciting the call for new behaviours, competences, and capabilities to tackle uncertainty and renewed resilience.

Putting aside the havoc caused to lives and human existence, there are a number of innovations that came to the fore because of the pandemic.

Many of these innovations have changed and continue to change our lives for the better. A number of things are now done virtually by adopting new effective tools, often digitally based, to manage works, relations, and contacts, which was not the case before the pandemic. COVID-19 prevented us from holding the Seventh GCGI Conference in 2020.

The conference held virtually in Lisbon in 2021 was a very successful conference despite the absence of the physical interaction and networking which are some of the essential features of international conference. Lisbon is hosting the 2024 face-to-face conference; we intend to catch up with what we missed face-to-face in Portugal in 2024.

This book has selected some of the best papers presented at the Seventh International Conference on CSR, Sustainability, Ethics and Governance held in Lisbon, Portugal, in June 2021. The conference was held under the auspices of the Global Corporate Governance Institute and hosted by the Lisbon Accounting and Business School at Lisbon Polytechnic Institute, being the seventh international gathering of this kind after successful events held in London, Nanjing, Cologne, Perth, Santander, and Abu Dhabi in previous years.

It was an excellent platform to deliberate on global themes of corporate social responsibility, sustainability, ethics, and governance in their various dimensions.

This event is one of the largest international conferences focusing on responsible business with delegates from over 20 nations who have presented their latest research and its practical implications for business, society, academia, and politics.

This publication collects 20 papers out of those presented at the conference that have been peer reviewed.

This book explores the current state of corporate social responsibility from an international perspective with the aim of sharing ideas and visions for a sustainable future and to provide useful tips for academics, practitioners, and policy-makers in the context of the 2030 Agenda for Sustainable Development of the United Nations.

The editors hope their readers would find the book interesting and that the networks it has created allow all of us to continue generating contributions to achieve responsible businesses in challenging times.

| | |
|---|---|
| Lisbon, Portugal | Irene Guia Arraiano |
| Santander, Spain | Belén Díaz Díaz |
| Urbino, Italy | Mara Del Baldo |
| Bern, Switzerland | René Schmidpeter |
| London, UK | Samuel O. Idowu |
| November 2022 | |

# Corporate Social Responsibility in a Dynamic Global Environment: Introduction

In the context of the 2030 Agenda for Sustainable Development of the United Nations, businesses should play an indispensable role alongside government and communities to solve complex global challenges such as poverty, inequality, unemployment, and climate change, contributing to the economic growth and international development.

There has been growing awareness of the need to consider the ESG (environmental, social, and governance) risks in the valuation of investments, not only because of the greater social demand for an inclusive economic growth for future generations but also because there is empirical evidence that shows a positive relationship between profitability and ESG criteria.

The research on corporate social responsibility (CSR) and ESG has evolved and become globalized over the last three decades. However, there are still many unanswered questions and further research studies are needed to understand the need for sustainability in business and society in general (Díaz et al., 2020).

The main objective of this book is to explore recent evidence about the importance of ESG on businesses, collecting experiences in different countries and considering the importance of reporting and evaluating CSR practices. The 20 chapters of the book from around the globe answer pertinent questions relevant to the title of the book—*Corporate Social Responsibility in a Dynamic Global Environment: Sustainable Management in Challenging Times.*

This book provides an insight into responsible business practices used by modern corporate entities in Europe, Asia, North and South America, and Africa, the continents where contributors of the 20 chapters included in the book are based. The book has been fortunate to join 39 authors from 15 different countries. They are all CSR scholars of repute and have worked in the field of CSR and its related disciplines for a number of years; they are members of the Global Corporate Governance Institute (GCGI) who attended the Seventh Conference in Lisbon, Portugal, in June 2021. The views many of them have expressed in their chapters are the results of their research studies on responsible business practices in the relevant countries featured in the book. We hope you will enjoy and find the chapters useful and understandable.

The book has been divided into four parts, each part focusing on issues relating to responsible business practices in the aforementioned continents. Part I looks at *Business and Society* in seven chapters. Part II focuses on *CSR Experiences in Different Countries* in six chapters. Part III covers *Corporate Governance* in three chapters. Part IV looks at *CSR and Performance: Reporting and Evaluation* in four chapters. Let us now look briefly at each of the 20 chapters.

The very first chapter by Baldarelli and Del Bado two Italian scholars entitled "*Women: Beyond Social Responsibility and toward social innovation for a sustainable society. Lesson from the past*" notes the contribution of women in developing social consciousness and responsibility and preserving the social and cultural framework of the society, through their direct involvement in managing activities aimed at solving and preventing economic and social plagues. The research undertaken by them adopts a historical perspective and is based on the analysis of an Italian organization—devoted to maternal help and infant care—that since the beginning of the nineteenth century was aimed at protecting illegitimate motherhood and children. Their empirical study contributes to understand if women's role was mainly "hidden" or externally recognized and appreciated for contributing to promote social change in the internal and external context.

The second chapter by Rodriguez, Abreu, and Fonseca, three Portuguese scholars of repute, analyses the development of Higher Professional and Technical Courses (HPTC) in Portugal. These new cycles of non-degree higher education were created and regulated by Decree-Law n°43/2014, 18 March, in response to the main higher education policy objective of the XIX Portuguese Government Programme and of the European Union 2020 Strategy, knowing that both promote economic growth and job creation in Europe and Portugal. Rodriguez, Abreu, and Fonseca rely on two different approaches. The first approach takes the form of a literature review, based on the legal regime of the HPTC. The second approach takes the form of a descriptive statistical analysis, using a questionnaire and the case study method in a Portuguese public polytechnic higher education institution. The authors find that these cycles improve professional education, both for the diversity of demand labour in Portuguese firms and for the supply of qualified professionals.

Goto and Yokoyama, Japanese scholars, in chapter three focus on social enterprises which are business entities that simultaneously pursue objectives of both economic goals necessary to continue the business such as sales and profitability and social goals of resolving social issues. However, tension between economic and social missions tends to occur; thus, multifaceted managerial solutions are required so that both these approaches can be pursued simultaneously. These authors comparatively analyse companies implementing employment transition support programmes, which are a welfare service for people with disabilities, to clarify the issues facing employment transition support businesses in Japan and what "sort" of management is being practised to simultaneously pursue both economic and social goals.

In the chapter entitled "Staying Legitimate in a Changing Climate: A Framework for Studying Corporate Climate Change Communication", Allam, a Romanian scholar, focuses on another important issue of the ESG, the environment, and climate change. The author states that it is imperative to deepen our knowledge of how oil and gas companies communicate their climate-related stance and performance. With this intention, Allam conducted an integrative literature review with the aim of defining and positioning the concept of corporate climate change communication and developing a conceptual framework that could be utilized to systematically examine the presence and hierarchy of climate information on the companies' websites, annual reports, and sustainability reports. This framework can be utilized to study the presentation of such information; whether it permits stakeholder dialogue; which themes, stakeholders, and external performance criteria are mentioned; and the extent to which the companies utilize symbolic and substantive legitimation strategies.

Bidarra, Abreu, and Rosa, Portuguese scholars, present their chapter on "Sustainable and smart living versus unworthy conditions". In the chapter, they focus on the legal concept of unworthy conditions for the citizen and rely on two analyses. The first is theoretical based on the literature review of the energy performance of building and the legal regime of the unworthy conditions of the sustainable living of the citizen. The second is an exploratory analysis, supported on statistics of the Agency of Energy. The results show that the European Union encourages the Member States to use European funds and to ensure transition between unworthy conditions and social inclusion, to improve the situation of housing for each citizen. It not only reduces the carbon emissions that impact on climate change, but also improves the energy efficiency of buildings that generates sustainable and smart living based on the EU Directives. Furthermore, the energy performance certificate demonstrates that more than one-third of citizens live in unworthy conditions. So, the challenge of this research is demanding the awareness of better housing and promoting the participation mechanisms, specifically emphasis on the (isolated) elderly, which need to be strengthened on rural and mountain areas.

In the sixth chapter of Part I, Kikwike from Tanzania presents his chapter on "Institutional Roles in promoting Corporate Social Responsibility in Tanzania". In the chapter, the author investigates the institutional roles played by the government, non-governmental organizations, and community members in the wake of promoting companies to indulge in corporate social responsibility in Tanzania. Kikwike employs the case study design approach using data extracted from semi-structured interviews with government officers and managers of non-governmental organizations and community members residing in places where companies operate. The data covered the period between 2018 and mid-2020. The findings suggest that companies are influenced by institutions through coercion from institutions and some other forms of provision of incentives, dialogue, awareness during registration, and some form of directives, stakeholder meeting, and requests of their most pressing need. Similarly, non-complying companies face some form of punishments to refrain them from avoiding practising corporate social responsibility. There exists a poor feedback mechanism between companies and community members residing in areas close to companies' operations.

In the last chapter of Part I, Saha, Dharod, and Janagam from India take a look at *"Implications of Fluidity in Compliance Regime for India's Compulsory CSR Law"*. In the chapter, the authors provide an in-depth review of the legal changes and analyse their impact in Indian CSR space. In 2013, India introduced compulsory CSR law. However, the authors state that, in reality, this law is yet to achieve its full potential. One of the primary factors is the ever-changing and heavy regulatory framework which shifted away from the original principles and acts as a structural impediment to innovation and effectiveness. CSR rules are in a state of constant flux with arbitrary (and at times) contradictory amendments and clarifications. These changes lack guiding philosophy and long-term objectives, reflecting only short-term political or policy imperatives. In fact, the entire framework of CSR laws was again changed on 22 January 2021, with significant changes in CSR definitions, implementation, reporting, and monitoring of projects. The authors conclude that corporates have adopted a risk-averse bureaucratic approach to CSR which has decelerated decision-making and impact. Over-regulation increases costs for companies, pushing them towards traditional, risk-averse programs. Overall, it creates a cautious environment where choices are made with the aim of avoiding bureaucratic complications than using CSR as a tool for sustainable impact.

The very first chapter of Part II the one on *CSR experiences in different industries* is focused on decent work (SDG 8) in the garment industry in Jordan. The global garment industry is notorious for recurring labour scandals, and the garment industry in Jordan has historically been no exception. In 2006, a report by an American NGO revealed severe incidents of forced labour, sexual assault and harassment, excessive overtime, and discriminatory wages. In 2016, international observers reported significant improvements and comparably favourable working conditions. Holzberg explores this progress conducting a case analysis of the Jordanian garment industry with a focus on the perspective of suppliers. Pertinent questions centred on the factors that enabled their decent work improvements and those that hindered further progress. A crossvergence perspective served as the theoretical basis for the analysis. This theory predicts that global and local influences interact at the factory level to shape the work practices of suppliers. Accordingly, the results illuminate the multiple-level developments needed to promote decent work in Jordan.

In the second chapter of Part II, Napiórkowska from Poland presents her chapter on "CSR activities of retail chains operating in Poland". In the chapter, the author notes that the development of social responsibility of retail chains in Poland is related to the rapid expansion of foreign enterprises in the Polish market, which started at the end of the 1990s. Gradually, new store formats began to emerge, and the growing competition was introduced by practices and rules of conduct adopted by foreign retail chains towards consumers, as well as local communities and suppliers. Literature shows that retailers' CSR activities focus on four main areas: environment, supply chain, employees, and society. However, these activities occur with different intensity in each of the areas mentioned. Napiórkowska identifies the intensity of CSR activities undertaken by retail chains operating in Poland in areas selected on the basis of literature. She also indicates significant differences in these activities between a group of retail chains with only Polish capital and a group of retail chains with foreign capital, analysing a sample of 259 medium and large retail chains operating in Poland.

Pham, a Japanese scholar, is the author of the tenth chapter of the book. His chapter entitled "University Social Responsibility: From the perspectives of Top 10 Universities in Japan" aims to (1) examine the USR's current situation of top universities in Japan and (2) assess the strengths and weaknesses of USR in these above universities. The author analyses top 10 Japanese universities and considers five dimensions of CSR: environmental, economic, social, stakeholder, and voluntariness. Pham concludes that to date Japanese universities, to some extent, have engaged properly in social responsibility and sustainable development. Moreover, these activities, projects, research, and implementations have also been announced and reported in their website contents.

The eleventh chapter of the book by two Colombian scholars, Carmona-Garces and Londono-Correa, focuses on "Institutionalization of CSR practices in nutrition". Sustainable development goals such as zero hunger and health and well-being set targets related to the control of excess weight. Within the food systems approach, there is an urgent need for the food producing industry to adopt responsible practices. The Global Strategy on Diet, Physical Activity and Health includes the action of the food producing industry among the strategies needed to support healthy diets. The chapter argues that it is of interest to explore how corporate social responsibility practices in nutrition could be approached from the new sociological institutionalism theory, to undertake empirical research paths in this regard.

Madera del Pozo and Cassinello Plaza, two Spanish scholars, present their chapter on "Do the ESG Factors affect a Bank Insolvency? A study applied to Spanish Credit Cooperatives (Cajas rurales) between 2010 and 2015". The authors think that ESG principles will be factors to consider in the definition of the future capital requirements of the financial sector although, previously, they consider necessary to quantify the relationship between ESG risks and probability of default. Madera del Pozo and Cassinello Plaza analyse the relationship between insolvency and several explanatory variables that measure ESG risks in Spanish Rural Credit Unions (credit cooperatives). The authors find that ESG variables related to equality between men and women, efficiency in the use of resources, salaries per employee, ATM network accessibility, and training expenses for employees have been significant for insolvency. On the other hand, equality within the board of director has not been significant. Their model assigns greater probabilities of default to the credit unions that make a worse management of ESG risks.

The final chapter of Part II entitled "Relation between internal social responsibility and job satisfaction in Serbian service companies" by Damnjanović from Serbia concludes that there is a strong positive correlation between internal social responsibility and employees' job satisfaction. Therefore, the chapter suggests managers in service companies to carefully plan and develop strategies for internal social responsibility as the service industry is often characterized by poor working conditions and workers' depression.

Moving swiftly on to the first chapter of Part III, the fourteenth chapter of the book by Wong from Malaysia focuses on the evolution of Chinese corporate governance. The chapter is crucial due to the lack of awareness in the foreign investor community concerning the transparency and reliability of the second largest economy's governance policy. The literature review offered in this study not only contributes to the expanding literature in relation to the Chinese CG model but also further deepens the understanding of the governance issues concerning the businesses operating in China among the foreign institutional investors and foreign corporate practitioners. This chapter sheds light on the latest development of Chinese capital market and regulatory framework as well as CG perspective for managers of multinational firms to evaluate their potential direct investment decision in China. Also, a comparison of the characteristics of Chinese, Anglo-American, and German models is furnished to the targeted audience. This comparison may be helpful for the decision-making process of Chinese policymakers and regulators in relation to the formulation of new codes, regulations, and policies in the future. The chapter finishes drawing several avenues in connection with the future development and improvement of Chinese CG.

Chapter 15 entitled "The Core Capital Approach to Board: A Case Study of People's Credit Banks in Indonesia" is authored by Rissy, an Indonesian scholar. The aims of this study are to analyse the significant contributions of the core capital approach as a theoretical framework towards People Credit Banks' (PCB) board, and practically, to evaluate the level of PCBs' compliance with their board structures based on their core capital, and to identify the barriers that prevent PCBs from complying with the required board structures. The author applies a mixture of legal doctrinal research and in-depth interviews with key participants from Indonesian Financial Services Authority (IFSA) and the chairperson of National PCBs Associations. This research revealed that theoretically, the core capital approach provides a useful framework as well as a strategic model that significantly contributes to the establishment of a PCB board. However, practically, this research also indicated that PCBs are not yet in compliance with their board structures due to several fundamental barriers. This study also sought to provide some recommendations to overcome the barriers.

In the last chapter of Part III, Sachs and Mugova from the USA take a look at *"Lessons in financial governance from Bowen's family systems model"*. In the chapter, the authors use Bowen's family systems theory to understand financial governance and governance relationships. Bowen's eight concepts of family systems are triangles, differentiation of self, nuclear family emotional process, family projection process, multigenerational transmission process, emotional cutoff, sibling position, and societal emotional process. These concepts are discussed in light of how financial information is shared in a business, how board composition affects the oversight of financial operations, how leadership balances financial and operational demands, and how the social interactions within the business interplay with the macro social-industrial environment. For example, the authors argue that a well-functioning family helps a family member develop into an independent adult. Likewise, the effective business leader enables the financial department to develop autonomy, allowing that department to provide meaningful, objective information for business decisions. When this does not occur, problems can arise which affect family stability and business efficiency.

The final part of the book commences with the seventeenth chapter of the book entitled "Mandatory non-financial disclosures: the state of art" by Eva Cerioni, Alessia D'Andrea, Marco Giuliani, and Stefano Marasca. The authors discuss the convergence in non-financial disclosures (NFD) regulatory provisions of different countries. The analysis is focused on ten countries with reporting rates of sustainability higher than 90%. The main results highlight that there are no similarities between the different legislations. Most of the regulations have a flexible approach leaving a wide margin of discretion to companies. Moreover, while it is true that the NFD today finds a regulatory base in most countries of the world, each legislation differs from the others. There are situations of convergence regarding the position of the report. There is wide flexibility on the choice of framework; despite this, there is de facto comparability in terms of adopted standards. The need for harmonization can therefore also be refuted at a regulatory level, except for specific sectors or groups of countries.

Moreno, Mañas, Montes-Pineda, and Fernández-Olit, scholars from Spain, authored Chapter 17 entitled "Towards a novel approach to companies-stakeholder relationship: Modelling IBEX35 long run value creation". The authors analyse the relationship between IBEX 35 companies with their stakeholders. Previous research points out two relevant aspects for the long-run value creation in organizations with CSR strategies. First, the existence of a stakeholder's dialogue which allows building a joint materiality matrix. Second, the ability to manage interest groups, where some relevant elements have been identified, such as the recommended proactivity of organizations. These elements are completed by carrying out a qualitative analysis of the best practices in the Dow Jones Sustainability Index (DJSI) companies. Moreover, the authors consider the stakeholder approach and the renewed accountability standards to model he IBEX 35 long-run value creation based on the relationship between companies and their stakeholders.

Chapter 19 by Matkevičienė and Jakučionienė from Lithuania analyses the role of communication professionals as social change agents in times of crisis to achieve an organization's CSR/sustainability goals in collaboration with various stakeholders. The authors argues that communication professionals, driven by the organization's values, take on the role of social change and CSR/sustainability "ambassadors" in the pursuit of the organization's corporate strategy, promoting and sharing sustainability ideas and encouraging responsible actions by the organizations and external stakeholders.

The actions of communication professionals are related to integrating the relevant values into the organization's activities and communication, promoting CSR/sustainability initiatives, motivating the launching of activities, and engaging actively in the implementation of CSR/sustainability activities. The authors conclude that the role of communication professionals in CSR/sustainability can be seen in the context of the organization's system and culture for initiating and implementing CSR/sustainability activities but is also determined by the communication professional's personal leadership and individual engagement in initiating sustainability activities, finding CSR/sustainability solutions, and implementing activities involving different stakeholders.

The final chapter of the book by Monteiro, Cepeda, and Barbosa from Portugal about the environmental information disclosure level aims to develop and analyse a theoretical model which seeks to measure the relationship between environmental certification, level of environmental information disclosure, and profitability of Portuguese companies listed on stock exchange. The authors develop two Environmental Disclosure Indices (IDA) to assess environmental information disclosure level in companies' annual reports and sustainability reports from 2015 to 2017. Based on a sample of 30 Portuguese listed companies, results show that companies are increasingly involved in responding to stakeholder pressures. In addition, they find that IDA is a mediating variable in relationship between environmental certification and profitability.

The 20 chapters of the book from across the globe in our view have made immense contributions to improving our understanding about CSR and ESG and have set up ideas for improving our world and achieving the sustainable development goals.

**Reference**

Díaz, B., Capaldi, N., Idowu, S. O., & Schmidpeter, R. (2020). Preface. In B. Díaz, N. Capaldi, S. O. Idowu, & R. Schmidpeter. Responsible business in a changing world. New management approaches for sustainable development (pp. IX–X). Springer.

Guildhall School of Business and Law, London Metropolitan University, London, UK — Samuel O. Idowu
Lisbon Polytechnic Institute (ISCAL), Lisbon, Portugal — Irene Guia Arraiano
University of Cantabria, Santander, Spain — Belen Diaz Diaz
University of Urbino, Urbino, Italy — Mara Del Baldo
BFH, Bern, Switzerland — René Schmidpeter

# Acknowledgements

We are eternally grateful to a number of people who have made the publication of this book on *Corporate Social Responsibility in a Dynamic Global Environment: Sustainable Management in Challenging Times* a reality. We express our thanks to all our contributors who have worked tirelessly despite their heavy professional commitments to put together their chapters. We also thank the anonymous peer reviewers of the chapters. Thank you everyone.

We would also like to thank our publishing team at Springer headed by the Executive Editor, Christian Rauscher, Barbara Bethke, Anette Weiss, and other members who have supported this project and all our other projects.

We are also grateful to our respective families for bearing with us during periods when we should have been with them but chose to spend the time meeting our obligations to completing the book.

Finally, we apologize for any errors or omissions that may appear anywhere in the text; please be assured that no harm was intended to anybody. Causing harm or discomfort to others is simply not the spirit of corporate social responsibility.

# Contents

**Part I Business and Society**

**Women: Beyond Social Responsibility and Toward Social Innovation for a Sustainable Society. Lesson from the Past** . . . . . 3
Maria-Gabriella Baldarelli and Mara Del Baldo

**Higher Professional Technical Course in Portugal: The Case Study of the Polytechnic Institute of Guarda** . . . . . 19
Joana Sá Rodrigues, Rute Abreu, and Cecília Fonseca

**A Study on Management Supporting Employment of Disabled in Japan** . . . . . 47
Yuichi Goto and Keiko Yokoyama

**Staying Legitimate in a Changing Climate: A Framework for Studying Corporate Climate Change Communication** . . . . . 65
Muhammad A. F. Allam

**Sustainable and Smart Living Versus Unworthy Conditions in Portugal** . . . . . 89
Sónia Bidarra, Rute Abreu, and Cecília Rosa

**Institutional Roles in Promoting Corporate Social Responsibility in Tanzania** . . . . . 107
Ibrahim Ramadhani Kikwiye

**Implications of Fluidity in Compliance Regime for India's Compulsory CSR Law** . . . . . 133
Debanjali Saha, Rinkesh Dharod, and Raj Janagam

**Part II CSR Experiences in Different Industries**

**Promoting Decent Work in Global Production: Lessons Learned from the Jordanian Garment Industry** . . . 157
Britta Holzberg

**CSR Activities of Retail Chains Operating in Poland** . . . 183
Anna Napiórkowska

**University Social Responsibility (USR): From the Perspective of Top 10 Universities in Japan** . . . 203
Long Tam Pham

**Institutionalization of CSR Practices in Nutrition** . . . 223
Carmona-Garces Isabel Cristina and Londono-Correa Diana

**Do the ESG Factors Affect Bank Insolvency? A Study Applied to Spanish Credit Cooperatives (Cajas Rurales) Between 2010 and 2015** . . . 235
Antonio Madera del Pozo and Natalia Cassinello Plaza

**Relation Between Internal Social Responsibility and Job Satisfaction in Serbian Service Companies** . . . 247
Milena Damnjanović

**Part III Corporate Governance**

**The Evolution of Chinese Corporate Governance** . . . 263
Shy Lih Wong

**The Core Capital Approach to Board: A Lesson from People's Credit Banks in Indonesia** . . . 285
Yafet Yosafet Wilben Rissy

**Lessons in Financial Governance from Bowen's Family Systems Model** . . . 305
Paul R. Sachs and Shame Mugova

**Part IV CSR and Performance: Reporting and Evaluation**

**Mandatory Nonfinancial Disclosures: The State of Art** . . . 321
Eva Cerioni, Alessia D'Andrea, Marco Giuliani, and Stefano Marasca

**Toward a Novel Approach to Companies-Stakeholder Relationship: Modeling IBEX35 Long Run Value Creation** . . . 337
Manuel Moreno, Elena Mañas, Oscar Montes-Pineda, and Beatriz Fernández-Olit

**Communication Professionals as Social Change Agents in Times of Crisis: How Pandemic Situation Has Changed Initiatives in CSR and Sustainability** . . . . . . 365
Renata Matkevičienė and Lina Jakučionienė

**Environmental Information Disclosure and Profitability: The Environmental Certification Influence** . . . . . . 381
Albertina Paula Monteiro, Catarina Cepêda, and Francisco Barbosa

**Index** . . . . . . 397

# About the Editors

**Irene Guia Arraiano**, Ph.D. in Financial Economics and Accounting, is Professor in Lisbon Accounting and Business School, Lisbon Polytechnic Institute (ISCAL), Lisbon, Portugal. She wrote her Ph.D. thesis on "Analysis of Socially Responsible Investing in European Stock Markets". She earned both Bachelor and Master's in Applied Mathematics, Statistics, and Operational Research at the Faculty of Sciences at Lisbon University. Irene Arraiano is Professor in the Mathematics Department and currently teaches Operational Research. She is a mobility coordinator in the Erasmus + Programme for the Business course. She is a member of the Research Centre of Évora University (CEFAGE), and her main research interests predominantly focus on the interaction between finance, mathematics, corporate social responsibility, and investing. Irene Arraiano has actively participated in the organization of National and International scientific events as the Chair of the Scientific and Organizational Committees. She participates in National and International Conferences as well as publishes in international academic journals. Throughout her academic career, Irene Arraiano has been part of various scientific and management bodies of ISCAL, such as scientific, directive, pedagogical councils and assembly of ISCAL representatives.

**Belén Díaz Díaz** is Ph.D. in Finance. She is Associate Professor of Finance at the University of Cantabria (Spain) and Academic Director at Santander Financial Institute (SANFI). Previously, she has been Vice-Dean of the Faculty of Economics and Business Administration and Director of Campus and Social Development at the University of Cantabria. She has been visiting researcher at the University of California in Berkeley, the London School of Economics, the University of Technology in Sydney, and different universities in Mexico, Chile, Brazil, Argentina, and Morocco. Professor Díaz is the author of several papers on CSR, corporate governance, banking, and corporate finance in academic journals and books. The quality of her research was recognized with the "Sanchis Alcover Award", awarded by the Scientific Association of Economy and Management in 2006. Since 1996, she has written several reports and developed more than 50 R&D projects about different financial and economic issues commissioned by different public and private entities.

**Mara Del Baldo** is Associate Professor (recognized as Full Professor by the Italian National Agency-ANVUR) of Business Administration and Economics of Sustainability and Accountability at the University of Urbino (Italy), Department of Economics, Society and Politics. She has extensive experience in collaborative research, in conjunction with universities and private industry. She visited several European Universities, i.e. the University of Vigo (Spain), the New Bulgarian University of Sofia (Bulgaria), the Corvinus University in Budapest (Hungary), and the University of Craiova (Romania). She is a member of the European Council for Small Business, the Centre for Social and Environmental Accounting Research (St. Andrews), the SPES Institute, and the European Business Ethics Network (EBEN), Italia, as well as of different Italian scientific associations. Mara is also a member of the academic advisory committee in the Global Corporate Governance Institute (USA). She is part of the editorial board, and she serves as a reviewer of several scientific journals, including Springer's *International Journal of Corporate Social Responsibility*. Among her main research interests are entrepreneurship and SMEs; corporate social responsibility; entrepreneurial leadership and business ethics; social accountability; integrated and sustainability reporting; and accounting and gender. She has published in several Italian and foreign journals as well as in national and international conference proceedings and books, including the Springer Series *CSR, Sustainability, Ethics and Governance*.

**René Schmidpeter** is Professor of Sustainable Management at the IU (International University) in Munich and Research Professor at the BFH in Bern. He is visiting professor at ISM in Vilnius as well as guest professor at various universities worldwide. He is a series editor for Springer's *CSR, Sustainability, Ethics and Governance* books, editor of the *Encyclopedia of Sustainable Management* (ESM), and editor of the *Dictionary of Corporate Social Responsibility* (DCSR). His research and teaching activities focus on the management of corporate social responsibility, international perspectives on CSR, social innovation, and sustainable entrepreneurship, as well as the relationship between business and society. He has published widely and is the editor of the German *management series on Corporate Social Responsibility* at Springer Gabler and is a member of various think-tanks, sustainability initiatives, and advisory boards in Europe and abroad. He has conducted research for the Federal Ministry of Education and Research, the European Union, the Bavarian Ministry of Social Affairs, the Anglo-German Foundation, the Bertelsmann Foundation, and the Deutsche Forschungsgesellschaft (DFG). He works closely with chambers of commerce and businesses in Europe to develop innovative strategies and entrepreneurial solutions for the sustainable transformation of business and society.

**Samuel O. Idowu** is Senior Lecturer in Accounting and Corporate Social Responsibility at London Guildhall School of Business and Law, London Metropolitan University, UK. He researches in the fields of corporate social responsibility (CSR), corporate governance, business ethics, and accounting and has published in both professional and academic journals since 1989. He is a freeman of the City of London and a liveryman of the Worshipful Company of Chartered Secretaries and Administrators. Samuel is the President of the Global Corporate Governance Institute (GCGI). He has led several edited books in CSR; he is the Editor-in-Chief of three Springer's reference books—the *Encyclopedia of Corporate Social Responsibility*, the *Dictionary of Corporate Social Responsibility*, and the *Encyclopaedia of Sustainable Management*—and the Editor-in-Chief of the *International Journal of Corporate Social Responsibility*. He is also Series Editor for Springer's books on *CSR, Sustainability, Ethics and Governance*. One of his edited books won the most Outstanding Business Reference Book Award of the American Library Association (ALA) in 2016 and another was ranked 18th in the 2010 Top 40 Sustainability Books by the *Cambridge University Sustainability Leadership Programme*. He is on the editorial boards of the *International Journal of Business Administration*, Canada, and *Amfiteatru Economic Journal*, Romania. Samuel has delivered a number of keynote speeches at national and international conferences and workshops on CSR and has on two occasions in 2008 and 2014 won Emerald's Highly Commended Literati Network Awards for Excellence. To date, Samuel has edited several books in the field of CSR, sustainability, and governance and has written seven forewords to books. Samuel has served as an external examiner to a number of UK universities and has examined Ph.D. theses in the UK, South Africa, Australia, the Netherlands, and New Zealand.

# Part I
# Business and Society

# Women: Beyond Social Responsibility and Toward Social Innovation for a Sustainable Society. Lesson from the Past

Maria-Gabriella Baldarelli and Mara Del Baldo

## 1 Introduction

This works aims to underline, using a historical perspective based on a research case (Naumes & Naumes, 2006), the fundamental role of women in promoting social consciousness, responsibility, and preserving the social and cultural framework of the society, through their direct involvement in managing activities (and covering different roles, such as entrepreneurs, managers, employees, and accountants) aimed to solve and prevent economic and social plagues.

The main research question is: "Which was (and still is) the role of women in orienting toward an authentic responsibility the mission, governance and accountability of organizations?" Accordingly, the empirical study helps to understand if women's role was mainly "hidden" or externally recognized and appreciated for contributing to promote social change in the internal and external environment.

The theme of women's contribution to CSR and sustainability, in terms of implementation of strategies, action, and accountability tools, is relatively new and now there are many gaps to fill in relative to several topics developed within

---

This paper is the work of a common research project and the result of common analysis with 50% contribution of Baldarelli Maria-Gabriella and 50% Del Baldo Mara for each paragraph.

M.-G. Baldarelli
Department of Business Administration, University of Bologna, Bologna, Italy

Istituto Universitario Sophia, Florence, Italy
e-mail: maria.baldarelli@unibo.it

M. Del Baldo (✉)
Department of Economics, Society and Politics, School of Economics, University of Urbino Carlo Bo, Urbino, Italy
e-mail: mara.delbaldo@uniurb.it

 
I. G. Arraiano et al. (eds.), *Corporate Social Responsibility in a Dynamic Global Environment*, CSR, Sustainability, Ethics & Governance,
https://doi.org/10.1007/978-3-031-24647-0_1

emerging research strands such as CSR and governance; board composition and gender diversity (Jamali et al., 2008); boards and strategy/decision-making and CSR decisions (Krüger, 2009; Rose, 2007); and boards' role and CSR reporting (Haniffa & Cooke, 2005). Specifically, there are still few studies on board diversity and CSR performance, board decision-making processes and CSR, and gender diversity and CSR. Most are "black box" empirical studies and there is a dearth of in-depth qualitative studies (Rao & Tilt, 2016).

Departing from these premises, the study is based on the analysis of an Italian organization (*Istituto S. Giuseppe per l'Aiuto Materno e Infantile di* Rimini) that from the beginning of the nineteenth century played a key role in the local and national context in protecting "illegitimate" motherhood and infancy ("illegitimate" children) which represented from centuries a social and cultural plague in the country. The mission, as well as the governance and accountability of this organization—still existent and currently named *S. Giuseppe ONLUS*—was mainly forged by women devoted "to take care" of the most fragile categories of people in society. The institution was found by a woman and managed by women, and the assistance provided by this institute was based on the voluntary and active participation and direct involvement of assisted mothers and women in general.

The selected institution represents a significant case (Yin, 1994) since for over a century its business model has been capable of undergoing cultural, economic, and social changes within the local (the province of Rimini) and national contexts (Italy). Indeed, the institute generated an *ante litteram* "spin-off" setting up an independent branch of health activity *(The Little Hospital Regina Elena").* The latter was among the first and innovative pediatric hospitals in Italy, and during the subsequent decades, it was merged into the hospital of the town where it is still "alive." Moreover, the activities of the S. Giuseppe continue even today through the two principal fields of social and social health assistance (residential communities, day centers for minors, or semi-residential communities for the seriously disabled). It created forms of care focused on the whole development of the individual (Fondazione San Giuseppe, 2020).

The research design is structured on a deductive and inductive level and includes a literature analysis about charismatic leadership, gender diversity, and CSR. The case analysis focuses on the women's role in acknowledging social trends and generating changes in institutions, thus providing evidence of the impact of the cultural (Baskerville, 2003; Hofstede, 1991) and social changes silently promoted by women.

The reminder of the chapter is organized as follows: Sect. 2 depicts the theoretical background, while Sect. 3 introduces methods and the case study. Finally, Sect. 4 discusses the main findings and draws up conclusion, in terms of insights and implications for future research.

# 2 Literature Framework

## *2.1 CSR and Gender: The Different "Universe" of Women*

Literature on CSR and women has mainly addressed the relationship between CSR and governance while focusing on the board composition, the board diversity, and gender diversity (Jamali et al., 2008). Other research streams have inquired into the link between boards and strategy/decision-making and CSR decisions (Krüger, 2009; Rose, 2007), while other contributions have deepened the boards' role and CSR reporting (Haniffa & Cooke, 2005). Other studies (Broadbent, 2016; Tilt, 2015) identified several gaps and, among these, the lack of studies on board diversity and CSR performance, the board decision-making process and CSR, and gender diversity and CSR.

Recently, a "black box" empirical study on board composition and CSR, aimed at inquiring into the role of diversity and gender in relation to strategy and decision-making (Rao & Tilt, 2016), has been addressed to 100 listed companies in Australia over 3 years (2009–2011), in order to analyze their reporting in annual reports and reports on CSR (governance, employees, environment, community, products, others) to consider ASX governance requirements. Moreover, the empirical study has been enriched through an in-depth qualitative analysis relative to a sample of chairpersons and board members (formed both by men and women) aimed to understand their perceptions, orientations, and the value assigned to the role of CSR, the board, and diversity. The differences that emerged from the interviews addressed to the four chairmen and the three female board directors were not so marked. However, "women told 'stories' to demonstrate commitment to CSR (...) and appear much more animated and passionate about CSR and their company's contribution to CSR issues" (Tilt, 2015: 17), while men seem to be more "objective" when it comes to their concerns about CSR. The same men assert that in their experiences, women are much more concerned about CSR than them and admit that "most men will do it because it's good for the company, and most women insist upon it because it's good for the world" (Tilt, 2015:13). Nevertheless, they affirm that in the participation of decision-making about CSR matters, they did not notice differences of opinion based on gender.

It does appear that the matter is not represented by the differences in women/men's opinions, but instead in the different approach; in other words, the men's approach to CSR seems to be "less robust." In particular, both men and women agree on the fact that CSR is necessary and important and that values and corporate culture should not be a separate reporting stream; accordingly, they agree on the centrality of a business culture, which permeates all sort of thinking and decision-making at some level. Nevertheless, for women, CSR never represents a compliance, because CSR is a "normal part of what they do" and this "female" perspective seems to be more rooted and authentic. Moreover, the views of men on gender diversity do not seem consistent because, although they consider it relevant, they also assert that the presence of independent directors is not important. "The values associated with the universal masculine and the universal feminine are different. The universal masculine is characterized by hard, rational and objective values, the universal feminine by emotion and more subjective values" (Broadbent, 2016: 6).

Broadbent (2016) emphasizes the need to introduce the universal feminine into accounting discourses and stresses the need for women's voices to be heard since the subordination of women's views (included heir "emotional" and relational approach) and voices is made apparent and real. Otherwise, the values of the universal feminine for which women, rather than men, are socialized will (and still) remain silenced (Broadbent, 1998).

The differences between the male and the female universe and its values (which underlie it) lead to different approaches of social responsibility and especially playing different roles in the development of projects and the concrete social innovations like the case presented in the subsequent parts of the work.

## 2.2 *Women's Charisma and Leadership*

The choice to use the frameworks about charisma and leadership is due to their usefulness in the investigation of the case from a historical perspective. The start-up of the Istituto S. Giuseppe was envisioned by two leaders who nurtured the embryonic idea to promote mothers' assistance and infant healthcare assistance as an independent activity. This idea derived from both Del Piano's charisma, a doctor who pushed him to help small children involving sanitary care, and Sister Soleri's charisma, a nun who decided to donate her family patrimony to provide money and assets for the foundation of the "Istituto S. Giuseppe."

The concept of charisma had been analyzed by Max Weber. Weber's theory of charisma refers to a certain quality of an individual personality by virtue of which he is set apart from ordinary men and is treated as endowed with a sort of supernatural, superhuman, or at least exceptional power or quality (Weber, 1947: 358–359). They are people capable of turning problems into opportunities and resources thanks to their ability "to see the world through different eyes." Many points in common for charismatic leaders are personal responsibility, vision, moral virtues, faith in personal commitment, shared social responsibility, and solidarity (Becker, 1998).

O'Brien (2013) puts forward interesting points of reflection through the analysis of three charismatic leaders (Eugene Debs, Martin Luther King Jr., and Dorothy Day) who have inspired and built social movements and institutions that played a fundamental role in the construction of American democracy. The author highlights the cultural and social impact of their charismatic message and addresses attention to the resistance or favor encountered by the institutions they set up. In the first case, Debs was an evangelist who promoted economic justice and democratic politics, a supporter of popular protests, and a critic of half-way reform measure. Debs was passionately committed to defending workers and overturning capitalist power through industrial relations. Debs' charisma arose from his ability to envision a future of democratic participation and shared responsibility beyond conflicts. His democratic idealism gave him solid ground to challenge prevailing class power, to critique reforms dependent on elites, and to bear witness to democratic responsibility by accepting jail rather than silence in the face of what he regarded as injustice.

O'Brien also emphasizes the charismatic leadership of Dorothy Day, an American radical who supported labor, worked for human suffrage, protested against World War I, and founded a catholic worker movement that in the 1930s resisted efforts during the Spanish Civil war, then the World War II, and the Vietnam War. Dorothy's courage and compassion expressed deep faith and an ever-changing spiritual life. Her love for the poor and the continuing search for constructive and creative initiatives to meet human needs without violence and exploitation all bear witness to the best aspirations of her country and her church. In her case, charismatic leadership was associated with personal conversion, creative discipleship, and the practice of hospitality. O'Brien highlights that all the analyzed leaders are characterized by common traits: personal responsibility, solidarity, and both grounded in visions of the belonging to the human family.

Moreover, the work of Walker (2006) concerns the empirical case below presented, as it deals with Octavia Hill who was among the promoters of social reform in the nineteenth century (1870) and active in philanthropy, during the Victorian period. Her charismatic personality stood out (like Sister Isabella Soleri) for the ability to propose innovational social reforms and for her methodical approach to the work, which focused on order, discipline, and charity. She had great business skills, precision, and expertise in organizational aspects of business management and governance as well as officiating as accountant, treasurer, and bursar. Such abilities and skills were used to improve the living conditions of poor and socially excluded people and provided an example for others who followed her path drawing on *stewardship and trust* (Walker, 2006: 175), fundamental qualities for the S. Giuseppe Institute.

Walker's work emphasizes the role of British women belonging to different social classes in philanthropic and accounting functions: "*During the nineteenth century women argued that their prescribed functions in caring, moral improvement and domestic management could be practiced in the public domain, thus offering legitimacy to the application of feminine skills beyond the home*" (Walker, 2006: 163 and 182). Indeed, it is precisely in the nineteenth century when the S. Giuseppe Institute was set up that women began to understand how their abilities, including their accounting skills, could be used "outside the home" and could therefore attain a significant role in the social and public arenas.

Among the authors who have contributed to enrich this thread of studies, Bruni and Sena (2013) have emphasized the role of charisma in social and civil life and observed how charisma is often considered an element of divine nature limited to a theological and Christian context and not appropriate to the economic and social sciences. By contrast, the authors highlight how even in the work of Schumpeter (1934) the root of charisma's role in economics can be found, in which he makes a distinction between innovators and imitators. The Schumpeterian message of innovation reveals the importance of charismatic traits as innovation is a social reality before being an economic *one* and the entrepreneur or company's private affair).

Likewise, reading the work of Schumacher (1973) in this key, it is possible to underline its influence on the cooperative movement, the social economy, and value-based organizations, "good corporations" which integrate responsible commitment into their own strategies (Opdebeeck, 2013).

# 3 The Role of Women: Social Innovation and Charismatic Leadership Through the History of the *Istituto S. Giuseppe* and the Little Hospital

## *3.1 Methodology*

The empirical research was based on the analysis of archival sources retrieved from the Foundation "S. Giuseppe" (Atkinson & Coffey, 1997). These included the correspondence written by the governing bodies addressed to congregations, banks, and public municipalities, extracts of reports and partners' meetings, and board minutes, corresponding to the period between 1914 and 1950. These sources were particularly interesting in order to understand the role of women at the various managerial levels and in corporate governance. Concurrently, the analysis includes the interpretation of financial statements and balance sheets, budget plans, and infra-annual reports as well as the analysis of accounting documentation (invoices, receipts of payment) and non-accounting data (technical reports).

## *3.2 The Historical and Social Context and the Pioneering Approach of "Maternal Help"*

The *Istituto S. Giuseppe per l'Aiuto Materno e Infantile di Rimini* played a pivotal role in fighting social injustice linked to female discrimination that represented a social plague and a form of poverty so widespread in the historical and social context of the time. It was established in 1910 in reply to the lack of identity of the women, mothers out of wedlock ("single mothers") especially belonging to the poorest social classes (Haynes, 2008a-b). This category of women was considered insignificant and kept on the margins of society together with old and sick people who were considered "to be cancelled," forcing them to entrust their children to "charity" institutions (especially of a religious nature). The *San Giuseppe* "intervened" to tackle this social plague.

The ability to "see," i.e., to understand the urge of women discrimination, is exactly the result of the sensitivity and intuition (Stein, 1996) of a woman (Sister Giuseppina Soleri) as well as of the professional competence and humanitarian dedication of a doctor (Prof. Antonio Del Piano) to protect "illegitimate" motherhood and "illegitimate" children (Del Piano, 1901).

To better understand the social role of S. Giuseppe in fighting this social injustice, one can mention that in 1890 over 21,800 charitable organizations (the so-called *Opere Pie*) existed in Italy. They were a result of the charitable nature of civil society through donations from wealthy families to the poor and needy (Servalli, 2013). Subsequently, they extended their care to children, in particular to illegitimate children and their mothers, like the *Opera Pia "Istituto San Giuseppe per l'Aiuto Matero e Infantile."*

Many of these charitable organizations had a religious matrix and were characterized by the presence of women, like lay charity institutions. However, women committed to charity and philanthropy and nuns, who were present in great numbers in monasteries and in the various religious orders dedicating themselves to the assistance of the sick and to the education of the children, did not have an external and social role and were therefore discriminated against. Hospitals in the modern meaning of the term were not widespread yet, nor was there specific attention to the care of pregnant women, of newborn babies, and of children.

In Rimini, indeed, the Congregation of the Daughters of Charity of *S. Vincenzo de' Paoli* to which Sister Isabella Soleri belonged for years contributed to the health and social policies of the city.

Under this national and local situation of social inequality toward women the *S. Giuseppe* bases its charitable action and its mission.

### 3.2.1 The Pioneering Approach of "Maternal Help"

The *Aiuto Materno* was characterized by a pioneering approach aimed at social causes and assistance compared to other similar and contemporary institutions. It was able to fill the gap in the needs of poor and alienated children and mothers who were so numerous and neglected. The prenatal and postnatal period, and the relative problem of "exposed" [1] especially in the poorest sectors of society, was extremely neglected and a source of prejudice and omissions. The charity and social health interventions carried out by the existing charity foundations were mainly addressed to education and only rarely to social and healthcare interventions. Before the constitution of *San Giuseppe Aiuto Materno* only 3% of institutions showed attention, albeit disorganized, toward maternal help, but nothing had been thought of helping mothers and recognizing illegitimate children. The effects of this gender discrimination undermine(d) the survival of society. Such injustice derived from a dominant culture (Kirkham, 1992; Lehman, 1992) which denied every social role to illegitimate mothers and their children (Haynes, 2008b, 2008c; Kirkham, 1992).

[1] The term "*Esposti*" ("the exposed") derived from a Roman custom which allowed fathers who refused to acknowledge a newborn child the rite of "removing him from the ground" (hence the word "raising") which implied exposing the newborn to the public at the "*columna lactaria*" (the Milk Column). For more detailed information about the history of the Esposti throughout the centuries, see Freddi (2007), "The 'Massacre of the Innocent' continues, a never ending story: The Tragedy of Illegitimate Children": 24–30.

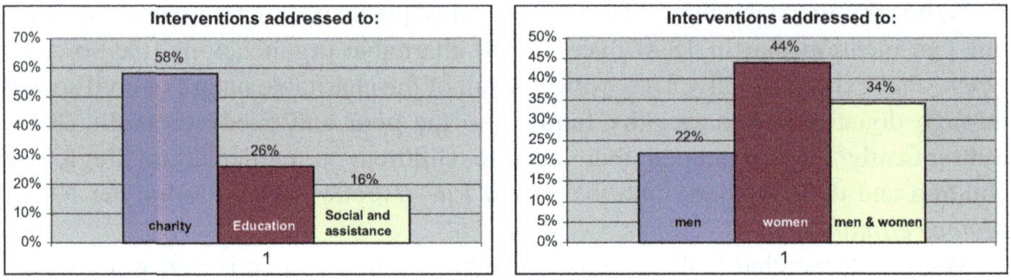


**Fig. 1** Types of interventions carried out by ("Opere Pie"). Source: Chiadini and Freddi (2008: 36)

**Fig. 2** Features and nature of the supporting institutions (1925). Sources: Chiadini and Freddi (2008: 37)

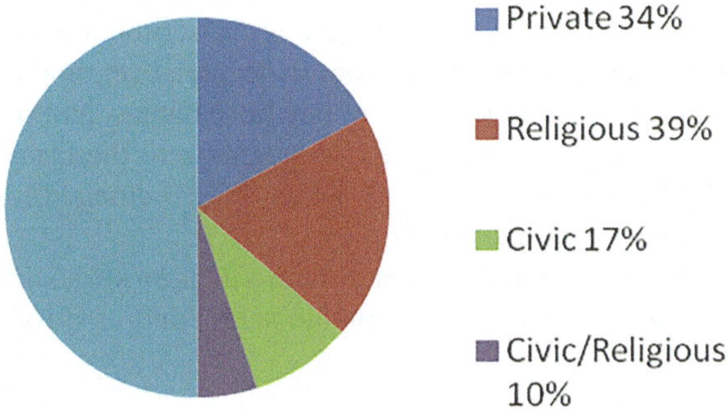


Such injustice was fought by the *S. Giuseppe* because a new identity was recognized for women. Women were the receivers of the activity of *the Istituto Opera Pia S. Giuseppe per l'Aiuto Materno e Infantile*, which helped them to raise their children and with the establishment of the *Ospedalino*, to treat them giving dignity to the condition of mother in her most critical period, before and after childbirth.

The graphs which follow (Fig. 1) highlight the charity and social health interventions carried out by the charity foundations ("Opere Pie") and their beneficiaries at the end of the 1800s.

The social scourge of the "vulnerable" took up 1.11% of the total number of interventions deriving in particular from the religious sphere (Fig. 2).

The issue of infant mortality was a direct consequence of the discrimination against women, since dignity and social legitimization were denied to the "single" mothers. Prevailing culture and society imposed the abandonment to their destiny of children upon the mothers, condemning the vast majority of them to malnutrition and sickness.

The medicine for children was confused with infant care and started as a medical specialty between the end of the 1800s and the beginning of the 1900s in Europe and in America. Before then, the old foundling hospitals (i.e., in Florence, the "*Spedale degl'Innocenti*") housed babies abandoned by their mothers but lacked a health structure.[2] Children got sick and died in large numbers. Doctor Del Piano in 1908 stated "*Perhaps we have not yet succeeded in establishing an adequate concept regarding the social benefit of infant care (...) however the child represents the future*" (Chiadini & Freddi, 2008: 52).

The *Opera Pia S. Giuseppe* was the first and only institution in Rimini to provide organized protection based on scientific principles for illegitimate mothers and vulnerable children by introducing a new approach to overcome this issue.

On the basis of data managed and supplied by the "Sisters of Charity"—who served as treasurer—Prof. Del Piano highlighted the "performance" attained by the *Aiuto Materno*: "*Illegitimate children recognition rose from 27% to 86%, that breastfeeding was possible for 90% of mothers assisted and that the mortality rate in the first two years of life had dropped from 43 to 20%.*"

On March 24, 1914, the President of *Aiuto Materno* wrote to the King of Italy to obtain the recognition of the institute as a Charity Foundation. In her petition, Sister Giuseppina Soleri explains that due to: "*...the commendable effort made to ensure the existence of the Aiuto Materno for the Country, expanding its activities to include the custody of children until the age of three, and with a small hospital for sick children, she was prepared to entrust to the Charity foundation her property of the Palace in Rimini and a part of her family possessions.*" A decree of the Kingdom of Italy of November 8, 1915, constituted it as a Charity Foundation.

The reading of the institutive and accounting documents permits us to acknowledge the role of the female members in its governance and management.

Firstly, women have been both ideally and actually the promoters as the *Opera Pia S. Giuseppe* were set up thanks to Sister Soleri's and other rich women's donations and to the fundraising activities held by women.

### 3.2.2 The Changing Role of Women During the Stages

Both Sister Soleri and Doctor Del Piano envisioned an innovative approach to women and illegitimate children. As a woman Soleri was able to interpret the needs of poorer and neglected women and children and provide an effective answer and anticipate serious social grievances, which other institutions and companies have not been able to express before. This social innovation was possible thanks to her strategic vision, which provided the idea with the means to realize it, by way of the donation of her family wealth that included buildings, goods, and financial resources necessary to establish the activities of the *S. Giuseppe*.

Secondly, women were entrusted with both managerial and operational roles—including that of accountants—and carried out numerous activities and functions. Women were, also, the recipients of the services provided. The *Opera Pia*

[2] The Italian Society of Pediatrics was founded in 1898 and the first Italian pediatric congress dates back to 1890. In 1886, the first university teachings of the specialization were introduced in Padua and Florence, where in 1891 the Meyer hospital for children began work.

*S. Giuseppe* statute, approved in 1915, states: *its main objective is to distribute food to the poor mothers of the city and suburbs of Rimini, to enable them to nourish their newborns, to supply milk, wet nurses, medical care, clothing and medicines to the children* (Art. 2).

During the first stage of *S. Giuseppe* the female presence in the roles of president and members of the board was fundamental: they served as patron ("patronesses," members of committees relative to different activities, responsible for internal and economic management) and assistant director and advisors. Women also served as external benefactresses and promoted fundraising through an intense relational activity directed at awakening civil society and institutions to the issue (Maltby & Rutterford, 2006; Walker, 2006; Wiskin, 2006).

In general terms, women had a strong impact on the institution's strategic and operative approach during the first phase of life of the institute (Baldarelli & Del Baldo, 2016). The spread of typically female values (ability to listen, assist and help, interact and communicate) oriented the investment choices and services qualification. Consequently, women acted as the driver of the elimination of social injustice.

Unfortunately, starting from 1916 they lost their visibility because they did not hold powerful roles any more, while maintaining "key" but subordinate roles. Women were swallowed up and absorbed by operative roles; assisting mothers acquire their dignity—upon which de facto all the activity of the *institute* was based, while they were replaced by men in governance roles, thus highlights a dichotomy: externally the assisted women acquire their dignity, while within *S. Giuseppe* women (founders, presidents, and those occupying managing positions) give way to men (Bell et al., 2002; Haynes, 2008a, 2008b). But despite women's ability to provide the necessary financial and human resources (Walker, 2006) they were not present in "the control room" and a glass ceiling (Bell et al., 2002; Ciancanelli et al., 1990) manifested. Indeed, from the end of 1916 the governance radically changed, and the female members were reduced. This is due to the social and cultural dominant construct, which generally discriminated against women with regard to men's power and social role (Hopwood, 1987). The board of directors became the expression of the institutional power of the city, including the bishop, the local bank, and the shareholders among whom few women were included, reducing them to a sort of "pink quota" presence.

However, despite not being directly involved in governing the *S. Giuseppe*, drawing from the writings and the reports of Doctor Del Piano we have analyzed, an agreement with the "silent" leadership of Sister Soleri emerges in veiled form in the reciprocal exchange and sharing of intuition and competence, which then finds synthesis in the accounting evidence (Chapman et al., 2009). They both had a profound reciprocal respect for each other. The traces of this fruitful synthesis between Del Piano's professionalism and Soleri's leadership emerged in the official recognition, which has been attributed to Sister Giuseppina Soleri, including the silver medal for her merits to public health.

Indeed, in 1929 Del Piano introduced the position of female inspectors who had an important role in the healthcare and assistance program, because they promote a new culture of "taking care" of children. Their activities included weekly consultations for mothers and newborns to provide them with notions of hygiene and child raising, pediatric check-ups, monitoring mothers' and children's environmental conditions, and supporting clothing and food.

Moreover, a key role was attributed to the nuns, being that the monastic order of the "Daughters of Charity of San Vincenzo de' Paoli" was historically devoted to assist the poor and the sick. They were entrusted with educational activities, administrative services, and infant care.

During the period of 1917–1918, the number of children and mothers assisted amounted to 350 and 76, respectively, and the mortality of children fell to 5.4%. These figures are a clear sign of the *institute*'s role in fighting the external discrimination suffered by women and fighting a social plague.

In 1920, the *Opera Pia S. Giuseppe* extended its activities to take custody of children from the orphanage and provide assistance and health services until July 1925, when the Italian Regina Elena hospital was officially inaugurated.

### 3.2.3 The Foundation of the "Little Hospital": From Assistance to Health Care

The idea of establishing a hospital for women giving birth and infancy dates back to those years of the constitution of *S. Giuseppe*: "As desired by Sister Isabella Soleri, and thanks to her donation (she decided to donate her family heritage to establish the Institute), the little hospital for sick children, called '*Ospedalino Regina Elena*,'" started its activities on July 29, 1925.[3] Assistance was supported by the nuns and the nursing and auxiliary staff. The hospital services included children up to the ages of six; birthing services for mothers of also legitimate children; and the maternity ward under the protection of the National Charity Foundation for mothers and children (Minutes of 23rd March 1927: 47).

The creation of this first gynecological and pediatric structure, which provided a concrete and innovative answer to the healthcare needs of mothers and children—that became part of the common sensitivity—deeply affected the evolution of the hospital history of Rimini province and promoted growth in social justice and civil society, thus guaranteeing the health of infants after having provided a social status to illegitimate mothers. Indeed, the *"Little Hospital"* also welcomed all mothers, giving healthcare dignity to all women giving birth, who at the time were paid little attention. This represented an important step in the path against "external" women's discrimination, transforming discrimination into social innovation. In 1939, the L obtained the qualification of specialized third category hospital both a pediatric and general hospital.

[3] Del Piano (1927: 9). See also *Minute of Board Meeting, of March 10, 1920*; book of reports from the Board of Director from September 24, 1915, to December 12, 1928, in the AUSL archive of Rimini: 19–21.

However, women's role—particularly that of nuns—remained veiled because the medical sector and more generally all professions were male-dominated (Walby, 1988). This discrimination stayed despite the fact that the consolidation of the first nucleus of competence of nursing and midwifery was developed by nuns. After the Second World War, they were the first in Italy to deal with the training of nursing staff and they contributed to opening the "school for professional nurses" next to the hospitals. The "Small Manual" written anonymously by a Sister of Charity was adopted in the training schools attended by generations of students for about 30 years. Other innovative services concerned the use of heliotherapy and thalassotherapy to prevent children's respiratory diseases. Such activities elicited the future "tourist vocation" of Rimini, which is, till today, famous for its seaside tourism.

To briefly complete the history of the *Little Hospital,* we underline that in 1970 it was classified as a provincial hospital specialized in pediatrics. It was then closed, in the midst of many controversial opinions, between 1985 and 1988. Currently, the San Giuseppe Foundation for Maternal and Child Aid Onlus continues the original activities by offering services addressed to children and young people hosting them in residential socio-educational and integrated semi-residential communities, services for people with disabilities hosted in a daytime socio-health structure, and residential service aimed at adolescents and neo-18-year-olds.

## 4 Discussion and Conclusion

The case of the S. Giuseppe Institute and the Little Hospital was analyzed to answer the following research questions: "Which is the role of woman in orienting toward an authentic responsibility: the mission, governance, and accountability of enterprises?" Is women's role formalized and externally recognized? How do women contribute to promoting change in the internal and external environment?

In the timeframes covered by the analysis, we highlighted various elements of female discrimination. Indeed, in the first phase of its foundation (until 1916) discrimination was the fruit of a social and cultural construct, still prevalent in the twentieth century, which denied dignity and social legitimization to single mothers, and it did not recognize (in the real sense of the term) the existence of their illegitimate children, who were destined to death and marginalization and "saved" just by "charity" association assistance (Haynes, 2008a, 2008b). Next to this "external" discrimination, inside the *S. Giuseppe*, women played a central role and participated in both governance and management functions. Almost paradoxically the institute was the fruit of the discrimination against women, since its mission was to tackle social injustice and the origins of the hospital are a consequence of this mission.

However, in the second phase, the role of women becomes more marginalized, in that they disappear "from the rooms of power," occupied by men (chairmen, directors, those in charge of accounting), even though they keep carrying on the majority of activities (of management and of accounting recording). This discrimination even goes on throughout the third phase, when the work of the *S. Giuseppe Institute* and of the *Ospedalino* was formally recognized by the local and national civil society for its struggle as regards the abandoning of children and infant mortality, as well as for having initiated pediatric and gynecological hospital services and triggered the profession of nursing.

Despite this, the *S. Giuseppe* and the *Ospedalino* gave rise to innovative social services (assistance to illegitimate mothers, education regarding breast-feeding children, infant healthcare). Behind these choices (which the accounting data summarize), the sensitivity and the competence of women are hidden (Stein, 1996). The history of *S. Giuseppe* demonstrates how women's dignity, recovered, and restored to its social dimension, contributes to the betterment of the whole society. It provides a strong example in *that*, more than other similar institutions of the time, it laid the foundations for a radical innovation in managing assistance and healthcare assistance for the benefit of women and children. Responding to a social injustice and leveraging on women's weaknesses, it developed internal competences which generated one of the first Italian pediatric hospital. It transformed a social injustice into social innovation based on the active participation of women (Bell et al., 2002).

Data, coming from accounting and extra-accounting documentation, demonstrate the economic and social value produced by the institution. They synthesize the effectiveness and the adherence to mission and resume the wealth produced by the economic activity, whose value is tied directly in with corporate values imbued with sensitivity, professionalism, tenacity, and female foresight, geared toward satisfying human needs. Through its activity, led by the female sensitivity, the S. Giuseppe and the Little Hospital triggered a new corporate culture (Baskerville, 2003; Hofstede, 1991; Hopwood, 1987), unusual for that time, orientated to the protection of childbirth and infancy and psychiatrically disabled. This generated an anthropological culture aimed at reinforcing and guaranteeing the same foundation of human survival and consequently corporate survival.

Women's "filigree" emerges with the leadership of Sister Soleri, capable to recognize the most urgent and deepest needs of others, using her personal resources and wealth to give substance to the promotion of humanity. Her role was not limited to philanthropy, but went beyond (Walker, 2006) anticipating essential social services and promoting an epoch-making cultural change (Walker, 2006). The innovations contributed to legislative changes, which allowed the setup of the *Ospedalino*'s activities and the pediatric healthcare directed by Doctor Del Piano.

In summary, drawing from the charismatic leadership framework (Bruni & Sena, 2013; Bruni & Smerilli, 2014; Meindl, 1990; O'Brien, 2013; Weber, 1947), we argue that the encounter between two charismatic personalities created a marriage between charity in its purest form and modern social medical science, giving rise to a range of innovative services that opened the way to a hybridization of public and private services—an utopia for that time—and putting into effect individual and collective well-being before the welfare state began.

About future perspectives of our research, attention should be redirected to charismatic women and their role in today's society. The need to tackle poverty and marginalization, as well as current crises, requires a radical rethink in the socio-economic system, lifestyles, and relationships in work dynamics that call for rethinking women leadership as a resource.

## References


Atkinson, P., & Coffey, A. (1997). Analyzing documentary realities. In D. Silverman (Ed.), *Qualitative research* (pp. 45–62). Sage.

Baldarelli, M. G., & Del Baldo, M. (2016). Accounting and charity. How to read the "pink" in accounting in the first half of the twentieth century to a real life case. Special issue: Accounting and charities in historical perspective. *Accounting History, 21*(2–3), 144–166.

Baskerville, R. F. (2003). Hofstede never studied culture. *Accounting, Organizations and Society, 28*(1), 1–14.

Becker, T. E. (1998). Integrity in organizations: Beyond honesty and conscientiousness. *Academy of Management Review, 23*, 111–137.

Bell, M. P., MacLaughlin, M. E., & Sequeira, J. M. (2002). Discrimination, harassment, and the glass ceiling: Women executives as change agents. *Journal of Business Ethics, 37*(1), 65–76.

Broadbent, J. (1998). The gendered nature of 'accounting logic': Pointers to an accounting that encompasses multiple values. *Critical Perspectives on Accounting, 9*, 267–297.

Broadbent, J. (2016). A gender agenda. *Meditari Accountancy Research, 24*(2), 169–181.

Bruni, L., & Sena, B. (2013). *The charismatic principle in social life*. Routledge.

Bruni, L., & Smerilli, A. (2014). *L'altra metà dell'economia*. Città Nuova.

Chapman, C. S., Cooper, D. J., & Miller, P. B. (2009). Linking accounting, organizations and institutions. In C. S. Chapman, D. J. Cooper, & P. B. Miller (Eds.), *Accounting, organizations, and institutions. Essays in honour of Anthony Hopwood* (pp. 1–29). Oxford University Press.

Chiadini, A., & Freddi, P. (2008; 2009; 2010). La storia di un valore. L'Istituto San Giuseppe per l'Aiuto Materno e Infantile di Rimini. Dagli inizi del Novecento agli anni Trenta, volume primo; Dagli anni Quaranta agli anni Settanta, volume secondo; Dagli anni Settanta al terzo millennio, volume terzo. FaraEditore.

Ciancanelli, P., Gallhofere, S., Humphrey, C., & Kirkham, L. (1990). Gender and accountancy: Some evidence from the UK. *Critical Perspectives on Accounting, 1*(2), 117–144.

Del Piano, A. (1901). *Pro Infantia*. Benzi.

Del Piano, A. (1927). *L'Opera dell'Aiuto Materno in Rimini dal 1910 al 1926*. Relazione Sanitaria del Dott. Prof. Del Piano, Tipografia Giuseppe Benzi.

Fondazione S. Giuseppe Rimini per l'Aiuto Materno e Infantile. (2020). *Crescere Responsabili, bilancio sociale*. FaraEditore. https://www.sangiuseppe.org/

Haniffa, R. M., & Cooke, T. E. (2005). The impact of culture and governance on corporate social reporting. *Journal of Accounting and Public Policy, 24*(5), 391–430.

Haynes, K. (2008a). Transforming identities: Accounting professionals and the transition to motherhood. *Critical Perspective on Accounting, 19*(5), 620–642.

Haynes, K. (2008b). (Re)figuring accounting and maternal bodies: The gendered embodiment of accounting professionals. *Accounting, Organizations and Society, 33*(4/5), 328–348.

Haynes, K. (2008c). Moving the gender agenda or stirring chicken's entrails? Where next for feminist methodologies in accounting? *Accounting, Auditing and Accountability Journal, 21*(4), 539–555.

Hofstede, G. (1991). *Cultures and organizations: Software of the mind*. McGraw-Hill.

Hopwood, A. G. (1987). Accounting and gender: An introduction. *Accounting, Organizations and Society, 12*(1), 65–69.

Jamali, D., Saffieddine, A. M., & Rabbath, M. (2008). Corporate governance and corporate social responsibility synergies and interrelationships. *Corporate Governance, 16*, 443–459.

Kirkham, L. M. (1992). Integrating herstory and history in accountancy. *Accounting, Organizations and Society, 17*(3–4), 287–297.
Krüger, P. (2009). *Corporate social responsibility and the board of directors. Job market paper.* Toulouse School of Economics.
Lehman, C. (1992). "Herstory" in accounting: The first eighty years. *Accounting Organization and Society, 17*(3–4), 261–285.
Maltby, J., & Rutterford, J. (2006). 'She possessed her own fortune': Women investors from the late nineteenth century to the early twentieth century. *Business History, 48*(2), 220–253.
Meindl, J. R. (1990). On leadership: An alternative to the conventional wisdom. In B. M. Straw & L. L. Cummings (Eds.), *Research in organizational behavior* (Vol. 12, pp. 159–203). JAI Press.
Naumes, W., & Naumes, M. J. (2006). *The art and craft of case writing* (2nd ed.). ME SHARPE Inc.
O'Brien, D. J. (2013). The charismatic principle in an American and democratic context. In L. Bruni & B. Sena (Eds.), *The charismatic principle in social life* (pp. 37–46). Routledge.
Opdebeeck, H. (2013). The role of the charismatic economist E.F. Schumacher in economic and civil life: CSR and beyond. In L. Bruni & B. Sena (Eds.), *The charismatic principle in social life* (pp. 151–170). Routledge.
Rao, K., & Tilt, C. (2016). Board composition and Corporate Social Responsibility: The role of diversity, gender, strategy and decision making. *Journal of Business Ethics, 138*(2), 327–347.
Rose, C. (2007). Does female board representation influence firm performance? The Danish evidence. *Corporate Governance, 15*(2), 404–413.
Schumpeter, J. A. (1934). *The theory of economic development.* Harvard University Press.
Schumacher, E. F. (1973). *Small is beautiful: Economics as if people mattered.* Harper & Row.
Servalli, S. (2013). The interface of power and charity in the government of poor. A case from the Italian context in the sixteenth-seventeenth centuries. *Accounting, Auditing and Accountability Journal, 26*(8), 1306–1341.
Stein, E. (1996). *Essays on woman* (2nd ed.). ICS Publications.
Tilt, C. (2015). Gender diversity on Corporate Boards and CSR. Paper presented at Paper for the SIDREA International Workshop Meditari Accountancy Research European Conference 2015, University of Bologna, Italy, Forlì, 2–3 July 2015 (pp. 1–19).
Walby, S. (Ed.). (1988). *Gender segregation at work* (pp. 1–13). Open University Press.
Walker, S. P. (2006). Philanthropic women and accounting. Octavia Hill and the exercise of 'quiet power and sympathy'. *Accounting, Business and Financial History, 16*(2), 163–194.
Weber, M. (1947). *The theory of social and economic organization.* Translated by A.M. Henderson and Talcot Pearsons. Edited with an introduction of Talcot Pearsons. The Free Press of Glencoe, Collier-Macmillan Limited (Copyright 1947 by Oxford University Press, New York, Inc. printed in the USA. First Free Press Paperback Edition 1964).
Wiskin, C. (2006). Businesswomen and financial management: Three eighteenth-century case studies. *Accounting, Business and Financial History, 16*(2), 143–161.
Yin, R. K. (1994). *Case study research: Design and methods* (2nd ed.). Sage.

**Maria-Gabriella Baldarelli**, PhD, is Associate Professor of Accounting at the University of Bologna, Department of Management, and Visiting Professor at the State University of Tirana, Albania (2017), and La Trobe, Melbourne, Australia (2015); Teaching staff mobility at the New Bulgarian University of Sofia, Bulgaria, from November 22 to 27, 2010; University of Sao Paulo, Brazil, from the end of May to June 1, 2011; University of Diocese of Buea (UDEB), Cameroon—from February 4 to 8, 2012; University of Pula-Croatia in May 2006; and University of Vlore (Albania) from May 12 to 15, 2009. Maria-Gabriella is a Partner of Editorial Board international Review "Economic Research" (UDK 338; ISSN 1331-677X). Research interests include financial statement in tour operator and travel agencies; corporate social responsibility; ethical, social, and environmental accounting and accountability; sustainability in tourist enterprises; responsible and accessible tourism for blind people; economy of communion enterprises; and gender (pink) accounting. Maria-Gabriella is a member of the board—Center of Advanced Studies on Tourism (CAST; http://www.turismo.unibo.it), as a delegate of the Department of Management, University of Bologna. She is also a member of SIDREA council from December 2009. Website: www.unibo.it/docenti/maria.baldarelli

**Mara Del Baldo** is Associate Professor (recognized as Full Professor by the Italian National Agency-ANVUR) of Business Administration and Economics of Sustainability and Accountability at the University of Urbino (Italy), Department of Economics, Society and Politics. She has extensive experience in collaborative research, in conjunction with universities and private industry. She visited several European Universities, i.e., the University of Vigo (Spain), the New Bulgarian University of Sofia (Bulgaria), the Corvinus University in Budapest (Hungary), and the University of Craiova (Romania). She is a member of the European Council for Small Business, the Centre for Social and Environmental Accounting Research (St. Andrews), the SPES Institute, and the European Business Ethics Network (EBEN) Italia, as well as of different Italian scientific associations. Mara is also a member of the academic advisory committee in the Global Corporate Governance Institute (USA). She is part of the editorial board, and she serves as a reviewer of several scientific journals, including the Springer's *International Journal of Corporate Social Responsibility*. Among her main research interests are entrepreneurship and SMEs; corporate social responsibility; entrepreneurial leadership and business ethics; social accountability; integrated and sustainability reporting; and accounting and gender. She published in several Italian and foreign journals as well as in national and international conferences proceedings and books, including the Springer Series *CSR, Sustainability, Ethics & Governance*.

# Higher Professional Technical Course in Portugal: The Case Study of the Polytechnic Institute of Guarda

**Joana Sá Rodrigues, Rute Abreu, and Cecília Fonseca**

## 1 Introduction

The chapter focuses on analysing students' situation from the perspective of Higher Professional Technical Courses (HPTC) (in Portuguese, *Cursos Técnicos Superiores Profissionais—cTeSP*), in a Portuguese public polytechnic higher education institution, based on a research hypothesis (Hasanefendic et al., 2016; Jin et al., 2019; Joensen & Nielsen, 2006). Professional and technical higher education, which can be considered "non-university" tertiary education, is taught at "Polytechnic" Institutes in Portugal. The research hypothesis refers to the understanding of how graduates of HPTC contribute to the increase of employability and how they meet "the needs of the economy and the regions in which they will be taught" and also "to the investment of polytechnic education in short-term training" (MEC, 2014). Thus, the aim of the chapter is to study the interaction between HPTC and the paths of its students within a certain social and economic context characterized by specific policies (UNESCO, 2016a, b). In this context, it is necessary to consider the "emergence of new audiences for higher education, not only young people but also adults, and young people from vocational secondary education", with a view to "increasing and diversifying the spectrum of higher education offer in Portugal and, thereby, increasing the number of citizens with higher qualifications, thus contributing to an increase in the national priority" (MEC, 2014). In this sense, the

J. S. Rodrigues · C. Fonseca
Instituto Politécnico da Guarda & UDI-IPG, Guarda, Portugal
e-mail: cfonseca@ipg.pt

R. Abreu (✉)
Instituto Politécnico da Guarda & UDI-IPG & CICF-IPCA & CISeD-IPV, Guarda, Portugal
e-mail: ra@ipg.pt

 
I. G. Arraiano et al. (eds.), *Corporate Social Responsibility in a Dynamic Global Environment*, CSR, Sustainability, Ethics & Governance,
https://doi.org/10.1007/978-3-031-24647-0_2

objective of this research is to present the case study of a polytechnic higher education institution in Portugal, seeking to answer the following research question:

- Are the HPTC responding to the objectives for which they were designed?

The answer to this question proves to be very important, as a way of understanding whether actions at national level by the Portuguese Member State adapt the three main priorities (EC, 2010) of the Europe 2020 Strategy, to their reality, as a way of, in this specific case, contributing to the creation of smart growth through investment in education.

Methodologically, this research relies on two different approaches. The first approach takes the form of a literature review, based on the legal regime of the *Higher Professional Technical Courses (HPTC)*, considering that, according to Decree-Law n° 43/2014 of March 18, the XIX Government's Programme promoted the HE policy to better adapt the educational offer both to the diversity of demand and to the needs of the country for qualified staff (ILO, 2019; Reinhardt et al., 2011); with a view to achieving the goals of the Europe 2020 Strategy, the percentage of young Europeans with a higher education diploma should reach 40%. In this sense, national data are collected on statistical studies at the level of HPTC, focusing mainly on the continuation of these studies that concluded the courses. The investigated data will also allow us to understand whether the goals outlined by the Europe 2020 Strategy were carried out by Portugal. The second approach takes the form of a descriptive statistical analysis, using a questionnaire and the case study method in a Portuguese public polytechnic higher education institution (Yin, 2018), based on the information provided by this institution, which shows the financed programmes and HPTC in general and the situation regarding the status of graduate students, in the 4 weeks, 6 months, and at the moment (2021) after the conclusion of the respective courses under analysis.

The chapter is organized as follows. Section 2 gives an overview of the Higher Professional Technical Courses (HPTC), as labour input needed for a country's economic growth and development. Section 3 presents the methodology that was used on this study. Section 4 introduces the empirical analysis based on a case study of a Portuguese public polytechnic higher education, to understand the real path of HPTC students. Section 5 offers an explanation about the obtained results. The last section summarizes main results and presents the conclusion.

## 2 Literature Review

Currently, higher education institutions (HEIs) are critical actors in a knowledge-based economy, being essential for the production, transmission, and dissemination of knowledge (Elena & Sánchez, 2013, pp. 48). According to Rytmeiter (2009), the centrality that knowledge and innovation assume in today's societies has been posing important challenges to Higher Education (HE) and its institutions, which are not restricted only to the national level, but rather, and each more and more, they tend to expand to the European and world context, where mobility and internationalization have been asserting themselves as essential pillars of development, both in terms of their competitiveness and in terms of their sustainability. In this context, some of these challenges are reflected in attracting more and better students, in the development of new forms of teaching and learning, in teachers with more and better qualifications, in the creation of highly relevant research structures, in the innovative character of research developed, in the ability to transfer knowledge to society, in improving the quality and performance of the institution, and in meeting the needs of stakeholders (Rytmeiter, 2009).

Recently, the concept of governance has been closely associated with corporate governance terminology. According to Lourenço (2017, p. 9), governance is a "terminology that derives from the English expression governance, which in the Portuguese lexicon has taken on different designations (*governo*, *governança*, and *governação*), represents a complex network of rules and instruments regarding management and the supervision of organizations, integrating the external perspective in a broader perspective than the market perspective (Williams, 2014). With a strong focus on the relationship between owners and executives, it has been the subject of an intense debate following the changes made in the top management of organizations, supporting conflict harmonization theories and regulations on governance principles and on control and supervision mechanisms". Thus, it is understood that governance is seen as the set of mechanisms and institutional practices that legitimize the functions exercised by the top management of organizations (Board of Directors), through the delegation of powers from the different entities that invest resources in them: investors equity capital (owners or shareholders), foreign capital investors (financial creditors), human capital investors (employees and managers), and other investors from other resources (Oman et al., 2003).

According to Schultz (1963), the fact that in recent decades there has been an increased awareness of human capital demonstrates the role of investment in education to improve economic and social achievements. This contributes to the strengthening of higher education and advanced training, which enables a quick response to market needs and an increase in employability.

In this way, the European Commission faced with a recessionary economic scenario and worsening social conditions instituted a new strategy: the Europe Union (EU) 2020 Strategy, for the period 2014–2020. This growth and employment strategy is based on policies whose main purpose is not only to restore the EU economy but also to increase employment, productivity, and social cohesion, with education and training being recognized as a key policy area in contributing to Europe's economic growth and social inclusion (Rodrigues et al., 2020). In general, the strategy, "consisting of measurable targets and underlying proposals, aimed to promote 'smart, sustainable, and inclusive' growth across the 28 EU member states" (at the time of writing, the UK was a member of the EU) (European Commission, 2010). In terms of increasing employability, "policy makers have realized the importance of investing in education and training as a way of improving the existing stock of skills" (Humburg et al., 2013). As regards education and training, the EU has set a double target to be achieved by 2020: one is to reduce early school leaving rates to less than 10% and the other is more targeted at HE attainment—at least 40% of 30–34-year-olds should have completed HE (EC, 2016; OECD, 2020). Thus, the interrelationship between HE and the labour market has been considerably reformulated over time. According to Tijdens et al. (2012), "the scarcity of skilled workers has often been attributed, among other things, to the considerable gap between educational systems and the needs of companies, or to the fact that learning and training profiles are not suitable for current industry configurations."

It should be noted that the key targets include increasing HE attainment so that, by 2020, 40% of young people have successfully completed not only their HE studies but also contributed to an increase in the population's employment levels, investments in research and development (R&D) and innovation, the reduction of greenhouse emissions, the reduction of school dropout rates, and also the reduction of the risk of poverty, which implies directly or indirectly a better knowledge base in the population. One of the flagship initiatives for the 2020 strategy is the "New Skills and Jobs" initiative. Through this initiative, the EU aims to stimulate key stakeholders to better anticipate changes in the skills needed for the future, to realize a better matching between available skills and those required in the labour market, and to bridge the gap between the worlds of education and work. Several practical measures have been devised to help achieve these aims, such as the European Vacancy Monitor, the Skill Forecasts undertaken by the European Centre for the Development of Vocational Training (in Portuguese, *Centro Europeu para o Desenvolvimento da Formação Profissional*—CEDEFOP), and the development of a European Framework for key competences for lifelong learning in which the key competences needed by all to succeed are outlined (EC, 2011).

This issue has already been underlined in the European Commission (2012) Communication on "Rethinking Education: Investing in skills for better socio-economic outcomes" (EC, 2012), warning that "investment in education and training for skills development is essential to boost growth and competitiveness: skills determine Europe's capacity to increase productivity. In the long-term, skills can trigger innovation and growth, move production up the value chain, stimulate the concentration of higher level skills in the EU and shape the future labour market" and that "European education and training systems continue to fall short in providing the right skills for employability, and are not working adequately with business or employers to bring the learning experience closer to the reality of the working environment" (EC, 2012).

Thus, this call to countries to promote the development of Vocational Education Training (VET) provision at higher levels. In accordance with EC (2016), member states were explicitly requested to "develop or maintain post-secondary or higher VET at EQF level 5 or higher, as appropriate, and contribute to achieving the EU headline target of 40% with tertiary or equivalent education", and to "promote flexible pathways between VET, general education and higher education, and enhance permeability by strengthening the links between them. To achieve this aim, as well as greater participation in lifelong learning, participating countries should accelerate the establishment and implementation of comprehensive national qualifications frameworks based on learning outcomes". As talent creation is largely dependent on HE as this supply of highly qualified human capital feeds our labour market and is of particular interest for knowledge-intensive sectors with a high technological profile and operating in the global market, it justifies the fact that VET at higher education level has been considered. In this sense, it has become necessary to act at the level of diversification of the higher education offer.

In the case of Portugal, Decree-Law No. 43/2014, of 18 March (MEC, 2014), established the creation of a new type of short higher education training not conferring a degree, short-cycle programmes of higher education/short-cycle tertiary education offer situated at level 5 of the European Qualifications Framework (EQF) for lifelong learning and also at level 5 of the International Standard Classification of Education (ISCED)—the *Higher Professional Technical Courses (HPTC)* (in Portuguese, *Cursos Técnicos Superiores Profissionais—cTeSP*) (MEC, 2014). According to the present legal diploma, these study cycles are taught within the framework of polytechnic HE and are composed of the following components: a general and scientific training component, technical training component, and a work-related training component, which is materialized through an internship—which allows us to prepare students to enter the labour market (MEC, 2014). However, these programmes may also provide a pathway to other tertiary education programmes. These courses have a strong regional insertion, and are taught in polytechnic HEIs, as well as in organic units of polytechnic HE integrated in university HEIs (MEC, 2014). "The aim is thus to broaden and diversify the spectrum of higher education on offer in Portugal and, in this way, increase the number of citizens with higher qualifications, thus contributing to an increase in national competitiveness. These courses will meet the needs of the economy and the regions where they will be taught, and may also attract new audiences to higher education, not only young people but also adults, and in particular young people coming from vocational secondary education" (MEC, 2014). As these courses were created to meet the needs of the economy and the regions where they are taught, they are obliged to establish partnerships with companies and business associations in the region. In addition, they can also attract new audiences to HE, not only young people but also adults and young people coming from vocational secondary education (MEC, 2014).

It should be noted that the training areas in which each HEI confers the professional higher technician diploma are defined by its legally and statutorily competent body, taking into consideration the professional training needs at qualification level 5, namely in the region where it is located. This means that these courses are created by the HEI in question and according to the needs of the region where they are located. They are submitted to a creation registration process, which is then sent to the General Directorate of Higher Education, which may promote visits to the HEIs to make an on-site assessment of the satisfaction of the conditions; as well as to listen to specialized entities in the area (MEC, 2014).

It is also informed that the following may apply to the HPTC: students who have completed secondary education course or legally equivalent qualification; students who have passed all the subjects of the 10th and 11th years of a secondary education course, or of legally equivalent qualification, and have not concluded the secondary

**Table 1** Number of students enrolled by training level, in Portugal, between the academic years 2014/15 and 2020/21

| | Academic year | | | | | | |
|---|---|---|---|---|---|---|---|
| Level of training | 2014/15 | 2015/16 | 2016/17 | 2017/18 | 2018/19 | 2019/20 | 2020/21 |
| *Total* | 349,263 | 349,969 | 350,895 | 372,753 | 385,247 | 396,909 | 411,995 |
| ISCED 5 | 395 | 6430 | 11,048 | 12,780 | 15,423 | 17,409 | 18,128 |
| ISCED 6 | 212,526 | 211,676 | 210,998 | 216,471 | 219,615 | 227,075 | 236,116 |
| ISCED 7 | 117,272 | 119,079 | 120,138 | 123,050 | 129,119 | 130,661 | 134,206 |
| ISCED 8 | 19,465 | 19,759 | 19,759 | 20,452 | 21,090 | 21,764 | 23,545 |

Source: DGEEC (2021)

education course, and are considered apt by means of a capacity evaluation test to be carried out by the HEI; those who have passed the especially adequate tests intended to evaluate the capacity to attend higher education of those older than 23 years old; and also holders of a technological specialization diploma, a professional higher technician diploma, or a higher education degree who wish to undergo professional requalification (MEC, 2014). In this context, it is important to analyse some national statistics in order to understand the role of HPTC as an alternative access to Portuguese HE. The data presented contribute to a better understanding of the trajectories in HE of the students who enrolled in and completed the HPTC, by analysing the variables academic years, areas of education and training, gender, and the status of the graduates after the conclusion of the courses.

Table 1 shows that although in the academic year 2014/2015, the demand for HPTC (ISCED 5) was practically inexpressive, because it was the first year of operation. In subsequent years, the growth in demand for this type of training level is visible from the 2015/2016 academic year.

Table 2 represents the status of graduates of HPTC courses in the following academic year, by education and training area of the HPTC, between the academic years 2014/2015 and 2019/2020.

The analysis of Table 2 leads to the general conclusion, based on the National Classification of Education and Training Areas (in Portuguese, *Classificação Nacional de Áreas de Educação e Formação*—CNAEF), that the education and training areas with the highest number of graduates, at national level, are "Business, administration, and law", "Engineering, manufacturing, and construction", and "Information and communication technologies". The lowest number of graduates, in all academic years under analysis, occurs in the areas of "Natural Sciences, mathematics and statistics" and "Social sciences, journalism and information." It should be noted that, in relation to the academic year 2015/2016, corresponding to the second year of operation of the HPTC, the areas of education and training were more aggregated, so there was a higher number of graduates in the areas such as "social sciences and law" (72) and "science, mathematics, and informatics" (51) and a lower number in the areas of "agriculture" (3), "services" (4), and "arts and humanities" (5). Regarding the academic year 2019/2020, to highlight that this

**Table 2** Status of graduates from HPTC in the following academic year, by education and training area of the HPTC, between the academic years 2014/15 and 2020/21

| Graduates in HPTC | | | Status of graduates in the following academic year | | |
|---|---|---|---|---|---|
| Academic year | CNAEF area | N.° Graduated | Enrolled in degree | Enrolled in another level of higher education | Not found in higher education |
| 2019/2020 | Agriculture, forestry, fisheries, and veterinary sciences | 361 | 74% | 1% | 25% |
| | Arts and humanities | 438 | 47% | 1% | 52% |
| | Business, administration, and law | 931 | 53% | 0% | 47% |
| | Natural sciences, mathematics, and statistics | 59 | 63% | 0% | 37% |
| | Social sciences, journalism, and information | 24 | 83% | 0% | 17% |
| | Education | 8 | 75% | 0% | 25% |
| | Engineering, manufacturing, and construction | 889 | 55% | 1% | 44% |
| | Health and social protection | 639 | 69% | 0% | 31% |
| | Services | 641 | 55% | 1% | 44% |
| | Information and communication technologies | 801 | 59% | 1% | 40% |
| 2018/2019 | Agriculture, forestry, fisheries, and veterinary sciences | 265 | 69% | 2% | 29% |
| | Arts and humanities | 437 | 43% | 0% | 56% |
| | Business, administration, and law | 824 | 53% | 0% | 46% |
| | Natural sciences, mathematics, and statistics | 80 | 69% | 0% | 31% |
| | Social sciences, journalism, and information | 11 | 73% | 0% | 27% |
| | Engineering, manufacturing, and construction | 808 | 55% | 1% | 45% |
| | Health and social protection | 463 | 64% | 1% | 35% |
| | Services | 615 | 53% | 1% | 46% |
| | Information and communication technologies | 678 | 59% | 0% | 40% |

(continued)

**Table 2** (continued)

| Graduates in HPTC | | | Status of graduates in the following academic year | | |
|---|---|---|---|---|---|
| Academic year | CNAEF area | N.° Graduated | Enrolled in degree | Enrolled in another level of higher education | Not found in higher education |
| 2017/2018 | Agriculture, forestry, fisheries, and veterinary sciences | 285 | 72% | 0% | 28% |
| | Arts and humanities | 375 | 49% | 1% | 50% |
| | Business, administration, and law | 690 | 57% | 0% | 43% |
| | Natural sciences, mathematics, and statistics | 50 | 68% | 0% | 32% |
| | Social sciences, journalism, and information | 14 | 93% | 0% | 7% |
| | Engineering, manufacturing, and construction | 727 | 56% | 1% | 43% |
| | Health and social protection | 475 | 56% | 1% | 43% |
| | Services | 465 | 54% | 1% | 45% |
| | Information and communication technologies | 675 | 62% | 0% | 38% |
| 2016/2017 | Agriculture, forestry, fisheries, and veterinary sciences | 289 | 71% | 1% | 29% |
| | Arts and humanities | 314 | 50% | 1% | 49% |
| | Business, administration, and law | 715 | 55% | 1% | 45% |
| | Natural sciences, mathematics, and statistics | 36 | 81% | 0% | 19% |
| | Social sciences, journalism, and information | 10 | 80% | 0% | 20% |
| | Engineering, manufacturing, and construction | 570 | 61% | 1% | 38% |
| | Health and social protection | 343 | 70% | 0% | 30% |
| | Services | 399 | 60% | 0% | 40% |
| | Information and communication technologies | 537 | 66% | 1% | 33% |

(continued)

**Table 2** (continued)

| Graduates in HPTC | | | Status of graduates in the following academic year | | |
|---|---|---|---|---|---|
| Academic year | CNAEF area | N.° Graduated | Enrolled in degree | Enrolled in another level of higher education | Not found in higher education |
| 2015/2016 | Agriculture | 3 | 33% | 0% | 67% |
| | Arts and humanities | 5 | 20% | 0% | 80% |
| | Social sciences, commerce, and law | 72 | 56% | 1% | 43% |
| | Science, mathematics, and informatics | 51 | 45% | 2% | 53% |
| | Engineering, manufacturing, and construction | 35 | 57% | 0% | 43% |
| | Services | 4 | 50% | 0% | 50% |

Source: DGEEC (2021)

**Table 3** Status of graduates from HPTC in the following academic year, by gender, between the academic years 2014/15 and 2020/21

| Graduates in HPTC | | Status of graduates in the following academic year | | | |
|---|---|---|---|---|---|
| Academic year | Sex | N.° Graduated | Enrolled in degree | Enrolled in another level of higher education | Not found in higher education |
| 2019/2020 | Male | 2701 | 57% | 1% | 42% |
| | Female | 2090 | 59% | 1% | 40% |
| 2018/2019 | Male | 2410 | 56% | 1% | 43% |
| | Female | 1771 | 56% | 1% | 44% |
| 2017/2018 | Male | 2176 | 59% | 1% | 41% |
| | Female | 1580 | 57% | 0% | 43% |
| 2016/2017 | Male | 1839 | 61% | 1% | 39% |
| | Female | 1374 | 63% | 0% | 37% |
| 2015/2016 | Male | 119 | 51% | 2% | 47% |
| | Female | 51 | 51% | 0% | 49% |

Source: DGEEC (2021)

was the first year with HPTC graduates, in the area of "education". Still at the level of CNAEF areas, it should be noted that the highest percentage of students registered in terms of continuing their studies, in the academic year 2019/2020, was in the areas of "social sciences, journalism, and information" (83%), "education" (75%), and "agriculture, forestry, fisheries, and veterinary sciences" (75%). In general, the previous academic years also recorded the same high percentages, in the above-mentioned areas.

**Table 4** Europe 2020 indicators - Portugal

| Europe 2020 indicators | | | | |
|---|---|---|---|---|
| | | Annual value | | |
| | | 2011 | 2019 | 2020 |
| Portugal | Goal 2020 | % | | |
| Employment rate, age group 20–64 | 75% | 68.8 | 76.1 | 74.7 |
| Early leavers' rate of education and training | 10% | 23.0 | 10.6 | 8.9 |
| Higher education rate | 40% | 26.7 | 36.2 | 39.6 |

Source: INE, Employment Survey—fourth quarter 2020 (2021)

Throughout the academic years under analysis, it is possible to verify the existence of a higher number of male CPETH graduates (Table 3).

Despite this, the percentage of those who continue their studies does not present major differences when analysed by gender, and this is a trend that has been maintained in recent years.

Finally, it is also important to analyse compliance with the Europe 2020 Strategy indicators. Under the Europe 2020 strategy, five headline targets have been set for employment, innovation, climate/energy, education, and social inclusion. These have been associated with measurable targets that reflect the diversity of situations between Member States, and progress is monitored through concrete and comparable indicators. The targets for employment (increasing the employment rate to 75% in the 20–64 age group) and for education (reducing the rate of early leavers from education and training to less than 10% and increasing to at least 40% the proportion of the population aged 30–34-year-olds with higher education) are monitored through indicators from the Employment Survey (INE, 2021).

Table 4 presents the three Europe 2020 indicators analysed for Portugal. The analysis of the progress indicators shows that, by 2019, Portugal had exceeded the target of an employment rate of at least 75% (it was 76.1%) for 20–64-year-olds. However, in 2020, because of the impact of the health crisis on the labour market, that employment rate decreased by 1.4 percentage points (p.p.) relative to 2019, to a total of 74.7%, failing to meet the target by 0.3 p.p. In 2011, Portugal was 6.2 p.p. away from that value (INE, 2021).

Regarding the objectives in education, Portugal has achieved one of the intended values and approached another. By 2020, the rate of early leavers from education and training was estimated at 8.9%, thus remaining below the maximum accepted value for this indicator (10%) and thus meeting the target with a margin of 1.1 p.p. a margin of 1.1 p.p. This value reveals a decrease of 1.7 p.p. in relation to 2019. In 2011, Portugal was 13.0 p.p. off target. In turn, the rate of enrolment in HE was estimated at 39.6%, 0.4 p.p. away from the target for 2020, which had been set at 40%. This estimate corresponds to an increase of 3.4 p.p. from the value observed in 2019.

## 3 Methodology

Methodologically, this research relies on two different approaches. The first approach takes the form of a literature review, based on the legal regime of the *Higher Professional Technical Courses (HPTC)*, considering that, according to Decree-Law n° 43/2014 of March 18, the XIX Government's Programme promoted the HE policy to better adapt the educational supply both to the diversity of demand and to the needs of the country for qualified staff; with a view to achieving the goals of the Europe 2020 Strategy, the percentage of young Europeans with a higher education diploma should reach 40%. In this sense, national data are collected on statistical studies at the level of HPTC, focusing mainly on the continuation of these studies that concluded the courses. The investigated data will also allow us to understand whether the goals outlined by the Europe 2020 Strategy were carried out by Portugal. The second approach takes the form of a descriptive statistical analysis, using a questionnaire and the case study method in a Portuguese public polytechnic higher education institution (Yin, 2018), based on the information provided by this institution, which shows the financed programmes and HPTC in general and the situation regarding the status of graduate students, in the 4 weeks, 6 months, and at the moment (2021) after the conclusion of the respective courses under analysis.

## 4 Case Study: Higher Professional Technical Courses of Polytechnic Institute of Guara

The Polytechnic Institute of Guarda (in Portuguese, *Instituto Politécnico da Guarda—IPG*), the entity under analysis in this investigation, stands out for being a polytechnic higher education institution under public law, consisting of four schools: School of Education, Communication and Sport (in Portuguese, *Escola Superior de Educação, Comunicação e Desporto—ESECD*), School of Technology and Management (in Portuguese, *Escola Superior de Tecnologia e Gestão—ESTG*), School of Health (in Portuguese, *Escola Superior de Saúde—ESS*), and School of Tourism and Hospitality (in Portuguese, *Escola Superior de Turismo e Hotelaria—ESTH*)—directly responsible for the development of training activities—and an organic training, research, and development unit (in Portuguese, *Unidade de Investigação para o Desenvolvimento do Interior—UDI*), which is responsible for the institution's research and development (R&D), knowledge transfer projects, and applications for national and community programmes (Rodrigues et al., 2020).

## 4.1 *Data Analysis*

Students enrolled in HPTC in public HEIs are considered within the framework of the application of the funding rules of these institutions. The information was collected to carry out the monitoring indicators to report to the European Commission, intrinsic to the projects applied for funding by the operational programmes:

1. *Achievement indicator:* characterizes the participant at the date of entry into the operation
2. *Immediate result indicator:* reports to the participant's status (at the level of qualifications or employment), 4 weeks after participation
3. *Long-term result indicator:* reports to the participant's status (at the level of qualifications or employment), 6 months after participation

### 4.1.1 Population

The population of this study was the students and graduates of 13 HPTC, which are financed by European funds, from IPG, who attended and/or completed their training between the academic years 2014/2016 to 2018/2020. From the first phase of collecting information, a population of 729 individuals was found who integrate and / or integrated 13 HPTC from three IPG schools (Table 5). The data allow us to conclude that the number of students decreased from 2015 (183) to 2016 (170) and 2017 (164), although it increased in 2019 (197). Regarding the student status in the course, it was observed that 323 students completed the HPTC and 209 decided by suspension and/or cancellation of registration (Table 5).

### 4.1.2 Sample

The sample consists of 202 students from 13 HPTC in IPG, labelled as (1) automobile industry; (2) cloud, network, and data centre infrastructures; (3) communication, protocol, and event organization; (4) computer applications development; (5) cooking and food production; (6) cybersecurity; (7) electromechanical industrial maintenance; (8) gerontology; (9) international trade and management; (10) mountain sports; (11) renewable energies and energy efficiency; (12) software testing; and (13) sound and image reporter. Therefore, it corresponds to a non-probabilistic sample with a sampling error of 6%.

In terms of participation the population was contacted, by e-mail and telephone, to take part in the study and to get informed of its goals. A voluntary participation of 202 students was obtained, representing a response rate of 27.7% (Table 6).

It should be noted that the response rate was influenced mainly by the existence of out-of-date telephone contacts and e-mail addresses; the students' lack of interest in participating, which may be related to the pandemic situation (COVID-19) and the period under analysis; as well as the extension of the questionnaire, but, in this case,

**Table 5** Distribution by operational programme, HPTC, school, year, and student status

| Operational programmes | HPTC | School | Year | Total number of students | State | | | | |
|---|---|---|---|---|---|---|---|---|---|
| | | | | | Completed | Cancellation of Registration/ Enrollment | Unregistered | Registered | Awaiting Payment |
| POCH-02-5368-FSE-000007 | Cloud, network, and data centre infrastructures | ESTG | 2014 | 15 | 9 | 5 | 1 | 0 | 0 |
| | Sound and image reporter | ESECD | 2015 | 13 | 8 | 4 | 1 | 0 | 0 |
| | Communication, proto-col, and event organization | ESECD | | 19 | 12 | 5 | 2 | 0 | 0 |
| | International trade and management | ESTG | | 24 | 13 | 6 | 5 | 0 | 0 |
| | Computer applications development | ESTG | | 31 | 12 | 15 | 4 | 0 | 0 |
| | Mountain sports | ESECD | | 26 | 13 | 10 | 3 | 0 | 0 |
| | Gerontology | ESECD | | 31 | 21 | 7 | 2 | 1 | 0 |
| CENTRO-03-5368-FSE-000002 | Cooking and food production | ESTH | | 23 | 9 | 7 | 4 | 3 | 0 |
| | Renewable energies and energy efficiency | ESTG | | 16 | 6 | 10 | 0 | 0 | 0 |
| POCH-02-5368-FSE-000035 | Computer applications development | ESTG | 2016 | 28 | 7 | 13 | 6 | 2 | 0 |
| | Gerontology | ESECD | | 35 | 20 | 10 | 2 | 3 | 0 |
| | International trade and management | ESTG | | 22 | 10 | 4 | 4 | 4 | 0 |
| | Communication, proto-col, and event organization | ESECD | | 27 | 12 | 8 | 1 | 6 | 0 |

(continued)

**Table 5** (continued)

| Operational programmes | HPTC | School | Year | Total number of students | State | | | | |
|---|---|---|---|---|---|---|---|---|---|
| | | | | | Completed | Cancellation of Registration/ Enrollment | Unregistered | Registered | Awaiting Payment |
| | Electromechanical industrial maintenance | ESTG | | 18 | 7 | 5 | 4 | 2 | 0 |
| | Software testing | ESTG | | 19 | 14 | 2 | 3 | 0 | 0 |
| CENTRO-03-5368-FSE-000007 | Cooking and food production | ESTH | | 21 | 9 | 3 | 4 | 5 | 0 |
| POCH-02-5368-FSE-000059 | Gerontology | ESECD | 2017 | 29 | 19 | 3 | 0 | 7 | 0 |
| | International trade and management | ESTG | | 23 | 5 | 6 | 2 | 11 | 0 |
| | Communication, protocol, and event organization | ESECD | | 18 | 4 | 6 | 0 | 6 | 2 |
| | Sound and image reporter | ESECD | | 16 | 11 | 0 | 1 | 2 | 2 |
| | Mountain sports | ESECD | | 20 | 13 | 5 | 0 | 2 | 0 |
| CENTRO-03-5368-FSE-000013 | Cooking and food production | ESTH | | 20 | 12 | 6 | 0 | 2 | 0 |
| | Computer applications development | ESTG | | 24 | 8 | 8 | 2 | 6 | 0 |
| | Cybersecurity | ESTG | | 14 | 6 | 4 | 0 | 4 | 0 |
| CENTRO-03-5368-FSE-000026 | Cybersecurity | ESTG | 2018 | 28 | 13 | 5 | 2 | 7 | 0 |
| | Cooking and food production | ESTH | | 20 | 11 | 6 | 0 | 3 | 0 |
| | Mountain sports | ESECD | | 22 | 12 | 2 | 0 | 8 | 0 |

| | | | | | | | | | |
|---|---|---|---|---|---|---|---|---|---|
| | Renewable energies and energy efficiency | ESTG | | 30 | 0 | 10 | 6 | 14 | 0 |
| | International trade and management | ESTG | | 31 | 4 | 10 | 2 | 15 | 0 |
| | Automobile industry | ESTG | | 24 | 9 | 11 | 1 | 3 | 0 |
| | Sound and image reporter | ESECD | | 21 | 7 | 6 | 1 | 7 | 0 |
| | Software testing | ESTG | | 21 | 7 | 7 | 0 | 7 | 0 |
| *Total* | | | | 729 | 323 | 209 | 63 | 130 | 4 |

Source: IPG (2021)

**Table 6** Distribution, by HPTC, of registered and graduate students

| | Population | | | Sample | | | Answer rate (%) | | |
|---|---|---|---|---|---|---|---|---|---|
| HPTC | M | F | MF | M | F | MF | M | F | MF |
| Cybersecurity | 40 | 2 | 42 | 9 | 0 | 9 | 22.5 | 0.0 | 21.4 |
| Communication, protocol, and event organization | 18 | 46 | 64 | 6 | 13 | 19 | 33.3 | 28.3 | 29.7 |
| Cooking and food production | 46 | 38 | 84 | 15 | 9 | 24 | 32.6 | 23.7 | 28.6 |
| Computer applications development | 78 | 5 | 83 | 30 | 0 | 30 | 38.5 | 0.0 | 36.1 |
| Mountain sports | 53 | 15 | 68 | 19 | 3 | 22 | 35.8 | 20.0 | 32.4 |
| Renewable energies and energy efficiency | 36 | 10 | 46 | 7 | 1 | 8 | 19.4 | 10.0 | 17.4 |
| Gerontology | 25 | 70 | 95 | 11 | 21 | 32 | 44.0 | 30.0 | 33.7 |
| International trade and management | 52 | 48 | 100 | 7 | 10 | 17 | 13.5 | 20.8 | 17.0 |
| Automobile industry | 20 | 4 | 24 | 4 | 1 | 5 | 20.0 | 25.0 | 20.8 |
| Cloud, network, and data centre infrastructures | 15 | 0 | 15 | 4 | 0 | 4 | 26.7 | 0.0 | 26.7 |
| Electromechanical industrial maintenance | 17 | 1 | 18 | 3 | 0 | 3 | 17.6 | 0.0 | 16.7 |
| Sound and image reporter | 28 | 22 | 50 | 7 | 8 | 15 | 25.0 | 36.4 | 30.0 |
| Software testing | 28 | 12 | 40 | 13 | 1 | 14 | 46.4 | 8.33 | 35.0 |
| *Total* | 456 | 273 | 729 | 135 | 67 | 202 | 29.6 | 24.5 | 27.7 |

Source: Self-elaboration

it was necessary to have all the information required under the operational programmes. Those factors were identified as important obstacles to be considered for the successful application of the questionnaire and, consequently, of the study. Considering the response rate by sex it can be concluded that it is balanced considering its representativeness in the courses. Furthermore, in the case of the response rate for the 13 courses, it highlights that there were three courses with less than 20%, namely electromechanical industrial maintenance, international trade and management, and renewable energies and energy efficiency (Table 6).

### 4.1.3 Results

About the sociodemographic characteristics of the 202 respondents, the data show that 24.5% of the students that participated in this study are female and 29.6% male (Table 6). Compared with the national data previously analysed (Table 3), the HPTC of the IPG are also mostly attended by male students. Table 7 represents the age of students at the date of entry into HPTC, and it is possible to observe that the 202 students who represent the sample under study have ages ranging from 18 to 53 years old, with the average of 21.9 years old.

To realize the status during the participants, information was organized by HPTC (Table 8). It could be concluded that the majority (164: 81.2%) concluded the course and similar registration, cancellation, and suspension occurred. In addition,

**Table 7** Distribution, by HPTC, of the age of students

| | Age (years) | | | | | | |
|---|---|---|---|---|---|---|---|
| HPTC | Minimum | Percentile 25 | Median | Percentile 75 | Average | Standard Deviation | Maximum |
| Cybersecurity | 18.00 | 19.00 | 20.00 | 21.00 | 20.89 | 4.31 | 32.00 |
| Communication. protocol, and event organization | 18.00 | 19.00 | 19.00 | 21.00 | 21.26 | 5.39 | 36.00 |
| Cooking and food production | 18.00 | 18.00 | 20.00 | 21.50 | 21.00 | 4.32 | 37.00 |
| Computer applications development | 18.00 | 19.00 | 19.50 | 20.00 | 20.57 | 3.99 | 38.00 |
| Mountain sports | 19.00 | 19.00 | 20.00 | 22.00 | 21.73 | 4.06 | 33.00 |
| Renewable energies and energy efficiency | 18.00 | 20.00 | 21.50 | 23.50 | 22.00 | 3.12 | 28.00 |
| Gerontology | 18.00 | 19.00 | 20.00 | 21.00 | 21.34 | 5.37 | 47.00 |
| International trade and management | 18.00 | 19.00 | 19.00 | 20.00 | 19.41 | 1.06 | 21.00 |
| Automobile industry | 20.00 | 22.00 | 22.00 | 39.00 | 29.00 | 10.58 | 42.00 |
| Cloud, network, and data centre infrastructures | 21.00 | 23.00 | 29.00 | 35.00 | 29.00 | 7.30 | 37.00 |
| Electromechanical industrial maintenance | 18.00 | 18.00 | 18.00 | 18.00 | 18.00 | 0.00 | 18.00 |
| Sound and image reporter | 18.00 | 18.00 | 19.00 | 20.00 | 20.93 | 7.81 | 49.00 |
| Software testing | 18.00 | 20.00 | 23.50 | 41.00 | 30.07 | 12.71 | 53.00 |

Source: Self-elaboration

**Table 8** Distribution, by HPTC, of students and graduate's situation

| HPTC | Situation in relation to the course | | | | | % Conclusion |
|---|---|---|---|---|---|---|
| | Concluded | Dropout (Registration Cancellation) | Registered | Registration Suspension | *Total* | |
| Automobile industry | 2 | 2 | 1 | 0 | 5 | 40.0 |
| Cloud, network, and data centre infrastructures | 3 | 0 | 0 | 1 | 4 | 75.0 |
| Communication, protocol, and event organization | 16 | 2 | 0 | 1 | 19 | 84.2 |
| Computer applications development | 22 | 6 | 1 | 1 | 30 | 73.3 |
| Cooking and food production | 17 | 3 | 1 | 3 | 24 | 70.8 |
| Cybersecurity | 5 | 0 | 2 | 2 | 9 | 55.6 |
| Electromechanical industrial maintenance | 3 | 0 | 0 | 0 | 3 | 100.0 |
| Gerontology | 30 | 1 | 0 | 1 | 32 | 93.8 |
| International trade and management | 15 | 1 | 1 | 0 | 17 | 88.2 |
| Mountain sports | 18 | 1 | 3 | 0 | 22 | 81.8 |
| Renewable energies and energy efficiency | 5 | 0 | 2 | 1 | 8 | 62.5 |
| Software testing | 13 | 0 | 1 | 0 | 14 | 92.9 |
| Sound and image reporter | 15 | 0 | 0 | 0 | 15 | 100.0 |
| *Total* | 164 | 16 | 12 | 10 | 202 | |

Source: Self-elaboration

considering the ratio between the number of students who completed and the total per course, values were observed to be higher than 80% in the following courses: sound and image reporter (100%), electromechanical industrial maintenance (100%), gerontology (93.8%), software testing (92.9%), international trade and management (88.2%), and communication, protocol, and event organization (84.2%), and only automobile industry course (40%) had a value less than 50%. Regarding the CNAEF areas, those with the highest percentage of completion in the analysed period were arts and humanities, engineering, manufacturing and construction, health and social protection, and social sciences, commerce, and law.

Based on the requirements inherent to each of the operational programmes under analysis (Human Capital Operational Program, in Portuguese *Programa Operacional Capital Humano*—PO CH and Central Region Operational Program, in Portuguese *Programa Operacional da Região Centro*—CENTRO), it is highlighted that only the students who do not hold a degree of academic qualification in higher education (ISCED level 5–6) are eligible for financing, since students who are already holders of this level of education (ISCED level 5–6) have already contributed, in this way, to the national goals of higher education graduates or equivalent, expected in the National Reform Program, therefore not showing a second participation in this offer supported by PO CH and CENTRO. In Table 9, it is possible to observe that most of the students in the present sample ($n = 115$) and ($n = 50$) entered the HPTC through level 4 (post-secondary non-higher education) and level 3 (secondary education), respectively. In this way, the authors confirm that there are 16 students with a HE level.

Considering the main objective of the HPTC, it was intended to understand the path taken by the participants in terms of their employment status at the entrance of the course and after its conclusion, considering 4 weeks and 6 months after finishing it (Graph 1). It should be noted that the following analysis after the end of the course considered the total of 164 participants who completed the course, for 4 weeks after finishing the course, and the 121 who completed and responded, for 6 months after finishing the course. At the beginning of the HPTC, most participants (166: 82.2%) were inactive; they were attending education, training, or in other situations. These values have decreased when the same analysis is done for 4 weeks (81: 49.4%) and 6 months (82: 67.8%) after completing the course. The number of employees, including self-employed and employee internal/external, increased from the entrance (16: 7.9%) to 4 weeks (46: 28.0%) and 6 months (18: 14.9%) after completing the course. Regarding the first job, it was observed that 10 (6.1%) participants obtained it 4 weeks after completing the course and 6 (5.0%) only achieved it after 6 months. Given the distribution of the unemployed in the sample, there was a decrease from 4 weeks (27: 16.5%) to 6 months (15: 12.4%) after completing the course.

Finally, the authors considered it equally pertinent to analyse the current professional situation of the students under analysis and, in this sense, the authors easily perceive, through the Graph 2.

In Graph 2, most students ($n = 79$) are studying. However, it is also worth noting the existence of student workers ($n = 10$), as well as the increase in the number of

**Table 9** Distribution, by HETC, of the academic qualifications of students

| HPTC | Academic qualifications | | | | | | | | |
|---|---|---|---|---|---|---|---|---|---|
| | Level 2 Basic Education | Level 3 Secondary Education | Level 4 Post-secondary non-higher education | Level 5 Short-term higher education (ex: *CET—Cursos de Especialização Tecnológica*) | Level 5 Short-term higher education (ex: *TeSP—Cursos Técnicos Superiores Profissionais*) | Level 6 Pre-Bologna degree | Nível 6 Graduation | Nível 7 Master's Degree | *Total* |
| Cybersecurity | 0 | 3 | 6 | 0 | 0 | 0 | 0 | 0 | 9 |
| Communication, protocol, and event organization | 3 | 4 | 11 | 1 | 0 | 0 | 0 | 0 | 19 |
| Cooking and food production | 1 | 0 | 21 | 0 | 0 | 0 | 1 | 1 | 24 |
| Computer applications development | 2 | 11 | 15 | 0 | 0 | 0 | 1 | 1 | 30 |
| Mountain sports | 3 | 5 | 14 | 0 | 0 | 0 | 0 | 0 | 22 |
| Renewable energies and energy efficiency | 2 | 4 | 2 | 0 | 0 | 0 | 0 | 0 | 8 |
| Gerontology | 7 | 9 | 15 | 0 | 0 | 0 | 1 | 0 | 32 |
| International trade and management | 2 | 5 | 9 | 0 | 1 | 0 | 0 | 0 | 17 |
| Automobile industry | 0 | 2 | 2 | 0 | 0 | 0 | 1 | 0 | 5 |
| Cloud, network, and data centre infrastructures | 0 | 2 | 0 | 1 | 0 | 1 | 0 | 0 | 4 |

| | | | | | | | | | |
|---|---|---|---|---|---|---|---|---|---|
| Electromechanical industrial maintenance | 0 | 1 | 2 | 0 | 0 | 0 | 0 | 0 | 3 |
| Sound and image reporter | 1 | 2 | 12 | 0 | 0 | 0 | 0 | 0 | 15 |
| Software testing | 0 | 2 | 6 | 1 | 0 | 2 | 2 | 1 | 14 |
| *Total* | 21 | 50 | 115 | 3 | 1 | 3 | 6 | 3 | 202 |

Source: Self-elaboration

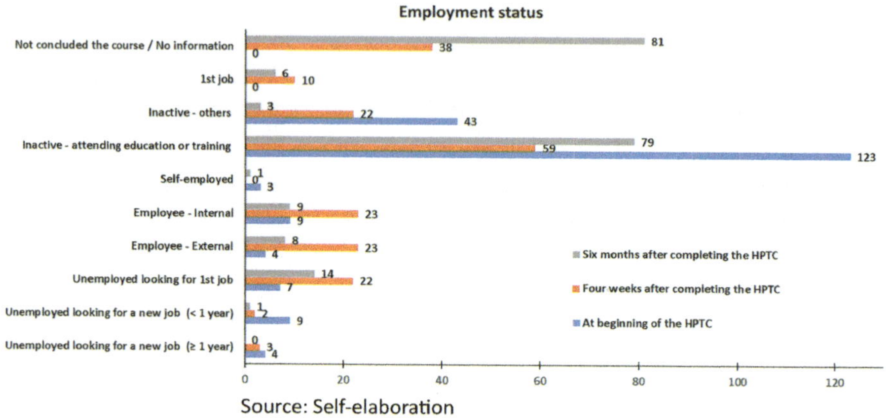


**Graph 1** Distribution, by HPTC, of employment status. Source: Self-elaboration

**Graph 2** Distribution by current professional status of HPTC students. Source: Self-elaboration

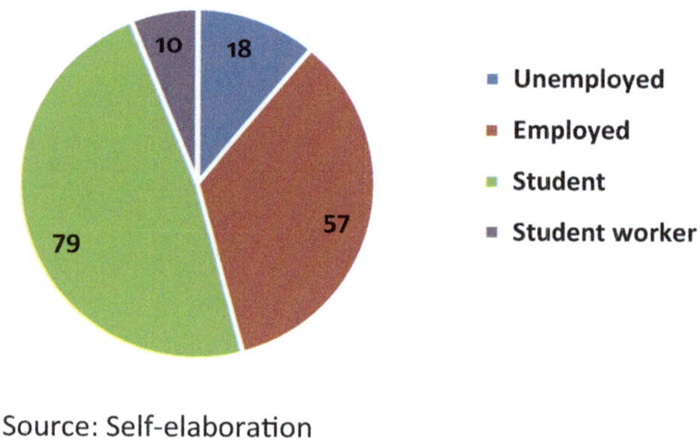


student employees ($n = 57$). Unfortunately, there is still a small proportion of students who are still unemployed ($n = 18$).

## 5 Discussion

The research discussion, in the last few years, shows that IPG increases the number of students in HPTC as a result of the growth of HPTC and evolution of the IPG training offer since 2014, in which the evolution of the number of HPTC is the main change in the last 2 years; HPTC diversifying its training action and collaboration with entities in the region, qualifying human resources in strategic areas and allowing their continuity for higher education level; and the increase of young adults who look for enriching their already acquired knowledge, improve professional performance, seek to increase higher qualification (in this case for student workers), and look for new job opportunities, in some cases in emerging areas, for students who are unemployed.

Although, and to respond to our research question: "*Are the HPTC courses responding to the objectives for which they were designed?*"—the data show that most students tend to continue their higher education studies, in the 4 weeks and 6 months after the completion of their training, this option is valid but only in the medium to long term, as they are increasing their knowledge and qualifications. The HE rate was very closed to reaching the target for 2020 (40%) as we could see in the analysis of Table 3. However, based on the questionnaire given to the students, the perception is that there is a lot of difficulty in getting a job after the conclusion. HPTC students are required, by law, to undertake internships in companies, specifically 750 h of on-the-job training. The feedback from the students surveyed is that companies in the region end up not hiring these students, which forces them to look for work in another region of the country and/or, given the difficulty of finding one, continue their studies to obtain a higher qualification. This makes the needs of companies persist and forces the purpose of these courses to be reformulated in terms of "responding to the needs of the region" and the supposed joint work that HPTC should develop with them, not only in the context of internship, but also to capture the human potential and young population.

Another aspect that makes us question whether the HPTC answer to needs of regions where they are operating comes from a news item from *Rádio Renascença*, from November 2021 (Radio Renascença, 2021), which alerts to the fact that "Several business associations in the interior of the country warn of the number of companies that cannot attract workers. There are many vacancies to be filled and there are more and more. In some cases, it is already holding back investments, preventing business expansion and causing a slippage in deliveries of orders and works". *Rádio Renascença* contacted almost a dozen business associations in the interior and the answer is always the same: "there is a lack of people to work." This problem, which was previously confined to a few sectors of activity, is now transversal, having worsened with the pandemic crisis. This is a problem that does not only depend on the HEIs but also on the government, since, according to Kosharskaya et al. (2021), "Institutional attitudes and public policies play a significant role in young people integration into the labor market."

These new cycles of non-degree higher education were created to increase the number of citizens with higher qualifications (OECD, 2012, 2015). Furthermore, they will meet the needs of the economy and regions in which they will be taught, aiming equally to attract new audiences for higher education, not only young people but also adults and young people coming from secondary vocational education.

## 6 Conclusion

In a society in which the capacity of human resources is one of the greatest riches, if not the decisive factor for their development, the way each one behaves and acts within the community becomes a determining element for individual achievement and for successful citizen participation in collective life. The importance and relevance of HE is attached to post-secondary training that allows students to acquire a solid and broad-spectrum training. It is relevant that universities and polytechnic institutes provide these broad-spectrum training courses, but at the same time develop courses with thinner bands and with a great relationship with the realities that are experienced in companies and in the world of the economy. It is in this context, as Grilo (2021) mentioned, also with this study it can be concluded that higher education institutions must have a remarkably close relationship with the business community, so that the training to be acquired is in line with the needs of employers, namely companies, which is where wealth is produced and the development of a country is promoted.

Despite the considerable development of research in the fields of business ethics and corporate governance, it should be noted that, to date, a comprehensive review and integration of these areas have not yet been carried out at the level of HEI. HEIs stand out for their predominant role as elite centres, responsible for sharing their mission between teaching and research, which evolves from an essentially cultural matrix, passing through a definition associated with the provision of services, as well as the link established between HEIs and society, through the transfer of knowledge and its role as a stakeholder. Thus, this chapter aims to contribute to calling for the need to focus future investigations on business ethics and corporate governance at the level of HEIs in general and, in this specific case, of Portuguese HEI.

Regarding the analysed case study, it is easy to observe the amount of training available in the IPG, but it is questioned to what extent the vast training offer corresponds to the needs of the labour market and to the region where the institution is located. This reality is transversal to all Portuguese polytechnic higher education institutions. So, based on the results, the need for a restructuring of these short-term courses stands out. They have a general and scientific component, a technical training component, and a training component in the work context (which takes place through an internship); they function in accordance with the objectives for which they are created. It is also important to emphasize that the training offer of a polytechnic HEI is not a static, immutable reality; it must not only accompany the evolution of the socio-economic fabric but also be an agent of social and economic transformation.

Finally, it can be concluded that one of the most important results is the incentive to the academic community, especially the Portuguese polytechnic HEIs, to promote studies and annual reports on this theme. Another result is that these cycles improve professional training, both because of the diversity of labour demand in Portuguese companies and because of the supply of skilled professionals. Other result is that these courses aim to substantially increase until 2030 the number of young people and adults with relevant qualifications, including technical and professional skills, for jobs, decent work, and entrepreneurship.

As limitations of the investigation, we highlight the fact that, at the national level, it was not possible to establish a comparison between these courses and their employability, since the statistical data available are aggregated by the level of education (higher education). It is urged that more studies be carried out on these courses and that there be the possibility of disseminating them, so that it is possible to analyse the employability rate, among other variables.

The analysis of Tables 2 and 3 highlights this problem, with the existence of the category "not found in HE", revealing that it is not sufficiently clarifying the situation of graduates who are not studying, as it does not become noticeable nor is there any indication that these students are employed or not. Even more so when in the analysed decree-law, in its article 40°, the obligation of HEIs to ensure the collection of information on the professional path of their graduates and the dissemination of summary information about it is highlighted.

**Acknowledgments** The authors wish to thank Professor David Crowther, Jose Angel Perez Lopez, and two anonymous referees for their helpful comments on earlier versions of this paper. Also, the organizing committee (Professor Irene Arraiano, in particular) and scientific committee (Samuel Olu Idowu, Réne Schmidpeter and Nicholas Capaldi of Global Corporate Governance Institute). The second author is financed by national funds through FCT - Foundation for Science and Technology, I.P., within the scope of multi-annual funding UIDB/04043/2020.

## References

Direção-Geral de Estatísticas da Educação e Ciência (DGEEC). (2021). Portal da Direção-Geral de Estatísticas da Educação e Ciência. Portugal. Available at: https://www.dgeec.mec.pt/np4/18/

Elena, S., & Sánchez, P. M. (2013). Autonomy and governance models: Emerging paradoxes in Spanish universities. *Perspectives: Policy and Practice in Higher Education, 17*(2), 48–56.

European Commission. (EC, 2010). *Europe 2020: A strategy for smart, sustainable and inclusive growth. Communication from the Commission*. Publications Office of the European Union.

European Commission. (2012). Communication from the Commission to the European Parliament, the Council, the European Economic and Social Committee and the Committee of Regions. Rethinking education: Investing in skills for better socio-economic outcomes.

European Commission, Directorate-General for Employment, Social Affairs and Inclusion. (EC, 2011). *An agenda for new skills and jobs: A European contribution towards full employment: Communication from the commission to the European parliament, the council, the European economic and social committee and the committee of the regions*. Publications Office. https://data.europa.eu/doi/10.2767/28479

European Commission, Directorate-General for Employment, Social Affairs and Inclusion. (EC, 2016). *Study on higher vocational education and training in the EU*. Publications Office of the European Union.

Grilo, E. M. (2021). *Que escola queremos?* Cadernos de Economia (134), 74. Polimeios-Produção de Meios, Lda.

Hasanefendic, S., Heitor, M., & Horta, H. (2016). Training students for new jobs: The role of technical and vocational higher education and implications for science policy in Portugal. *Technological Forecasting and Social Change, 113*(B), 328–340.

Humburg, M., Velden, R., & Verhagen, A. (2013). *The employability of higher education graduates: The employers' perspective*. Publications Office of the European Union.
Instituto Nacional de Estatística (INE). (2021). Estatísticas do Emprego: 4° Trimestre de 2020. Available at: https://www.ine.pt/xportal/xmain?xpid=INE&xpgid=ine_destaques&DESTAQUESdest_boui=415270523&DESTAQUESmodo=2&xlang=pt
Instituto Politécnico da Guarda (IPG). (2021). *Sistema de informação dos Cursos do IPG*. IPG.
International Labour Organization (ILO). (2019). *Decent work country programmes*. International Labour Office. Available online: https://www.ilo.org/global/about-the-ilo/how-the-ilo-works/departments-and-offices/program/dwcp/lang%2D%2Den/index.htm.
Jin, X., Li, T., Meirink, J., van der Want, A., & Admiraal, W. (2019). Learning from novice–expert interaction in teachers' continuing professional development. *Professional Development in Education, 47*(5), 745–762. https://doi.org/10.1080/19415257.2019.1651752
Joensen, H., & Nielsen, H. (2006). *Is there a causa effect of high school math on labor market outcomes?* IZA Discussion Papers, No. 2357, Institute for the Study of Labor (IZA).
Kosharskaya, E., Lapshova, E., & Menshenina, S. (2021). Higher professional education and vocational training based on dual education model: A stable factor in differentiating young people position in the labor market. In proceedings of the international scientific and practical conference on sustainable development of regional infrastructure (ISSDRI, 2021). https://doi.org/10.5220/0010596607240729
Lourenço, R. T. (2017). *Governação e Desempenho Organizacional nas Instituições de Ensino Superior Públicas Portuguesas*. Universidade de Coimbra.
Ministério da Educação e Ciência. (MEC, 2014). [Decree-Law N° 43/2014]. Decreto-Lei n.° 43/2014, cria os cursos técnicos superiores profissionais, como formação superior de curta duração não conferente de grau. *Diário da República, I Série, 18*(03), 2074–2081.
OECD (2012). OECD reviews of vocational education and training: A skills beyond school review of Denmark. Available at: http://dx.doi/org/10.1787/9789264173668-en
OECD (2015). ISCED 2011 Level 5: Short-cycle tertiary education. In *ISCED 2011 Operational Manual: Guidelines for Classifying National Education Programmes and Related Qualifications*. OECD Publishing. Available at: https://doi.org/10.1787/9789264228368-10-en
OECD. (2020). Education policy outlook: Portugal. Available at: https://www.oecd.org/education/policy-outlook/country-profile-Portugal-2020.pdf
Oman, C., Fries, S., & Buiter, W. (2003). *La gouvernance d'entreprise dans les pays en développement, en transition et les économies émergentes*. OECD, Center de Développment.
Radio Renascença. (2021). Precisamos de pessoas no interior que agoniza com a desertificação de mão-de-obra. Available at: https://rr.sapo.pt/especial/economia/2021/11/10/precisamos-de-pessoas-interior-agoniza-com-desertificacao-de-mao-de-obra/260196/
Reinhardt, W., Schmidt, B., Sloep, P., & Drachsler, H. (2011). Knowledge worker roles and actions—Results of two empirical studies. *Knowledge and Process Management, 18*(3), 150–174.
Rodrigues, J., Abreu, R., Fonseca, C. (2020). *O Controlo Interno nos Cursos Técnicos Superiores Profissionais.* ICAFI 2020: Proceedings of the 1st International Conference in Accounting and Finance Innovation – Business Innovation and Digital Transformation, Aveiro.
Rytmeiter, C. (2009). Governing university: Perceptions and practice of governance and managements roles. *Tertiary Education and Management, 15*, 137–156.
Schultz, T. W. (1963). *The economic value of education*. Columbia University.
Tijdens, K. G., De Ruijter, J., & De Ruijter, E. (2012). Measuring work activities and skill requirements of occupations: Experiences from a European pilot study with a web survey. *European Journal of Training and Development, 36*(7), 751–763.

UNESCO. (2016a). *Global education monitoring report 2016: Education for people and planet.* UNESCO.

UNESCO. (2016b). *UNESCO technical and vocational education strategy, 2016–21.* UNESCO.

Williams, P. (2014). Rethinking university assessment. *International Journal of Technology and Inclusive Education, 3*(1), 257–264.

Yin, R. (2018). *Case study research: Design and methods.* Sage Publications.

**Joana Sá Rodrigues** has a degree in Human Resources Management from the Polytechnic Institute of Guarda, Portugal. Presently, she is finishing her dissertation of the master's degree of Integrated Management Systems (Environment, Quality, Safety, and Social Responsibility) at the Polytechnic Institute of Guarda. After doing her internship at the Research Unit for the Development of the Interior of the Polytechnic Institute of Guarda, Joana worked as a research fellow between May 2020 and 2021. In addition, Joana is the promoter of the sustainability project "Rumex Induratus: The discovery for New Village", participated as co-promoter of the "Hydroponic Evolution Farm" Project, which was distinguished in the 16th Regional and National Competition of Poliempreende 2019, and collaborated in the Project "UNDERGUARDA" in the 17th Edition of Poliempreende—Projects of Entrepreneurial Vocation. Finally, note that she has also developed articles of scientific research presented and published under the scope of the Digital Conference ICAFI; IV Jornadas de Educação: "Educar para o Futuro" and XXIV SEMEAD—Seminars on Administration of the University of São Paulo, Brazil.

**Rute Abreu** is Professor since 1990 in Accounting and Finance at the Polytechnic Institute of Guarda, Portugal. Rute teaches financial accounting, auditing, quality, environment and safety standards, sustainability, and social responsibility management. Rute has a master's degree in Industrial Engineering from the Universidade Nova de Lisboa, Portugal, and a Ph.D. in Accounting and Finance from the Universidad de Salamanca, Spain. Actually, she is Accounting Degree Director and Director of master's degree in Integrated Management System. Usually, she develops activities with Social Responsibility Research Network and Global Corporate Governance Institute. Rute researches at the Center for Research in Accounting and Taxation of the Polytechnic Institute of Cávado and Ave (CICF-IPCA), Portugal, on CSR, corporate governance, accounting, and firm valuation, and it has been published in several journals. Also, she participates in the blind review process for several journals and helps as Editor on *The Social Responsibility Journal* and *Accounting and Management Review of Portuguese Chartered Accountants.* Regularly, she develops scientific activities related to participation and organization, all over the world, conferences, and meetings. Additionally, Rute promotes research projects related to accounting education, tax citizenship, management quality systems in social economy, and performance analysis at beauty firms that transfer knowledge to society.

**Cecília Fonseca** holds a Ph.D. in Applied Mathematics from the University of Beira Interior. She is a professor at IPG since 1997 at the Math Department and mainly teaches subjects such as data analysis, probabilities, and statistics. She has research and scientific publication related to dependence measures for spatial extreme events and data analysis for health and management areas. She has developed work in the lifelong education of teachers in IPG, since 2008, with an emphasis on training activities. She was part of multidisciplinary teams that carried out work within the scope of IPG's technical-scientific service to the regional community. Regarding participation in ERASMUS mobility programmes, she gave lectures at universities in Hungry, Croatia, and Czech Republic and did an international mobility programme in Belarus.

# A Study on Management Supporting Employment of Disabled in Japan

**Yuichi Goto and Keiko Yokoyama**

## 1 Introduction

In recent years, interest in social enterprises, which are businesses involved in solving social issues in many areas, has been increasing. Social enterprises are business entities that simultaneously pursue objectives of both economic goals necessary to continue the business such as sales and profitability and social goals of resolving social issues. However, tension between economic and social missions tends to occur (Doherty et al., 2014); thus, multifaceted managerial solutions are required so that both these approaches can be pursued simultaneously.

This study comparatively analyzes companies implementing employment transition support programs (i.e., a type of work integration social enterprise [WISE]), which are a welfare service for people with disabilities, to clarify the issues facing employment transition support businesses in Japan and what "sort" of management is being practiced to simultaneously pursue both economic and social goals.

Y. Goto (✉)
Onomichi City University, Hiroshima, Japan
e-mail: gotou@onomichi-u.ac.jp

K. Yokoyama
Kansai University, Osaka, Japan
e-mail: yokokei@kansai-u.ac.jp

 
I. G. Arraiano et al. (eds.), *Corporate Social Responsibility in a Dynamic Global Environment*, CSR, Sustainability, Ethics & Governance,
https://doi.org/10.1007/978-3-031-24647-0_3

## 2 Review of Prior Research

Prior research on social enterprises has been accumulated, particularly in the USA and Europe. However, as noted, there is no clear definition of a social enterprise as of yet, and the definition continues to be debated (Wry & York, 2017). Thus, we cannot say that the research to date has reached any particular consensus.

For example, the research conducted in the USA has focused on discussions about the commercialization of nonprofit organizations (NPOs) and the solving of social issues by corporate-type NPOs (e.g., Dees, 1998). Conversely, the research done in Europe has developed along the lines of discussing how cooperative associations and the like are grappling with the problem of social exclusion (e.g., Borzaga & Defoury, 2001). In the field of management studies in Japan as well, research into social enterprises and social entrepreneurs has been proceeding with social innovation as the keyword (e.g., Suzuki, 2014; Takahashi et al., 2018; Tanimoto, 2006; Tanimoto et al., 2013). However, the studies in Japan cannot be said to have been conducted sufficiently compared with the research done in the USA and Europe on this topic.

With such background, on the assumption that tension exists between the economic and social purposes of social enterprises, studies are being accumulated in the USA and Europe with WISEs as study subjects to look at the optimal management format for pursuing both purposes. In Japan, however, although WISEs have been surveyed and researched in terms of social welfare, social policy, and sociology, there has been almost no exploration in the field of management studies with the exception of Fujii et al. (2013).[1]

Organizing the prior US and European research on WISEs by what they focused on, we came up with the following four elements: (1) the tension between economic purpose and social purpose, (2) the degree of recognition of economic purpose and social purpose within the organization, (3) how management can surmount such tension, and (4) economic outcomes and social outcomes. Prior studies include theoretical and empirical research on one or more of these four elements. Specifically, they are the elucidation of the structure of (1) (e.g., Stevens, Moray, & Bruneel, 2015), quantitative and qualitative studies of the relationship between (2) and (3) (e.g., Battilana et al., 2015), quantitative study of the relationship between (2) and (4) (e.g., Stevens, Moray, Bruneel, & Clarysse, 2015b), and theoretical studies and case studies regarding (1) and (3) (e.g., Battilana & Lee, 2014; Cooney, 2011; Doherty et al., 2014; Pache & Santos, 2013; Smith et al., 2010), among others.[2] These studies have made partial clarification on the relationship between

---

[1]The study by Fujii et al. (2013) is limited to workers' collectives, and it does not discuss management practices.

[2]In addition, some studies examined WISEs or social enterprises from the perspective of institutional logic (e.g., Jay, 2013; Mair et al., 2015; Santos et al., 2015; Smith & Besharov, 2019). In the research from the perspective of institutional logics, research focuses on the sort of management, organizational change, and the relationships between management and outcomes from the perspective of responding to multiple institutional logics.

the degree of recognition of management's economic and social purposes in social enterprises and organizational outcomes, verification that they tend to fall into mission drifts, and identification of ways for management to prevent them.

However, no case study has attempted to analyze all four of these elements. In other words, there has been no adequate identification of specifically how management should be for the simultaneous pursuit of both economic and social purposes at social enterprises.

Therefore, this study makes comparative analysis focusing on all four of these elements in Japanese WISEs, a topic that has never before been explored, and clarifies the current status of managements that are simultaneously pursuing economic and social purposes at Japanese social enterprises and the issues they face and considers options for the future.

# 3 Research Methodology

## 3.1 Research Targets

As mentioned above, the employment transition support business in Japan is relatively new and has not been sufficiently examined from the perspective of management studies. Therefore, through comparative case studies, we would like to clarify the way of management that pursued both economic and social purposes in Japanese companies engaged in the transition support business for disabled persons.

This study compares two for-profit corporations: X and Y (hereinafter, Company X and Company Y), which are involved in the employment transition support business in Japan. Both companies focus primarily on employment transition support programs for people with mental disabilities and developmental disabilities.[3] Table 1 provides an outline of both companies.

We decided to compare Company X and Company Y because we thought that in considering the best management format for simultaneously pursuing both economic and social purposes, a comparison and examination of these companies was appropriate for the following three reasons.

First, both Company X and Company Y started their employment transition support businesses in 2012. As of March 2021, Company X was operating 80 employment transition support offices, while Company Y had 8 offices. While Company X is executing a multi-branch strategy, Company Y is limiting its number of locations. In other words, we can regard Company X and Company Y as having different operating policies.

Second, both companies have above-average employment rates and job retention rates after 1 year of employment; these are outcome indicators in the employment

[3] Although Company X accepts people with intellectual and physical disabilities, 70% of its users are those with developmental and mental disabilities.

**Table 1** Outline of Company X and Company Y

| | Company X | Company Y |
|---|---|---|
| Organizational type | Listed corporation | Unlisted corporation |
| Main businesses | Employment transition support | Employment transition support |
| Establish date | December 2011 | September 2009 |
| Launch date of business | April 2012 | October 2012 |
| Number of offices | 80 | 10 |
| Office management | Direct management | Direct management |

transition support business. In 2016, the national average transition rate to general employment for people of developmentally disabled persons was 25.1%, and also, developmentally disabled persons with a 1-year retention rate of 71.5%.[4] Conversely, in the same year, Company X had an employment rate of 55% for its users,[5] with a 1-year retention rate of 76%. Additionally, Company Y's user employment rate was over 100% in 2016, with a 1-year retention rate of over 90%.[6] Compared with the national average, both companies are achieving relatively high employment rates and 1-year retention rates. Between them, Company Y, particularly, realizes a higher outcome in terms of its social purpose of employment of disabled persons by general corporations, which is the mission of the employment transition support business.

Third, both companies are running their employment transition support businesses as corporations. Among the companies offering employment transition support in Japan, only two are listed on the stock exchange, one of which is Company X. Conversely, although Company Y is a corporation, it is not listed on any stock exchange. Both are corporations, but their business operation policies are considered different in that one is a public company, and the other is not.

Moreover, looking at their economic outcome, as of the end of March 2021, Company X had total sales of 8.17 billion yen and an operating margin of 24.9%. The situation was somewhat different for the other listed company engaged in the same business, which had sales of 16.1 billion yen and an operating margin of 10.5%. Company X's operating margin is thus very high.[7] Meanwhile, although Company Y is not open to the public, thorough investigation, we were able to presume that its outcome is almost the industry average.

Both companies are achieving a certain level of outcome within the employment transition support business. However, they differ in their business-operating format and in the degree of their economic and social accomplishments. Thus, in our

[4] National Institute of Vocational Rehabilitation ed. (2017).

[5] This is calculated based on 20 users per day at each office, excluding offices opened in 2016.

[6] Company Y's one-year retention rate in 2015 was 95%, and its retention rate has been at least 90% every year since then.

[7] The other listed company also conducts programming education and Internet businesses, besides providing employment support for people with disabilities and education for children with developmental disabilities. These other businesses constitute about 12% of the company's sales.

opinion, by performing a comparative analysis of the two companies, through identification of the optimal management format for pursuing economic and social purposes at a corporation and where differences between them occur, we should be able to obtain new knowledge about the type of management that can simultaneously pursue both economic and social purposes.

### *3.2 Analytical Viewpoint and Data Collection*

As mentioned above, to the extent of our knowledge, there are no studies that comprehensively and thoroughly analyze all the following four elements: (1) the tension between economic purpose and social purpose, (2) the degree of recognition of economic purpose and social purpose within the organization, (3) how management can surmount such tension, and (4) economic outcomes and social outcomes.

This study, therefore, comparatively analyzed Company X and Company Y by focusing on the aforementioned four elements.

Incidentally, a number of methods were used in this study to gather the data for the analysis. Specifically, data were collected through conducting semi-structured interviews of the companies' executive teams and middle management and by reviewing the companies' websites, newspaper and magazine articles, and reference materials provided. The interviews were conducted in 2018 and 2019, and each interview lasted approximately 90 min. The data obtained from interviews were transcribed into text data and used for analysis. And also, additional surveys were conducted via email during the analysis.

The analysis of the data was carried out using the following procedure. First, each author carefully read the textual data, focusing on the four elements mentioned above, and extracted the relevant key concepts. Second, authors discussed the interpretation of the text data and the validity of the kye concepts with each other. In case of disagreement, the process of re-reading and discussing the textual data was repeated until a consensus was formed among authors.

## 4 Case Overview

### *4.1 Employment Transition Support in Japan*

In Japan, the employment transition support business is a welfare service for disabled persons aged 18 and over who want to work at general corporations and so on. This type of program was first set up under *Services and Supports for Persons with Disabilities Act* and went into effect in 2006.[8]

[8] *Services and Supports for Persons with Disabilities Act* was revised to become *the Act for Comprehensive Welfare for Persons with Disabilities* in 2012.

Traditionally, the main providers of welfare services to persons with disabilities were organizations such as social welfare corporations, but putting in effect the *Services and Supports for Persons with Disabilities Act* allowed for-profit corporations to provide these services. However, since employment transition support is a type of welfare service, the business is subject to various restrictions, including a general 2-year limit on the total length of time that users can avail themselves of these services, variations in basic remuneration depending on the capacity of daily users per office, and Ministry of Health, Labour and Welfare regulations on pay standards and the like.[9]

Since the 2006 enactment of *Services and Supports for Persons with Disabilities Act,* many organizations have entered the employment transition support business. The number of places of business offering employment transition support rose approximately 5.5 times, from 603 in 2007 to 3301 in 2020. Moreover, as shown in Fig. 1, in 2007, about 75% of the places of business engaged in employment transition support were social welfare corporations, while 3% were for-profit corporations, and 13.6% were specified NPOs.[10] However, by 2017, for-profit corporations had expanded rapidly compared with other organizational types to comprise about 30% of the places of business engaged in employment transition support, while specified NPOs grew somewhat marginally to account for 19%, while the share of this business done by social welfare corporations fell to about 40%. In other words, for-profit corporations have come to assume a key role in providing employment transition support services.

Since the enactment of *Services and Supports for Persons with Disabilities Act*, many types of organizations have started offering employment transition support services. However, not all of these organizations have been successful. The social impact of the employment transition support business can be ascertained by the employment rates and retention rates of disabled persons at general corporations, which are the main objectives of these businesses.

As stated previously, the national average transition rate to general employment for people of developmentally disabled person was 25.1% in 2016 and 26.4% in 2017, which was higher than in the past. However, the percentage of establishments that have a 0% transition rate of people with disabilities to the general workforce is between 30% and about 40% of the total, almost unchanged for nearly 10 years.[11]

---

[9]Basic remuneration in the employment transition support business varies depending on the capacity of daily users. For businesses with 20 or fewer users, the rate is 804 units/day, and for businesses with 21–40 users, the rate is 711 units/day (see Ministry of Health, Labour and Welfare (2015c), Shogaisha no shuro shien ni tsuite [Employment support for persons with disabilities], and Ministry of Health, Labour and Welfare (2018a), Heisei 30 nendo teichaku shien jigyo, hoshu kaitei ni tsuite [FY 2018 revision of retention support businesses and remuneration]).

[10]The percentages by management type are not published after 2018, so the data until 2017 are used Figure 1 is prepared based on Ministry of Health, Labour and Welfare (2008, 2010a, b, 2011, 2012, 2014, 2015a, b, 2016, 2017a, 2018b).

[11]Ministry of Health, Labour and Welfare (2017b), Shuro iko shien ni kakaru hoshu, kijun ni tsuite [Employment transition support remuneration and standards].

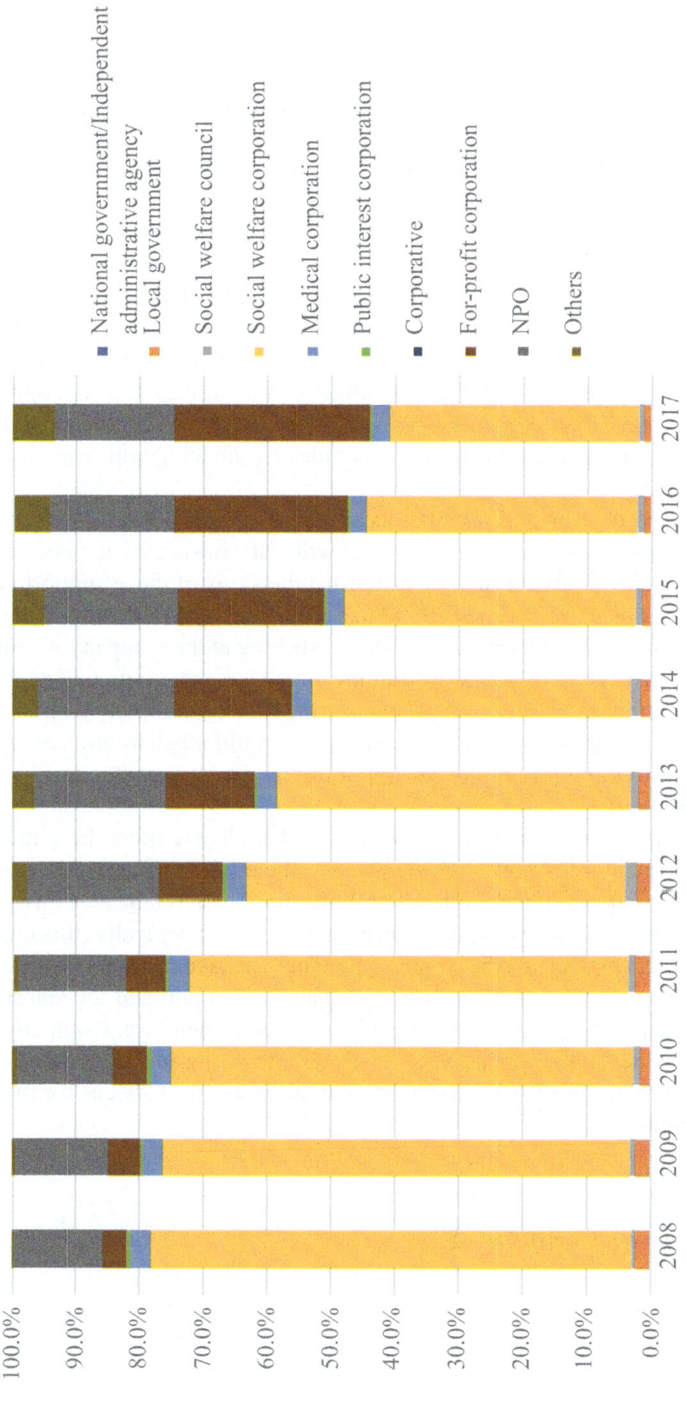


Source: Prepared from Surveys on General Conditions at Social Welfare Facilities, etc. (2007–2017).

**Fig. 1** Changes in the breakdown of employment transition support businesses by management type. Source: Prepared from Surveys on General Conditions at Social Welfare Facilities, etc. (2007–2017)

## 4.2 Summary of Company X

Company X was founded in December 2011 for engaging in employment transition support for disabled persons. Company X launched its employment transition support business in April 2012, and by the end of March 2021, it was operating 80 employment transition support offices. Moreover, the company went public in October 2017.

Company X was established because its founding members were looking for an opportunity to create a new business by zeroing in on the employment transition support business to overcome the issues that had confronted NPOs and social welfare corporations. They proactively adopted a corporate perspective in operating this business and started the business with a view to developing it rapidly, efficiently, and effectively. Although Company X's founding members consisted of people with management experience as entrepreneurs and executives, as well as people who had work experience at companies, they did not include anyone with work experience in the welfare sector.

Company X directly manages all 80 of its employment transition support offices. When the company decided to expand nationwide, it considered it important to standardize the quality of its services and improve the skills of those supporting the employment transition and thus endeavored to develop manuals and other materials. Moreover, to improve the support skills of those working at the company, Company X implemented two types of training: training based on job grade and technical (skill) training. Furthermore, it took steps in ensuring that the company's philosophy was stated during the training so that its employees would all share the company's philosophy and providing all employees opportunities to get together and increase their sense of belonging.

Company X applies three numerical objectives that closely track the remuneration of its employment transition support business: the number of users, the number of people who were successfully employed, and the number of retainers. Conversely, Company X thoroughly cuts its business operations costs. Specifically, this consists of reducing indirect costs such as personnel and administrative costs at headquarters, which do not directly affect user services. Company X has adopted the stance that providing good support will enable it to gain new users. Therefore, Company X is managed in a manner that its employees, instead of focusing on economic aspects such as sales and profits, will strive to ensure that the quality of services for its users does not deteriorate.

## 4.3 Summary of Company Y

Company Y was founded in September 2009, and it launched its employment transition support business in October 2012. As of March 31, 2021, it was operating 10 employment transition support offices, mostly in the Tokyo metropolitan area. Although Company Y is a corporation like Company X, it is not public.

Until shortly before its starting employment transition support business, Company Y was engaged in the cities' model businesses for the support of developmentally disabled.

As Company Y went through the process of implementing its model businesses for the cities, it found out about the employment transition support system, so it then embarked on developing an employment transition support business. Company Y's founder has a child who has been diagnosed with a developmental disability, so the founder became interested in developmental disabilities, and wanting to do something about it, decided to establish this business. According to the founder, at that time, there was no awareness that this business was a social business. Moreover, the founder of Company Y, although had experience working at a company, had no management experience, work experience in the welfare sector, or acquaintances involved in welfare-related endeavors at the time the business was established.

Company Y operates its business with the primary objective of providing good support services that would lead to employment for persons with developmental disabilities. Specifically, the company offers a wide range of elaborately prepared programs, including the in-house development of more than 300 videos that teach computer skills and other things necessary for employment and more than 100 types of job training. Thus, Company Y has achieved high employment rates for its users. However, Company Y has not set any specific numerical objectives for its business. As stated previously, the employment transition support business is subject to many restrictions. Therefore, the founder of Company Y thought that it would be meaningless to set any specific numerical economic outcome objectives. Thus, the company has been managed with an emphasis on getting jobs at general corporations for persons with disabilities, which is the purpose of the employment transition services business.

Like Company X, Company Y has various programs, such as off-site meetings and company-wide social gatherings, to promote a shared awareness among its employees. In recent years, Company Y began trying to reorganize itself as the scale of its operations grew. However, the founder is placing more emphasis on helping disabled persons, stating that it is more fun to be on-site to support than to build management structures.

## 5 Comparative Analysis and Results

In this study, we conducted a comparative analysis of Company X and Company Y, taking into account all of the above factors (1) through (4). As a result, six elements that should be focused on social enterprises research, especially WISEs, were clarified. Six elements are (1) industry characteristics, (2) entrepreneur team, (3) value pursued, (4) strategy, (5) organizational management, and (6) economic and social outcomes which are drawn in Fig. 2. Next, we follow the framework in Fig. 2 to trace the interpretations of the case studies that led to the creation of this figure.

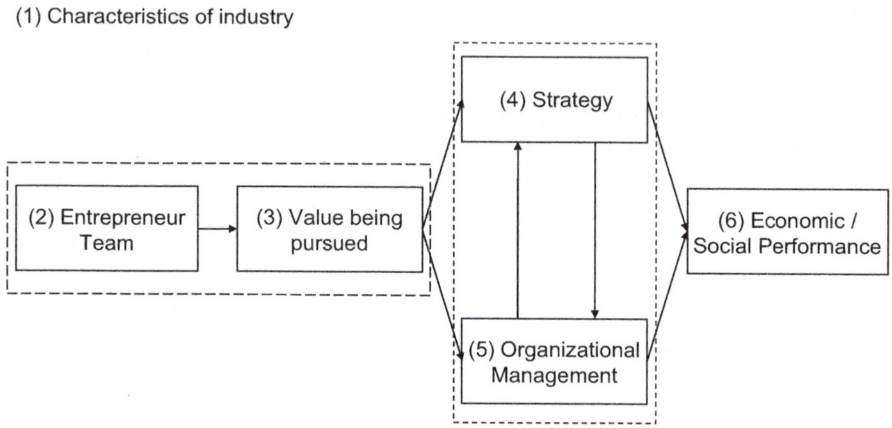


**Fig. 2** Six elements and relationships

## 5.1 Characteristics of Industry

Two points can be mentioned as industry characteristics of the employment transition support business: (a) growth in the size of the market and (b) the presence of structural tension between economic and social purposes.

(a) Growth in the size of the market

In Japan, the statutory employment rate of persons with disabilities is gradually being raised,[12] so that hiring of disabled persons is a managerial consideration for many organizations, including for-profit corporations and the national and local governments. Additionally, the number of people using employment transition support services is increasing, with the base at 100 as of April 2008: 486 for persons with mental disabilities, 160 for persons with physical disabilities, and 139 for persons with intellectual disabilities, all at February 2015.[13] As the hiring of persons with disabilities is accelerating throughout Japan and more disabled persons are seeking employment at general corporations and so on, which means that the market for the employment transition support business is expanding.

[12]The statutory employment rate of persons with disabilities by for-profit corporation is being increased incrementally, from 1.8% in 2008 to 2.0% in 2013 and then to 2.2% in 2018. Furthermore, starting in 2018, this rate is being set at a higher level for governmental bodies, with national and local governments at 2.5% and prefectural boards of education at 2.4%.

[13]Ministry of Health, Labour and Welfare (2015c), "Shogaisha no shuro shien ni tsuite" [Employment support for persons with disabilities]

(b) The presence of structural tension between the economic and social purposes

The employment transition support business is a so-called benefit provision business, in that basic remuneration varies according to the capacity of daily users and the remuneration units, the remuneration standards, and so on stipulated by the Ministry of Health, Labour and Welfare. In other words, there is a maximum amount of sales per office, so continuously securing the users will bolster sales and profit will be determined by the level of costs. Therefore, the economic outcome (sales and profitability) of an employment transition support business depends both on its ability to attract and maintain users until the limit is reached (sales) and on its costs (profitability). Therefore, getting users translates into getting sales, which can easily lead to the moral hazard of cutting some corners on providing services to disabled persons to increase the company's profitability.

Meanwhile, the social outcome of the employment transition support business can be ascertained by the employment rates and retention rates of disabled persons at general corporations and other employers, which are the main purposes of this business. As stated above, although the employment rate of people using employment transition support services has increased, many offices still do not achieve disabled persons getting job. That is, many offices are receiving remuneration but are not generating social outcomes.

From the preceding, we see that Japan's employment transition support sector is characterized by an increasing number of users and a growing market size. However, since it is systemically difficult to increase sales and as profits are determined by the cost of the services being provided, increasing profitability requires cutting costs that could end up sacrificing the quality of the services being provided. This results in a trade-off between profits (i.e., economic outcome) and service quality (i.e., social outcome), so that a structural tension between economic and social purposes exists in this industry.

## 5.2 The Entrepreneur Team and the Values Being Pursued

Both Company X and Company Y were founded by people with no practical experience in establishing a company or working in the welfare sector. While employment transition support is now the main business line of both companies, the way that their business was founded was very different.

Company X's founders consisted of people with corporate experience as founders or management or with work experience at companies in industries outside the welfare sector. That is, they had in-depth corporate management experience and understood human resources requirements from the corporate perspective. These people believed that they could adopt a corporate approach while offering employment transition support by employing people with practical experience in the welfare sector to look at companies' human resources needs from the perspective of a person with a disability. Moreover, they regarded the fact that few of these businesses existed as a business opportunity. In other words, Company X's perception of the employment transition support business was as a business opportunity that could reap economic rewards while fulfilling a social need. Since the company was established as an economic business opportunity, it is positioned as an economic value-driven business concept.

Company Y, conversely, was founded by a person with no experience of starting or managing a company. Company Y was not established with the idea of being an economic business opportunity. Instead, it was launched when model businesses were being established to serve people with borderline developmental disabilities. The founder recognized the employment transition support business when exploring ways to connect individuals with developmental disabilities with jobs and searching for a format that is possible within the company's own managerial resources. In other words, it is positioned as a social value-driven business concept.

## *5.3 Strategy*

Besides the differences in Company X's and Company Y's business organization formats, there were also differences in their operating formats.

Company X is primarily geared toward economic value while it also clearly pursues social values. Specifically, Company X has developed general support curricula for mass people with developmental disabilities and has lean operations by keeping down its administrative costs. This can be observed by its efforts not only to lower the operating costs of each of its offices but also to reduce its shared costs by conducting multi-office operation. Moreover, although the company tries not to let its employees be overly focused on economic aspects with numeric objects, it is still run based on certain quantitative targets.

Meanwhile, Company Y is primarily geared toward social values. Its services are targeted at persons with developmental disabilities, and it offers high-quality support curricula. With this support curricula, Company Y has found employment for many users over a short period. Thus, it has not set any numerical targets of economic purposes such as the number of users, but it is instead being run with its major focus on social outcomes.

The differences in the business management formats of Company X and Company Y are considered owing to the differences in the emphasis that each company places on the values it is pursuing. As Company X regards the employment transition support business as an economic opportunity, it places more emphasis on the economic value of utilizing external resources, while Company Y has made a business out of its concern for persons with disabilities, so it places more emphasis on social value. Both companies are simultaneously pursuing both economic and social purposes in employment transition support, but their value emphasis is different. This emphasis on different values prescribes their operating formats.

### 5.4 Organizational Management

The knowledge and experience of the two companies' founders have influenced the differences in the organizational management at the two companies. Since the founding members of Company X had knowledge of and experience in launching and managing companies, they established an organizational management structure that is both rational and systematic. The founder of Company Y, however, did not have any prior experience founding or managing a company. Moreover, since Company Y has placed more emphasis on pursuing its social purpose of helping persons with disabilities, assembling an organizational management structure was of secondary importance. Furthermore, as the management team had not been composed, we can say that the organization has been run in a somewhat ad hoc manner. Thus, the knowledge and experience of the founders and the conditions whereby the management teams were assembled have led to differences in the two organizations' management formats.

### 5.5 Economic Outcomes and Social Outcomes

As stated previously, both Company X and Company Y continue to do business while realizing the social purpose of finding jobs for the users of their services. Of the two, Company X has realized a superb economic outcome, with an operating margin of 26%. Meanwhile, Company Y has exhibited excellent social outcome, with a placement rate of over 100% and a 1-year retention rate of over 90%. In other words, in the employment transition support business, which is subject to institutional restrictions, both companies are simultaneously pursuing both economic and social purposes, but their emphasis is different.

## 6 Conclusion

### 6.1 The Significance of this Study

This study made a comparative analysis of Japanese WISEs on the four elements of (1) the tension between economic purpose and social purpose, (2) the degree of recognition of economic purpose and social purpose within the organization, (3) how management can surmount such tension, and (4) economic outcomes and social outcomes. This research topic has not been previously explored. Moreover, this study identified the current condition of managements at Japanese WISEs that are simultaneously pursuing economic and social purposes and considered options for the future. Due to our analysis, we derived the characteristics of Japanese WISEs that should be considered and their relationship as the framework in Fig. 2.

The theoretical implication of this framework is that when considering management that simultaneously pursues economic and social purposes, we need to go back and consider the composition of the entrepreneurial team and their perception of business opportunities. In other words, the diagram shows that the emphasis given to the values being pursued by the organization (i.e., economic and social values) depends on the management team's experience and expertise and how they perceive business opportunities and that applying this to strategy and organizational management can lead to unique outcomes. Although a relationship between entrepreneurial teams and organizational outcome has been discussed in the field of entrepreneurship research (e.g., Kobayashi, 2015), it has not been adequately covered in research on social corporations or WISEs.

Additionally, through this study, we identified the following two points with respect to the current state of and issues facing the employment transition support businesses in Japan. First, the Japanese employment transition support business is characterized by a susceptibility to moral hazard, so reconsideration needs to be given to the optimal management format. Through the employment transition support business, more than 20% of places of business are achieving a high level of social outcomes in that they have transition rates to employment at general corporations in excess of 50%. Meanwhile, many offices still have transition rates to employment at general corporations of 0%. Japan's employment transition support business is a benefit provision business; thus, structurally, it can earn revenue even if it does not perform well. The ideal format for this industry and its management needs to be reconsidered.

Second, we found that a variety of formats exist in the employment transition support business. The subjects of this study—Company X and Company Y—managed their businesses differently. Specifically, Company X is oriented toward economic value, while Company Y's operations are oriented toward social value, so each company has developed its own strategy and organizational management. Although employment transition support is a benefit provision business and there are differences within this sector as to how much emphasis is placed on the economic and social purposes, there are many ways for managements to simultaneously pursue and achieve both purposes to a certain degree. Many organizations with various organizational types, including companies, NPOs, and social welfare corporations, have entered Japan's employment transition support business, and the competition in this business has become fierce. Therefore, organizations oriented toward economic value need to regard their management skills as an inherent strength and work toward enhancing their social purposes (i.e., social outcome). Conversely, organizations oriented toward social value need to build management teams that recognize management skills and to consider taking steps that will enhance their economic purposes, bearing in mind that high-quality curricula will either be imitated or standardized some time. To this end, they will need either to gain synergies by expanding their scale or developing other businesses or to build a more innovative business model.

## 6.2 Future Issues

Several limitations were observed in this study. First, this study focused only two companies involved in the employment transition support business in Japan. In order to generalize and sophisticate the results obtained through this study, it will be necessary to examine the increase in the number of companies. In addition, NPOs, social welfare corporations, and other types of organizations are also active in the employment transition support business. Therefore, there needs to be a comparison of organizations that have developed this business using other organizational form or a comparison that includes employment transition support organizations in other countries and regions. Second, social issues are varied, and thus, the management format that is best for the simultaneous pursuit of both economic and social purposes probably varies as well. Therefore, studying and comparing organizations involved with social issues such as poverty and the environment would probably be useful in deepening our understanding of managements that simultaneously pursue both economic and social purposes.

## References

Battilana, J., & Lee, M. (2014). Advancing research on hybrid organizing – Insights from the study of social enterprises. *Academy of Management Annals, 8*(1), 397–441.

Battilana, J., Sengul, M., Pache, A., & Model, J. (2015). Harnessing productive tensions in hybrid organizations: The case of work integration social enterprises. *Academy of Management Journal, 58*(6), 1658–1685.

Borzaga, C., & Defoury, J. (Eds.). (2001). *The emergence of social enterprise*. Routledge.

Cooney, K. (2011). An exploratory study of social purpose business models in the United States. *Nonprofit and Voluntary Sector Quarterly, 40*(1), 185–196.

Dees, J. G. (1998). Enterprising nonprofits. *Harvard Business Review, January-February*, 54–57.

Doherty, B., Haugh, H., & Lyon, F. (2014). Social enterprises as hybrid organizations: A review and research agenda. *International Journal of Management Reviews, 16*, 417–436.

Fujii, A., Harada, K., & Otaka, K. (Eds.). (2013). *Tatakau shakaiteki kigyo: Komyuniti enpawamento no ninaite [Social enterprise tackling social exclusion]*. Keiso-Shobo. (in Japanese).

Jay, J. (2013). Navigating paradox as a mechanism of change and innovation in hybrid organizations. *Academy of Management Journal, 56*(1), 137–159.

Kobayashi, H. (2015). Bencha sougyou chi-mu no keisei purosesu: Tsuushin bencha no jirei hikaku [A study on formation process of founding teams: A comparative case analysis of telecom start-up]. *Soshiki kagaku [Organizational Science], 48*(3), 41–54. (in Japanese).

Mair, J., Mayer, J., & Lutz, E. (2015). Navigating institutional plurality: Organizational governance in hybrid organizations. *Organization Studies, 36*(6), 713–739.

Ministry of Health, Labour and Welfare. (2008). Heisei 19 nen Shakai fukushi shisetsu-tou chousa kekka no gaikyo [Overview of Survey of Social Welfare Facilities in 2007] (in Japanese).

Ministry of Health, Labour and Welfare. (2010a). Heisei 20 nen Shakai fukushi shisetsu-tou chousa kekka no gaikyo [Overview of Survey of Social Welfare Facilities in 2008] (in Japanese).

Ministry of Health, Labour and Welfare. (2010b). Heisei 21 nen Shakai fukushi shisetsu-tou chousa kekka no gaikyo [Overview of Survey of Social Welfare Facilities in 2009] (in Japanese).

Ministry of Health, Labour and Welfare. (2011). Heisei 22 nen Shakai fukushi shisetsu-tou chousa kekka no gaikyo [Overview of Survey of Social Welfare Facilities in 2010] (in Japanese).
Ministry of Health, Labour and Welfare. (2012). Heisei 23 nen Shakai fukushi shisetsu-tou chousa no gaikyo [Overview of Survey of Social Welfare Facilities in 2011] (in Japanese).
Ministry of Health, Labour and Welfare. (2014). Heisei 24 nen Shakai fukushi shisetsu-tou chousa no gaikyo [Overview of Survey of Social Welfare Facilities in 2012] (in Japanese).
Ministry of Health, Labour and Welfare. (2015a). Heisei 25 nen Shakai fukushi shisetsu-tou chousa no gaikyo [Overview of Survey of Social Welfare Facilities in 2013] (in Japanese).
Ministry of Health, Labour and Welfare. (2015b). Heisei 26 nen Shakai fukushi shisetsu-tou chousa no gaikyo [Overview of Survey of Social Welfare Facilities in 2014] (in Japanese).
Ministry of Health, Labour and Welfare. (2015c). Shogaisha no shuro shien ni tsuite [Employment support for persons with disabilities] (in Japanese).
Ministry of Health, Labour and Welfare. (2016). Heisei 27 nen Shakai fukushi shisetsu-tou chousa no gaikyo [Overview of Survey of Social Welfare Facilities in 2015] (in Japanese).
Ministry of Health, Labour and Welfare. (2017a). Heisei 28 nen Shakai fukushi shisetsu-tou chousa no gaikyo [Overview of Survey of Social Welfare Facilities in 2016] (in Japanese).
Ministry of Health, Labour and Welfare. (2017b). Shuro iko shien ni kakaru hoshu, kijun ni tsuite [Employment transition support remuneration and standards] (in Japanese).
Ministry of Health, Labour and Welfare. (2018a). Heisei 30 nendo teichaku shien jigyo, houshu kaitei ni tsuite [FY 2018 revision of retention support businesses and remuneration] (in Japanese).
Ministry of Health, Labour and Welfare. (2018b). Heisei 29 nen Shakai fukushi shisetsu-tou chousa no gaikyo [Overview of Survey of Social Welfare Facilities in 2017] (in Japanese).
National Institute of Vocational Rehabilitation ed. (2017). Shougaisha no shugyoujyoukyou-tou ni kansuru chousa kenkyu [Surveys and research on the employment status of people with disabilities] (in Japanese).
Pache, A., & Santos, F. (2013). Inside the hybrid organization: Selective coupling as a response to competing institutional logics. *Academy of Management Journal, 56*(4), 972–1001.
Santos, F., Pache, A. C., & Birkholz, C. (2015). Making hybrids work: Aligning business models and organizational design for social enterprises. *California Management Review, 57*(3), 36–58.
Smith, W. K., & Besharov, M. L. (2019). Bowing before dual gods: How structured flexibility sustains organizational hybridity. *Administrative Science Quarterly, 64*(1), 1–44.
Smith, W. K., Binns, A., & Tushman, M. L. (2010). Complex business models: Managing strategic paradoxes simultaneously. *Long Range Planning, 43*, 448–461.
Stevens, R., Moray, N., & Bruneel, J. (2015a). The social and economic mission of social enterprises: Dimensions, measurement, validation, and relation. *Entrepreneurship Theory and Practice, 39*(5), 1051–1082.
Stevens, R., Moray, N., Bruneel, J., & Clarysse, B. (2015b). Attention allocation to multiple goals: The case of for-profit social enterprises. *Strategic Management Journal, 36*, 1006–1016.
Suzuki, Y. (Ed.). (2014). *Sosharu entapuraizu ron: Jiritsu wo mezasu jigyo no kakushin [On social enterprise]*. Yuhikaku. (in Japanese).
Takahashi, M., Kimura, T., & Ishiguro, T. (2018). *Sosharu inobeshon wo rironka suru: kiri hirakareru shakai kigyouka no aratana jissen [Theorizing social innovation]*. Bunshin-do. (in Japanese).
Tanimoto, K. (Ed.). (2006). *Sosharu entapuraizu: shakaiteki kigyou no taitou [Social enterprises]*. Chuoukeizai-sha. (in Japanese).
Tanimoto, K., Omuro, N., Ohira, S., Doi, M., & Komura, K. (2013). *Sosharu inobeshon no soushutsu to fukyu [Creation and diffusion of social innovation]*. NTT-shuppan. (in Japanese).
Wry, T., & York, J. G. (2017). An identity-based approach to social enterprise. *Academy of Management Review, 42*(3), 437–460.

**Yuichi Goto** is Associate Professor, Faculty of Economics, Management and Information Science, Onomichi City University, Japan. He graduated from the Graduate School of Economics, Hokkaido University, is a Ph.D. in Business Administration, is interested in social business, social entrepreneurs, and multi-sector collaboration, and conducts field research. He has edited several books: *Management of Strategic Collaboration* and *Collective Impact in Japan.*

**Keiko Yokoyama** is Professor at the Faculty of Commerce, Kansai University, and Representative Director of Sobakuri Labo, a general incorporated association. She graduated from the Graduate School of Economics, Hokkaido University, and is a Ph.D. in Business Administration. While working in the field of research, she came in contact with people who were boldly working to create new value by breaking out of the existing framework and became fascinated by the way of life of entrepreneurship. She has edited several books, including *Corporate Social Strategy and NPOs, Ethical Entrepreneurship* and *Collective Impact in Japan.*

# Staying Legitimate in a Changing Climate: A Framework for Studying Corporate Climate Change Communication

**Muhammad A. F. Allam**

## 1 Introduction

Climate change and its many risks, most notably extreme weather events and rising sea levels, is one of the most overarching issues of our times. It is a multifaceted environmental issue that has roots and implications in the social, institutional, economic, and technological aspects of human life (Pachauri et al., 2014). A problem of this magnitude has the potential to disrupt established norms and societal systems, including business practice and discourse.

With a great role in today's economy, businesses fall under increased pressure from internal and external stakeholders that expect higher levels of corporate social responsibility (CSR) towards the society and the environment. In response, companies, especially those in polluting industries such as the oil and gas sector, tend to increase their level of information disclosure to communicate their stance on environmental issues (Frynas, 2009). This sector has presumably been granted a license to operate, hence legitimized by stakeholders such as governments, other business sectors, non-governmental organizations (NGOs), the media, and the general public. On the other hand, some of these stakeholders have been recently challenging the legitimacy of the oil and gas industry. Accusations included the irresponsible business actions that harm the environment and contribute to the deteriorating of earth's climate, hence putting human lives at risk (Cai et al., 2012).

To respond to such pressure, counter negative image, and attain and retain legitimacy required for long-term survival, oil and gas companies are more than ever engaged in CSR activities on local and global levels. This is coupled with communication strategies that promote these activities through different channels

M. A. F. Allam (✉)
Radboud University, Nijmegen, The Netherlands
e-mail: muhammad.allam@ru.nl



I. G. Arraiano et al. (eds.), *Corporate Social Responsibility in a Dynamic Global Environment*, CSR, Sustainability, Ethics & Governance,
https://doi.org/10.1007/978-3-031-24647-0_4

such as corporate websites, annual reports, public relations, CSR advertising, and social media (Du & Vieira, 2012). Yet, it has been proven in some instances that oil and gas companies are practising green-washing: a strategy adopted "to engage in symbolic communications of environmental issues without substantially addressing them in actions" (Walker & Wan, 2012).

## *1.1 Research Problem*

The survival and ongoing operations of oil and gas companies indicate a certain level of legitimacy, that is, societal acceptance of their existence and operations. However, products of these companies, when burned, are the main sources of greenhouse gas (GHG) emissions, the main cause of global warming (NASA, 2017). Operating in such a polluting and controversial industry, oil and gas companies are a main target of criticism and must deal with an increased internal and external stakeholder sensitivity (Du & Vieira, 2012) that presumably challenges their legitimacy and thus threatens their survival. These stakeholders range from employees to media, civil society, and the general public. From the corporate perspective, managing these challenges might require resorting to substantive, symbolic, or a mix of these two approaches to managing legitimacy (Woolfson & Beck, 2005). While substantive management of legitimacy involves actual changes in structures or operations to meet societal expectations, symbolic management is merely appearing to conform to societal expectations.

The focus of many researchers was the study of the symbolic approach to managing legitimacy of organizations in the oil and gas sector or in relation to addressing climate change (e.g. Ferraro et al., 2015; Van Halderen et al., 2016; Whiteman et al., 2013). While only a small number of studies have integrated both substantive and symbolic management of legitimacy (e.g. Matejek & Gössling, 2014; Walker & Wan, 2012), this research answers the call for more studies in this direction (Deephouse et al., 2016, p. 28). In addition, when communicating their climate-related stance and performance, it is argued that companies do not make use of the potential offered by the available digital tools.

## *1.2 Research Aim and Objectives*

The aim of this study is to provide a framework through which we can deepen our understanding of the extent and nature of climate change communication by oil and gas companies on websites and in corporate reports.

Towards this end, this research seeks to achieve the following objectives:

1. To position corporate climate change communication within the sustainability communication literature.

2. To identify the relevant constructs underpinning corporate climate change communication and utilize them to develop a conceptual framework by which it can be systematically studied.
3. To develop a multi-level analysis approach to differentiate between the symbolic and the substantive legitimation tactics and strategies in climate change communication.

## 1.3 Research Method and Question

Climate change communication and legitimation of oil and gas companies are well researched topics. Nevertheless, the celebrated deal on climate action in 2015, known as the Paris Agreement, calls for new research to gain new insights on the issue at hand, especially that the private sector, including oil and gas companies, had a major role in "orchestrating" the agreement, which was criticized for having no mention of GHG sources, nor how to stop exploration and use of fossil fuel (Spash, 2016). Hence the proposition that oil and gas companies' engagement in the global climate action efforts is aimed merely at legitimacy.

Thus, we identified the need for a conceptual framework through which the aforementioned proposition can be tested and for studies to be replicated. To achieve this, our contribution is the development of a framework to investigate how oil and gas companies utilize websites, annual reports, and sustainability reports as tools of climate change communication.

More specifically, this research answers the following question:

How can we systematically examine climate change communication by oil and gas companies on their websites and corporate reports?

To answer this question, we utilize the integrative review method of inquiry for its appropriateness to the aim of this study, i.e. generating a new framework (Torraco, 2005) and providing a definition of concepts (Souza et al., 2010). Integrative literature review can address a mature topic, or an emerging one. Since corporate climate change communication is a relatively new sub-field of research, our study can be associated with the second type; that is, it will result in a "preliminary conceptualization of the topic" (Torraco, 2005), p. 357).

The study's research question has already provided certain limitation to which literature can be included in the review. Accordingly, we took the following steps:

First, since this review is focused on a relatively new business practice, that is, utilizing websites and corporate reports, we decided to initially include literature from the past 10 years (2008–2017). After starting the review, it seemed we were missing important earlier contributions, and so the period was expanded to 20 years (1998–2017).

Due to the multidisciplinary nature of the topic of corporate climate change communication, it was decided not to search on a specific journal or publisher, but to utilize an academic search engine, that is Google Scholar.

To decide on which keywords should be utilized for the search, more inclusion criteria were deemed necessary. The criteria selection was guided chiefly by the research question but also the initial readings done by the author to identify the specific research problem. Thus, we chose to first include academic articles and book chapters on corporate climate change communication. Then, we included literature on corporate sustainability disclosure or CSR in the oil and gas industry. Lastly, we chose to include literature on the legitimacy theory that tackles legitimation strategies by organizations in controversial industries.

Second, based on the criteria identified above, we started the search process by utilizing search filters for the specified dates and various combinations of search terms that were further finely adjusted as the search process continued. Some of the queries utilized are "corporate climate communication", "corporate climate change communication", "corporate communication" + "climate change" + "oil and gas", and "legitimacy theory" + "legitimation strategies".

Third, the author utilized purposeful sampling technique (Patton, 2002) to select the relevant and important literature to be synthesized. The result was organized in a spreadsheet highlighting the literature attributes (e.g. year of publication, title, industry discussed, country, and topic).

Finally, a thematic analysis was conducted on the literature with a critical lens. The analysis and findings are organized in the following sections from the general to the specific, i.e. starting with the definitions of the broader discipline of CSR and sustainability communication moving to one of its elements and the core focus of this research, that is, corporate climate change communication. Then, the legitimacy theory literature is discussed and integrated.

## *1.4 Relevance of the Study*

This study has important implications for both research and practice. Communication practitioners can make use of the findings to better understand how companies in controversial industries and of several sizes, ownership structures, and nationalities deal with stakeholder pressure. More specifically, the study provides them with a way of systematically studying how the oil and gas sector utilizes websites, annual reports, and sustainability reports to communicate about climate change. This study comes at a time when climate action is gaining momentum, with an unprecedented recognition of the critical role that the corporate sector needs to play (Ernst & Young, 2015), hence the necessity to understand corporate communication regarding this issue.

It also offers insights on an under-researched area of corporate CSR, sustainability, and environmental communication, that is, the use of corporate websites for climate change communication by controversial industries. Identifying the position of climate communication within the whole CSR, sustainability, or environmental communication of the company will offer uncovered knowledge.

## 2 A Conceptual Framework for Corporate Climate Change Communication

The following section highlights the importance of reporting and communicating corporate social and environmental stances and performances. It focuses on the relations between corporates, the society, communication channels and their audiences, the challenges they face, and finally the motivations for corporate sustainability communication.

We then provide a detailed analytical review of the position of the climate change issue within the reviewed literature and introduce a definition of corporate climate change communication. Next, we reflect on the oil and gas industry mentions with regard to their sustainability communication motivations and challenges. Finally, the legitimacy theory is integrated as a useful approach to understand corporates' motivations to communicate about sustainability in general and climate change in particular.

### *2.1 Theoretical and Practical Definitions of Corporate Sustainability*

Studies of corporate social responsibility (CSR), corporate sustainability, and corporate citizenship have often used the three terms to refer to the same practice. In theory, it is beneficial for scholars to disagree and provide multiple terms and definitions to the same concept with the purpose of enriching and advancing the discipline, while in practice, the presence and usage of several terms to refer to the same concept can result in challenging implications.

Providing an attempt to utilize a joint concept of CSR, corporate citizenship, and sustainable development (CSR/CC/SD), Moreno and Capriotti (2009) suggest differences but also stress similarities in the three concepts. Most studies, however, have opted to use only one term to refer to the increased phenomenon of corporate social, environmental, and economic (financial) reporting. For example, Adams and Frost (2006a) have used sustainability, Ziek (2009) used CSR, albeit integrating sustainability and citizenship as indicators of CSR communication, and to a much lesser extent scholars have used corporate citizenship (e.g. Maignan & Ferrell, 2001).

Maignan and Ferrell (2001, p. 457) defined corporate citizenship as "the extent to which businesses assume the economic, legal, ethical and discretionary responsibilities imposed on them by their various stakeholders." Considering the pressure "imposed" by stakeholders, and the multifaceted perceived responsibility of corporates, this definition aligns and becomes a part of CSR, especially when we consider the widely cited broad definition by Carroll (1991, p. 43), as he postulates that "the total corporate social responsibility of business entails the simultaneous fulfilment of the firm's economic, legal, ethical, and philanthropic responsibilities".

On the other hand, the term "sustainability" provides new dimensions. In 1987, the World Commission on Environment and Development (WCED) coined the term "sustainable development" which refers to "development that meets the needs of the present without compromising the ability of future generations to meet their own needs" (Brundtland, 1987). In a business context, the term sustainability is often used to encompass the environmental, social, and financial performance of a company that later came to be known as triple bottom line (TBL) (Elkington, 1997). Borrowed from accounting, this refers to the last line in companies' financial statements usually representing profits, implying profitability from social and environmental performances, in addition to the financial one. These three dimensions mirror the three pillars of sustainability, known as the 3Ps: people, planet, and profit (Elkington, 1997).

For the purpose of this study, the concept of sustainability is preferred to CSR for the following reason. CSR is evidenced to provide companies with sought-after eco-efficiency, through lowering polluting emissions, or recycling, practices that have been proven to positively corelate with financial gains (Frynas, 2009; Walker & Wan, 2012). On the other hand, CSR practices fail to provide sustainable development, that is, managing not to compromise the health, safety, and well-being of future generations (Frynas, 2009). This aspect of considering future needs is of great importance for the debate around the urgency of climate action and energy transition. Sustainability communication is as well more relevant than environmental communication, for climate change is considered a sustainability problem, not merely an environmental (i.e. ecological) one (Newig, 2011).

In summary, disagreements and a variety of definitions of concepts in academic literature are implicitly encouraged to advance the knowledge of certain disciplines. However, in practice, it can create confusion and result in lack of standards, hence possible loss of credibility. This is especially true when looking at the practice of corporate sustainability. The blurry definitions may result in three challenging implications: first, from the corporate internal perspective, companies' confusion not knowing which term refers to which practice might contribute to local and global lack of reporting standards (e.g. in Finland; see Laine, 2005). It is then even more difficult for management to define and communicate their sustainable performance (Herzig & Schaltegger, 2011). Second, because of lack of standards, assurances, and accountability, and due to the voluntary nature of corporate sustainability reporting, corporates might resort to constructing their own preferred versions of sustainability, which might lead to the socially undesired business-as-usual scenario (Laine, 2005). Third, from a communication point of view, and from a stakeholders' perspective, receivers of the corporate sustainability messages might stand confused by the variety of terms and definitions. This adds to their potential lack of technical knowledge to understand, for instance, oil companies' emission results (Cox & Pezzullo, 2016).

## 2.2 *Motivations and Channels of Corporate Sustainability Communication*

Sustainability communication was not always as integrative as it is nowadays (Herzig & Schaltegger, 2011). Corporate social reporting emerged in the 1970s, following pressure on corporates by social movements, a decade later environmental reporting came to existence in the late 1980s and early 1990s to tackle the environmental problems perceived to be mainly caused by corporates (Herzig & Schaltegger, 2011). In the 1990s, sustainability reports started to gain momentum (Kolk, 2004). Today, communication is a "hot topic" within CSR and sustainability research; its challenging nature has prompted Dawkins (2005, p. 108) to describe it as "the missing link in the practice of corporate responsibility".

Whether discussing CSR or sustainability, corporate communication is seen as a vital element (Du & Vieira, 2012). When used effectively, sustainability communication informs corporate stakeholders of its CSR performance, likely resulting in enhancing its credibility, reputation, and legitimacy, in addition to assisting in the process of stakeholder identification (Deegan, 2002; Du et al., 2010; Maignan & Ferrell, 2004).

Sustainability communication importance has not only been recognized by corporates but can also be noticed in the increasing sustainability reporting by "industry bodies and associations, government institutions, consulting firms, non-governmental organisations and research institutions" (Herzig & Schaltegger, 2011, pp. 151–152).

Four recurrent themes found in the literature reviewed were the relationship between corporations and the society, sustainability communication channels, challenges, and motivations behind such practice.

Companies do not exist in isolation from other members of the society, as one conceptualization of CSR is "the construct describing the relationship between companies and society" (D'Aprile & Mannarini, 2012, p. 48). This relationship is often described to be metaphorically formalized in a "social contract"; that is, society expects certain requirements from businesses, in exchange for granting them legitimacy or a "licence to operate" (Deegan, 2002; Du & Vieira, 2012; Steiner, 1972). While corporations are positioned on one side of the relationship, the society at large is on the other side; this can refer to individuals, e.g. consumers, investors, and employees, or entities, e.g. governments, non-profits, and media (Maignan & Ferrell, 2004).

When discussing the roles of different types of stakeholders in the CSR domain, the following two perspectives are usually considered. From the society perspective, partnerships and cooperation are praised as tools of renegotiation of the social contract with corporations (Hamann & Acutt, 2003). On the other hand, corporates are collaborating with NGOs in developing their environmental policies (Pulver, 2007) while using collaboration with legitimate entities as a legitimation strategy (Oliver, 1990).

The role played by NGOs, especially environmental, has been vital in the sustainability agenda. Mostly assuming a critical stance against corporate unsustainable actions, they reported on various issues ranging from "environmental violations, human rights abuses, detrimental impact on local communities, to breaches of labour and safety standards" (Du & Vieira, 2012, pp. 413–414). Also, they criticized and exposed green-washing practices (Pulver, 2007). According to Dawkins (2005), NGOs are always looking for evidence of companies' sustainability claims. Therefore, NGOs are one of the most powerful stakeholders a corporate must consider, when communicating their social and environmental performance.

Honouring their end of the social contract is not the only reason companies engage in sustainability disclosure and communication. On the same side of the spectrum of motivations lies management acceptance of accountability or responsibility, that is, acknowledging the stakeholders right to know (Vaccaro & Madsen, 2009), thus gaining their support (De Roeck & Delobbe, 2012).

On the other hand, CSR and sustainability actions are often attributed to the corporate need of attaining, maintaining, and repairing legitimacy, that is, future survival (Dowling & Pfeffer, 1975) and long-term prosperity (Du & Vieira, 2012). This motivation is usually associated with controversial industries such as oil and gas, alcohol, and tobacco (Bhattacharya & Sen, 2004). These industries are under external pressure by different stakeholders, mainly the media, NGOs, regulatory institutions, and the public. In a strategic effort to counter and manage this pressure, corporates opt to act socially and environmentally responsible, or at least appear to do so (Ramus & Montiel, 2005), to allow them to continue their business as usual (Palazzo & Richter, 2005), or for their new structures or operations to appeal to the sociocultural norms and societal expectations (Dacin, 1997; Deegan, 2002). Reflecting on the motive of gaining stakeholders' support, this interplay of motivations is consistent with what Gray et al. (1995) postulated, which is the need to consider the legitimacy theory and stakeholder theory as overlapping perspectives, not competing ones.

Turning to corporate sustainability communication channels, the literature reviewed provides accounts on the following facets: the link between the channel, message, and receiver, in addition to the advantages, limitations, and challenges faced when using web technologies for corporate sustainability communication.

As a crucial step in any communication strategy, identifying and utilizing a variety of channels is advised for CSR messages to reach the target audience (Dawkins, 2005; Du et al., 2010). Corporates have abundance of channels to choose from; these include annual reports, social, environmental, and sustainability reports (Gray, 2001), corporate websites (Moreno & Capriotti, 2009), CSR advertising (Yoon et al., 2006), public relations, and social media platforms (Colleoni, 2013; Dawkins, 2005).

Looking at practice, Esrock and Leichty (1998) found that 82% of their sample of US Fortune 500 index used corporate websites to communicate CSR activities. This and other studies suggest a growing utilization of the Internet and World Wide Web (WWW) technologies for corporate sustainability communication, in contrast to the formerly dominant form of using annual reports (Herzig & Godemann, 2010). Some

**Table 1** Functional characteristics of websites for corporate sustainability communication

| Functionality | Description and literature |
|---|---|
| Accessibility | A medium to communicate to an international audience (Moreno & Capriotti, 2009) |
| | Use of the HTML format (Herzig & Schaltegger, 2011) |
| | 24-h accessibility (Adams & Frost, 2006b; Herzig & Schaltegger, 2011) |
| | Increasing information accessibility and comprehensibility (Adams & Frost, 2006b) |
| Interactivity | Permits dialogue (two-way communication), feedback (to respond to questions, concerns, and problems) (Esrock & Leichty, 1998; Kent & Taylor, 1998) |
| | More active interaction with wide and diverse public (Adams & Frost, 2006b) |
| | Mutual asynchronous forms such as mail-to functions or discussion forums (Herzig & Schaltegger, 2011) |
| | Mutual synchronous forms such as chats, audio, or video-conferencing (Herzig & Schaltegger, 2011) |
| Value | Exert agenda-setting function, thus avoiding gate-keeper function of the mass media (Esrock & Leichty, 1998) |
| | Reduced information costs (Herzig & Godemann, 2010) |
| | Environmentally friendly (Adams & Frost, 2006b) |
| Multimedia | Links to other information sources on the company or other organizations (Herzig & Schaltegger, 2011) |
| | Combination of different media elements such as words, figures, images, or videos (Herzig & Schaltegger, 2011) |
| Capacity | Permits large quantities of information than print reports (reporting is for example no longer limited by the number of printed pages) (Herzig & Schaltegger, 2011) |
| | Provision of historical information (archives) (Herzig & Schaltegger, 2011) |
| Velocity | Provision of real-time data and immediate updates on events and information (Adams & Frost, 2006a, 2006b; Esrock & Leichty, 1998) |
| Personalization | Integrated view for audience to select from (Herzig & Schaltegger, 2011) |
| | Addressee-specific information tailoring and distribution (to meet various technical and information needs) (Herzig & Schaltegger, 2011) |
| | Individual access for stakeholders (Herzig & Schaltegger, 2011) |

companies (e.g. Adidas and E. ON UK) have even abandoned the printed reports to focus solely on e-reporting (Herzig & Schaltegger, 2011). Annual reports have also evolved; McQueen (2001) observed that they now have two functions. The traditional function of regulated financial reporting and a new unregulated function of public relations. In the same context, he argues that the audience of these reports have expanded from merely shareholders to include employees, potential investors, and the general public.

While reviewing literature, some patterns have emerged when discussing corporate websites as the main vehicle for sustainability reporting. These include functionality characteristics, advantages, and opportunities offered by using the Internet for corporate sustainability communication. For better presentation, and easier further reference, we grouped these into seven themes (Table 1): interactivity, velocity, accessibility, multimedia, capacity, personalization, and value.

Besides, some researchers highlighted the challenges that might face companies trying to utilize web technologies for communicating their sustainable actions. These include the lack of standards, regulations, and assurance of web content, which might result in a lower level of credibility (Herzig & Schaltegger, 2011). This challenge results from the possibility to change web content very easily. Another challenge being the digital divide (Herzig & Schaltegger, 2011), which, in my opinion, is one of the most important factors considering it might largely affect stakeholder engagement in countries where there is a high level of divide. Finally, some stakeholders might still prefer to consume information the traditional way, that is, printed reports (Herzig & Schaltegger, 2011).

Even though the potential and possibilities offered by the Internet seem to far outweigh the limitations, in practice companies should resort to corporate websites as one medium of sustainability communication that complements, or is rather complemented by, using traditional means such as printed annual and sustainability reports (Herzig & Schaltegger, 2011). In addition to their original content, websites have the potential of widening the audience of these reports. Whereas printed reports reach a limited audience, digital versions of these reports can reach an international audience.

Beyond the challenges facing corporate sustainability pertinent to channels and audience, corporate sustainability challenges arise mainly due to organizations failing to achieve the goals of their economic, social, and environmental practice. There is a great mounting pressure imposed by societal actors on corporations, whether it is a governmental requirement, industry standard, NGO scrutiny, media criticism, or internal powerful stakeholder engagement. Tackling the conflicting interests of the widely diverse stakeholder groups is a challenge that requires an integrative, not only communicative, approach. Herzig and Schaltegger (2011, p. 166) offer a recommendation to corporations as they note that:

> A well-managed, interdisciplinary team-based process seems to be required, one that involves different departments, external stakeholders and possibly communication agencies, as well as diverse competencies in identifying the sustainability issues that are most relevant to both the company and society. Likewise, communicating these issues in a comprehensible way and integrating sustainability reporting with other sustainability communication media and the company's more general corporate communications concept appears to be vital if sustainability communication is to move to a higher level.

Thus, a corporate that utilizes several media for sustainability communication, integrates it into the corporate wider communication strategy, and utilizes the same integrative approach for identification and engagement with internal and external stakeholders can expect better organizational performance and results.

## 2.3 Corporate Climate Change Communication

When discussing climate change communication, one of two general perspectives is usually considered, communication *of* climate change, or communication *about* climate change (Newig, 2011). While the former takes an instrumental or managerial view to inform and educate the audience about climate change, its implications, and actions, the latter takes an analytical view and is more concerned with how societal actors communicate about the issue. Although Newig (2011) did not mention corporate role in either, we argue that corporate climate change communication belongs to the *about* category, that is, corporations are societal actors that do not hold the scientific evidence, as do scientists and environmental groups, which legitimize and qualify them as a source of objective knowledge. Furthermore, corporates, as political figures, utilize the power of interpretation when communicating their stance and performance regarding climate change, therefore using their own logic and frames instead of merely mirroring that of scientists.

> ...communication is a key element in societal strategies to cope with climate change. (Newig, 2011, p. 119)

As mentioned briefly before, this chapter assumes the stance that climate change communication is part of sustainability communication; thus, corporate climate change communication is part of corporate sustainability communication. This notion, although implicitly postulated throughout the CSR and sustainability literature reviewed (e.g. Escobar & Vredenburg, 2011; Livesey, 2002; Walker & Wan, 2012), has only been made explicit by Newig (2011), as he based his argument on three characteristics of climate change that call for societal communication:

1. High level of complexity and uncertainty
2. No global consensus on goals to be achieved
3. Widely distributed efforts and capacities

This need for a global consensus on goals and call for societal communication has resonated widely in the international political and governance forums. The highest level of these is the United Nations (UN) that historically led the coordination of global efforts to tackle the negative effects of climate change. As early as 1987, the WCED report on sustainable development mentioned climate change as a global threat requiring strategies for minimizing damage and coping with the risks imposed (Brundtland, 1987). The following few years marked several UN resolutions addressing multiple aspects of the phenomenon (Table 2).

Furthermore, the year 1988 witnessed the establishment of the Intergovernmental Panel on Climate Change (IPCC), the world's leading authority on climate change (IPCC, 2013; The Royal Society, 2007). On the other hand, this period marked not only the climate change debate gaining momentum in the international political forums (Paterson, 1996) but also a first substantial intervention from the corporate side in the form of establishing the Climate Change Working Group (CCWG) at the International Petroleum Industry Environmental Conservation Association

**Table 2** Some climate change-related UN resolutions (United Nations, 1992)

| Year | Resolution number | Addressing the issue of |
|---|---|---|
| 1988 | 43/53 | Global warming |
| 1989 | 44/206 | Rising sea levels |
| | 44/172 | Combating desertification |
| | 44/207 | Protection of global climate for present and future generations of humankind |
| 1990 | 45/212 | |
| 1991 | 46/169 | |

(IPIECA) in 1988 (Gale, 2015), and the Global Climate Coalition (GCC) in 1989. While the former was concerned with voicing the industry stance to the UN forums, the latter attempted to influence the public opinion by questioning the science of climate change and challenging the need for action (IPIECA, 2007; Mooney, 2007).

More recently, the need for global climate action is listed as number 13 out of 17 UN Sustainable Development Goals (SDGs) of 2016 (United Nations, 2015a). The SDGs give the issue more salience than the preceding Millennium Development Goals (MDGs) for 2000–2015 that included climate change as a sub-topic under goal number seven, which was to "Ensure Environmental Sustainability" (United Nations, 2015b). In the SDGs, climate action is not only tackled in goal 13, but it is also central to other goals as well, including the following:

Goal 7: Ensure access to affordable, reliable, sustainable, and modern energy for all.
Goal 12: Ensure sustainable consumption and production patterns.
Goal 14: Conserve and sustainably use the oceans, seas, and marine resources.
Goal 15: Sustainably manage forests, combat desertification, halt and reverse land degradation, and halt biodiversity loss.

In addition to politics and science, the role of the media in the climate change debate is often highlighted. Weingart et al. (2000) discussed the development of the discourse on climate change in three spheres: science, politics, and media. One more sphere, or rather actor that Cox and Pezzullo (2016) highlighted as having a great role in the broader environmental communication arena, is universities. For instance, they noted that climate change communication is incorporated today in some universities' curricula.

Moving on to look at the role of corporates, or how the climate discourse is taking place in the business sphere, it does not seem to be mentioned as much as other actors or spheres. For example, Weingart et al. (2000) briefly highlighted how the oil industry financed climate-sceptic scientists. Moreover, corporates, especially in polluting industries, are heavily pressured by the new sustainability discourse (Escobar & Vredenburg, 2011; Frynas, 2009). Consequently, they engage in symbolic or substantive actions to address such pressure and manage their legitimacy (Walker & Wan, 2012). Regardless of the real intentions behind these actions, we argue that it is of great importance to consider the business sphere more often when discussing climate change communication, and not only focus on the media, politics, and science. This is of special importance given the increasing role of corporates in shaping our experiences (Allen & Craig, 2016), and since the often hailed "Paris Climate Agreement" adopted in 2015 involved unprecedented activity from the private sector (Grossman, 2017).

Even though corporate climate change communication was not discussed extensively in the literature reviewed, one report was completely dedicated to how companies report on their climate-related performance, that is, "The Corporate Climate Communications Report 2007" published in 2008 by Corporate Register, a portal of non-financial reports from different sectors and company types. In this report, they study how Global FT500 companies disclose climate change data. The report starts with a general discussion and moves towards performance disclosure, activity disclosure, target-setting disclosure, assurances, and guidelines disclosure. The report covers a wide range of sectors categorized into heavy, light, and service industries while also covering a wide range of regions including North America, South America, Europe, Africa and Middle East, Asia, Japan, and Australasia (Corporate Register, 2008).

Interestingly, out of 39 oil and gas companies in the sample, 28 (72%) published CSR reports. Of these, 71% had a climate-specific section in their reports. Of these, 64% had climate issues covered in the CEO introduction, and 18% stated management responsibility for corporate climate action. While most industries were proven likely to align with Greenhouse Gas Protocol of the Global Reporting Initiative (GRI), the oil and gas were the second least likely to align with the protocol. When it comes to setting SMART (specific, measurable, achievable, realistic, and time-scaled) targets (Doran, 1981) to reduce emissions, less than 25% of oil and gas companies set SMART targets, while 50% set no targets at all. About 10% of the oil and gas companies had specific climate change assurance for their reports (Corporate Register, 2008).

In summary, the report suggests that climate reporting has become a mainstream business practice. The report also is one of the first to recognize the beginning of "corporate climate activism". This claim is backed by the fact that "companies have the means in terms of global reach and capital, consumer loyalty and international networks to effect change on a significant scale" (Corporate Register, 2008, p. 44).

Turning to another recurrent theme when discussing climate change communication, or the wider umbrella of sustainability communication, we now discuss the presence, or lack thereof, of reporting standards, regulations, and external assurance. Herzig and Schaltegger (2011) echo the differences between corporates in their reporting motivations; following a certain level of guidelines can demonstrate accountability or play a legitimizing role if minimum guidelines are followed. In the same context, going green is distinguished from green-washing (Cox & Pezzullo, 2016). These two motives, we argue, cannot be seen in isolation from each other, but rather as two ends of a continuum. An organization can achieve a certain level of both goals by reporting their social, environmental, and financial performance following guidelines and standards.

Sustainability reporting can be perceived as a green-washing activity. Thus, to move away from such accusation, and for effective reporting, corporates need to reflect an outside-in as well as an inside-out perspective (Herzig & Schaltegger, 2011), that is, explaining activities contributing to sustainable development and corporate strategy, while responding to external expectations. Although Herzig and Schaltegger (2011) did not specify what he means by external expectations, and how it is different from the other perspective, he echoes Bebbington et al.'s (2007) call for a true dialogue with stakeholders. Thus, stakeholder dialogue is considered a measurement of corporate willingness to meet societal expectations, hence acknowledging accountability.

Tackling a global issue such as climate change, corporates need to consider two different types of pressures; these are global and local (Escobar & Vredenburg, 2011). Since there are no enforceable standards—on corporates—agreed upon globally, issues like climate change are tackled mostly by local and regional regulations and standards; thus, multinational companies would respond only to powerful local and regional stakeholder pressure, giving little to no real attention to global concerns (Escobar & Vredenburg, 2011; Oliver, 1991).

However, there have been many global voluntary standards tackling greenhouse gas emissions, energy efficiency, and environmental communication, mainly developed by global standardization organizations such as the Global Reporting Initiative (GRI) and the International Organization for Standardization (ISO). But very few have developed a comprehensive framework for reporting on climate action and related performances. Two of these are the Climate Change Reporting Framework developed by the Climate Disclosure Standards Board (CDSB) and the Science Based Targets (SBT) initiative developed by Carbon Disclosure Project (CDP), World Resources Institute (WRI), the World Wide Fund for Nature (WWF), and the United Nations Global Compact (UNGC).

In addition to standards and regulations, external criteria that corporates seek to conform to include rankings and competitions (Herzig & Schaltegger, 2011). They often aim to improve the practice of sustainability reporting and contribute to standardization. Although a voluntary procedure, having a good ranking or winning a reporting award in sustainability is a symbolic measure that contributes to companies' enhanced image and reputation (Deegan & Carroll, 1993). As the competitions often reflect stakeholders' expectations (Herzig & Schaltegger, 2011), it is also likely to influence organizational legitimacy. With regard to climate change, there is only one award that is given to companies in special recognition of their "Corporate Climate Change Communication". In 2012, this recognition awarded by the Taiwan Institute for Sustainable Energy was given to Taiwan Power Company for their sustainability report (Taiwan Power Company, 2013).

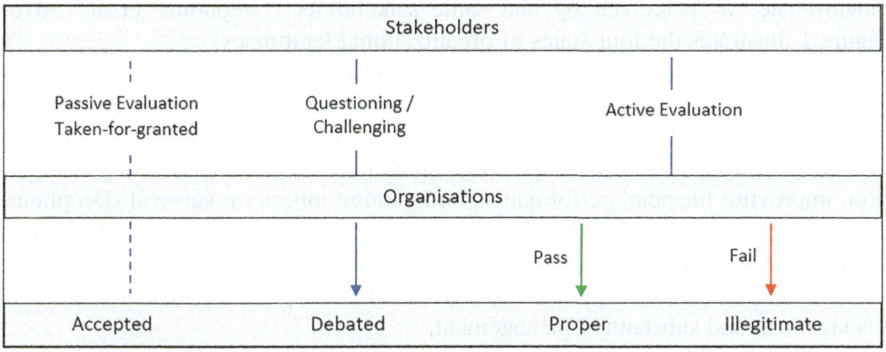


**Fig. 1** States of organizational legitimacy, based on Deephouse et al. (2016), pp. 33–34

## *2.4 The Legitimacy Theory: Symbolic and Substantive Climate Communication*

The legitimacy theory has been one of the dominant theoretical lenses, through which researchers examined CSR, sustainability (e.g. Deegan, 2002; Du & Vieira, 2012; Matejek & Gössling, 2014; Walker & Wan, 2012; Welbeck et al., 2017), and recently corporate climate change communication (e.g. Giannarakis et al., 2017; Qian & Schaltegger, 2017; Van Halderen et al., 2016). In their meta-review of legitimacy literature, Deephouse et al. (2016, p. 9) suggest the following definition:

> Organizational legitimacy is the perceived appropriateness of an organization to a social system in terms of rules, values, norms, and definitions.

They affirm the importance of societal actors' perception of an organization in assessing its legitimacy, hence the notion that it is not a universal concept, but rather context-dependent (Barkemeyer, 2007). This is especially true when considering the different outcomes of legitimacy. Historically, legitimacy has been characterized as having a fundamental dichotomous nature (Deephouse & Suchman, 2008). Therefore, an organization can be considered either legitimate or illegitimate, while recently, organizational legitimacy can be assessed and placed on a continuum.

Most prominently, Deephouse et al. (2016) put forward four basic outcomes and states of organizational legitimacy; these are accepted, proper, debated, and illegitimate. The main difference between the first two states is active evaluation by societal actors. Accepted organizations that have been actively inspected by regulators are deemed "proper" from the regulator and similar institutions' perspective. They could, however, be deemed "accepted" from the perspective of the consumers who take the organization's existence and operation for granted (Deephouse et al., 2016). On the other hand, organizations can have the "debated" state when their values or operations are actively questioned or challenged by some stakeholders. Finally, entities that do not "pass" the active evaluation by a stakeholder, i.e. not meeting stakeholder expectations, will be deemed inappropriate, hence "illegitimate" as perceived by that same stakeholder (Deephouse et al., 2016). Figure 1 illustrates the four states of organizational legitimacy.

In this study, we adopt Deephouse and Suchman's (2008) strategic perspective to legitimacy, that is, emphasizing organizations' ability to manage legitimacy towards achieving their goals. Examples of these goals or "consequences of legitimacy" include gaining market access, gaining stakeholder support, eliminating competition, improving financial performance, and gaining long-term survival (Deephouse & Suchman, 2008). In managing their relationship with different stakeholder groups, organizations use different approaches to extend, maintain, or defend their legitimacy. Ashforth and Gibbs (1990) highlight two main types of approaches: symbolic management and substantive management.

Symbolic management, on the one hand, comprises the organizational practices that aim at making acts "appear" consistent with stakeholder expectations. Substantive management, however, represents actual changes in organizational goals, structures, or processes (Deephouse et al., 2016). Symbolic management has been criticized because of its ability to "deceive" and "distract" from substantive issues (Bundy & Pfarrer, 2015); nevertheless, both approaches are valuable, especially when used together (Deephouse et al., 2016). We assume this approach and through this research answer to their call for more consideration to the interplay between the two management approaches for organizational legitimacy.

Communication strategy differs greatly according to the management strategy chosen by the organization. If green-washing is only appearing to be responsible by means of symbolic management, green-highlighting is utilizing both symbolic and substantive approaches; "green highlighting combines talking the talk with walking the walk" (Walker & Wan, 2012, p. 232).

Legitimating accounts are usually presented in the form of text (Phillips et al., 2004), hence the central role of communication in managing organizational legitimacy. This aligns with our earlier characterization of legitimacy as a stakeholder perception of the organization. While symbolism enables an organization to be multiple things to multiple stakeholders (Kraatz & Block, 2008), it is the embodiment of commitments that trigger the reciprocal commitment needed for organizational survival (Stinchcombe, 1997).

The text produced and distributed by organizations to manage legitimacy usually takes the form of stories or narratives (Lounsbury & Glynn, 2001), through which they strategically deploy rhetoric that aims at increasing the impact of the messages conveyed (Suddaby & Greenwood, 2005). Thus, researchers investigating organizational legitimation approaches have used methods of rhetorical analysis (Erkama & Vaara, 2010), narrative analysis (Brown, 1998), discourse analysis (Phillips et al., 2004), and framing analysis (Garcia & Greenwood, 2015).

Finally, based on the literature review, we offer the following definition for corporate climate change communication: *the production and distribution of information on corporate performance and stance regarding climate change causes, consequences, and solutions, to relevant stakeholders with the aim of gaining or maintaining their support.*

## 2.5 The Conceptual Framework

Following the extensive literature review above, we developed a legitimacy-based conceptual framework that can be organized as follows. Oil and gas companies are operating in a controversial industry that is a major contributor of GHG emissions, one of the main causes of climate change. They are facing increasing stakeholder pressure, presumably challenging, and questioning their operations and existence; hence their legitimacy state being "debated". Motivated to survive, some companies are driven by their acceptance of accountability and responsibility towards societal actors, while others are driven by the mere aspiration to continue to exist and stay relevant, that is, being legitimate.

Those accepting accountability adopt a business strategy involving substantive management. This comes in forms of real changes in operations or structures of the company that reflect stakeholder expectations. On the other hand, those merely seeking legitimacy adopt a business strategy that involves symbolic management. In doing so, they appear to be conforming to stakeholder expectations, with no actual changes. A mixed strategy is possible; hence, a company can seek both accountability and legitimacy with different degrees.

This study utilizes the legitimacy theory as a lens through which we can develop a better understanding of how oil and gas companies communicate their climate-related performance and stance. To achieve this, the following four dimensions should be considered.

Presence, Hierarchy, and Accessibility of Climate-Related Information

First, the presence of climate-related information on the companies' websites, annual reports, and sustainability reports should be identified.

Presentation, Format, and Interactivity of Climate-Related Information

Next, how climate information is presented should be examined.

Content Characteristics of Climate Information

To have a more in-depth analysis and meaningful findings, climate content should be examined based on the following: first, conformity to external criteria, including regulations, standards, ranking, awards, and agreements. Second, the explicit mention of stakeholders. Third, themes and topics discussed.

Legitimation Strategies

Finally, to identify the legitimation approaches utilized by the oil and gas industry, the following is suggested. The content characteristics revealed in earlier stages, the conformity to external criteria, the explicit mention of specific stakeholders, and the inclusion of specific themes, will all act as the categories by which the legitimation approach is deduced (Fig. 2).

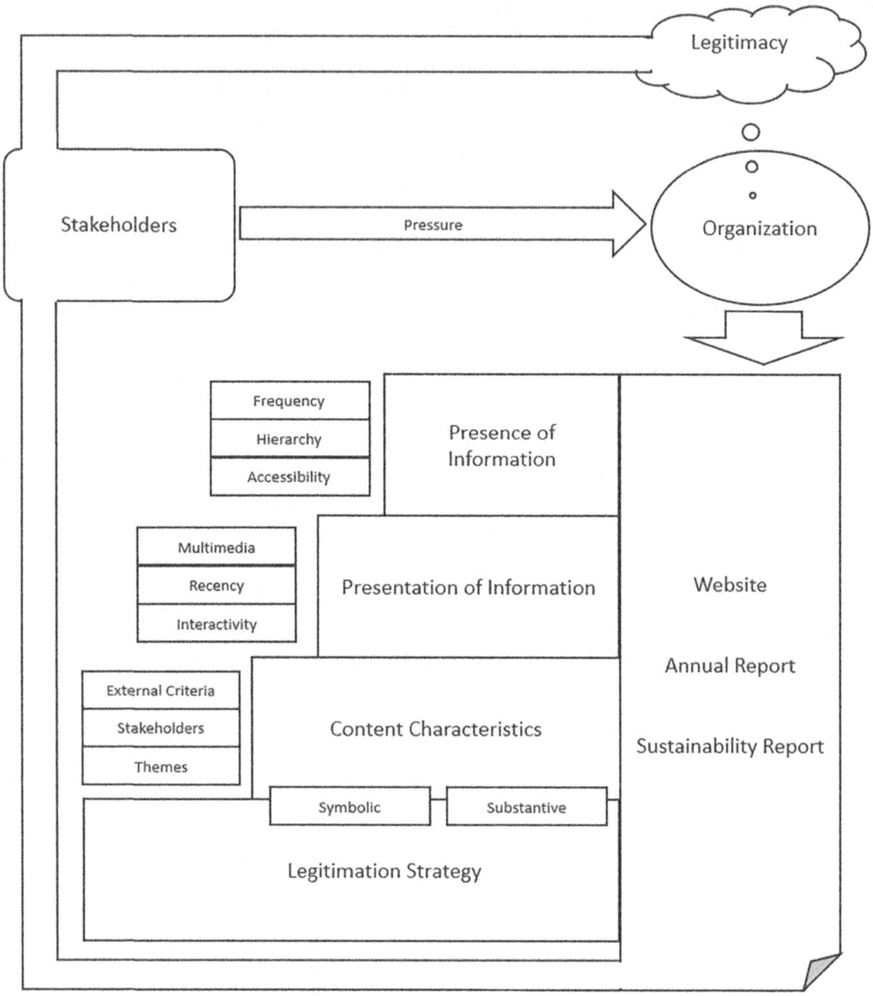


**Fig. 2** Conceptual framework developed for the study of corporate climate change communication, based on literature review

## 3 Conclusion

At a time when anthropogenic climate change is one of the greatest challenges facing our planet, it is crucial for societal actors to engage in a meaningful and honest discussion that is built on common understanding, two-way communication, and willingness to change. This has recently materialized in a global momentum of climate action. Towards this end, this chapter contributes new and important knowledge to the climate debate. Theoretically, we offered, for the first time, a definition and a conceptual framework for corporate climate change communication. The concept was also positioned within sustainability communication after critically discussing the differences with CSR.

We performed an integrative review of a purposeful sample of the literature published during two decades on CSR, sustainability communication, and their link to organizational legitimacy. The developed conceptual framework is a preliminary contribution to the emerging sub-field of corporate climate change communication.

Future studies are encouraged to build on this initial attempt and empirically utilize the framework to provide a deeper understanding of how corporates communicate about this overarching issue; whether organizations operating in controversial industries merely disclose information on performance for symbolic legitimacy or communicate their position to various groups of stakeholders with the underlying belief of a social contract that ties them with multiple actors who can influence these organizations' survival.

**Acknowledgements** The author thanks the anonymous reviewer for their insightful comments and feedback. The author would also like to thank Roel van Veen, Dr. Maryja Šupa, and Professor Wim Elving, for their valuable feedback on an earlier version of this paper. The author would like to acknowledge as well that this paper is based on a broader research project carried out as partial fulfilment of a master's degree from Hanze University of Applied Sciences, The Netherlands, and Vilnius University, Lithuania.

## References

Adams, C. A., & Frost, G. R. (2006a). Accessibility and functionality of the corporate web site: Implications for sustainability reporting. *Business Strategy and the Environment, 15*(4), 275–287. https://doi.org/10.1002/bse.531

Adams, C. A., & Frost, G. R. (2006b). The internet and change in corporate stakeholder engagement and communication strategies on social and environmental performance. *Journal of Accounting & Organizational Change, 2*(3), 281–303. https://doi.org/10.1108/18325910610690090

Allen, M. W., & Craig, C. A. (2016). Rethinking corporate social responsibility in the age of climate change: A communication perspective. *International Journal of Corporate Social Responsibility, 1*(1), 1.

Ashforth, B. E., & Gibbs, B. W. (1990). The double-edge of organizational legitimation. *Organization Science, 1*(2), 177–194. https://doi.org/10.1287/orsc.1.2.177

Barkemeyer, R. (2007). Legitimacy as a key driver and determinant of CSR in developing countries. Paper presented at "The 2007 Marie Curie Summer School on Earth System Governance", Amsterdam.

Bebbington, J., Brown, J., & Frame, B. (2007). Accounting technologies and sustainability assessment models. *Ecological Economics, 61*(2), 224–236. https://doi.org/10.1016/j.ecolecon.2006.10.021

Bhattacharya, C. B., & Sen, S. (2004). Doing better at doing good: When, why, and how consumers respond to corporate social initiatives. *California Management Review, 47*(1), 9–24. https://doi.org/10.2307/41166284

Brown, A. D. (1998). Narrative, politics and legitimacy in an IT implementation. *Journal of Management Studies, 35*(1), 35–58. https://doi.org/10.1111/1467-6486.00083

Brundtland, G. H. (1987). Report of the world commission on environment and development: Our common future. United Nations.

Bundy, J., & Pfarrer, M. D. (2015). A burden of responsibility: The role of social approval at the onset of a crisis. *Academy of Management Review, 40*(3), 345–369. https://doi.org/10.5465/amr.2013.0027

Cai, Y., Jo, H., & Pan, C. (2012). Doing well while doing bad? CSR in controversial industry sectors. *Journal of Business Ethics, 108*(4), 467–480. https://doi.org/10.1007/s10551-011-1103-7

Carroll, A. B. (1991). The pyramid of corporate social responsibility: Toward the moral management of organizational stakeholders. *Business Horizons, 34*(4), 39–48. https://doi.org/10.1016/0007-6813(91)90005-g

Colleoni, E. (2013). CSR communication strategies for organizational legitimacy in social media. *Corporate Communications: An International Journal, 18*(2), 228–248. https://doi.org/10.1108/13563281311319508

Corporate Register. (2008). The Corporate Climate Communications Report 2007 (Rep.). Retrieved November 20, 2017, from https://www.corporateregister.com/pdf/CCCReport_07.pdf

Cox, J. R., & Pezzullo, P. C. (2016). *Environmental communication and the public sphere* (4th ed.). Sage Publications.

D'Aprile, G., & Mannarini, T. (2012). Corporate social responsibility: A psychosocial multidimensional construct. *Journal of Global Responsibility, 3*(1), 48–65.

Dacin, M. T. (1997). Isomorphism in context: The power and prescription of institutional norms. *Academy of Management Journal, 40*(1), 46–81.

Dawkins, J. (2005). Corporate responsibility: The communication challenge. *Journal of Communication Management, 9*(2), 108–119. https://doi.org/10.1108/13632540510621362

De Roeck, K., & Delobbe, N. (2012). Do environmental CSR initiatives serve organizations' legitimacy in the oil industry? Exploring employees' reactions through organizational identification theory. *Journal of Business Ethics, 110*(4), 397–412. https://doi.org/10.1007/s10551-012-1489-x

Deegan, C. (2002). Introduction: The legitimising effect of social and environmental disclosures – A theoretical foundation. *Accounting, Auditing & Accountability Journal, 15*(3), 282–311. https://doi.org/10.1108/09513570210435852

Deegan, C., & Carroll, G. (1993). An analysis of incentives for Australian firms to apply for reporting excellence awards. *Accounting and Business Research, 23*(91), 219–227. https://doi.org/10.1080/00014788.1993.9729881

Deephouse, D. L., Bundy, J., Tost, L. P., & Suchman, M. C. (2016). Organizational legitimacy: Six key questions. In R. Greenwood, C. Oliver, T. Lawrence, & R. E. Meyer (Eds.), *The Sage handbook of organizational institutionalism* (2nd ed., pp. 27–54). Sage.

Deephouse, D. L., & Suchman, M. C. (2008). Legitimacy in organizational institutionalism. In R. Greenwood, C. Oliver, R. Suddaby, & K. Sahlin-Andersson (Eds.), *The Sage handbook of organizational institutionalism* (pp. 49–77). Sage.

Doran, G. T. (1981). There's a SMART way to write management's goals and objectives. *Management Review, 70*(11), 35–36.

Dowling, J., & Pfeffer, J. (1975). Organizational legitimacy: Social values and organizational behavior. *Pacific Sociological Review, 18*(1), 122–136. https://doi.org/10.2307/1388226

Du, S., Bhattacharya, C. B., & Sen, S. (2010). Maximizing business returns to corporate social responsibility (CSR): The role of CSR communication. *International Journal of Management Reviews, 12*(1), 8–19. https://doi.org/10.1111/j.1468-2370.2009.00276.x

Du, S., & Vieira, E. T. (2012). Striving for legitimacy through corporate social responsibility: Insights from oil companies. *Journal of Business Ethics, 110*(4), 413–427. https://doi.org/10.1007/s10551-012-1490-4

Elkington, J. (1997). *Cannibals with forks: The triple bottom line of 21st century business*. Capstone Publishing.

Erkama, N., & Vaara, E. (2010). Struggles over legitimacy in global organizational restructuring: A rhetorical perspective on legitimation strategies and dynamics in a shutdown case. *Organization Studies, 31*(7), 813–839. https://doi.org/10.1177/0170840609346924

Ernst & Young. (2015). COP21 – What it means for the Australian Government and businesses (Rep.). Retrieved April 1, 2018, from http://www.ey.com/Publication/vwLUAssets/EY-cop21-what-it-means-for-the-australian-government-and-businesses/$FILE/EY-cop21-what-it-means-for-the-australian-government-and-businesses.pdf

Escobar, L. F., & Vredenburg, H. (2011). Multinational oil companies and the adoption of sustainable development: A resource-based and institutional theory interpretation of adoption heterogeneity. *Journal of Business Ethics, 98*(1), 39–65. https://doi.org/10.1007/s10551-010-0534-x

Esrock, S. L., & Leichty, G. B. (1998). Social responsibility and corporate web pages: Self-presentation or agenda-setting? *Public Relations Review, 24*(3), 305–319. https://doi.org/10.1016/S0363-8111(99)80142-8

Ferraro, F., Etzion, D., & Gehman, J. (2015). Tackling grand challenges pragmatically: Robust action revisited. *Organization Studies, 36*(3), 363–390.

Frynas, J. G. (2009). *Beyond corporate social responsibility: Oil multinationals and social challenges*. Cambridge University Press.

Gale, J. (2015). IEAGHG information paper: 2015-IP12; Exploring methane emissions with IPIECA. Retrieved October 17, 2017, from http://ieaghg.org/docs/General_Docs/Publications/Information_Papers/2015-IP12.pdf

Garcia, M. M., & Greenwood, K. (2015). Visualizing CSR: A visual framing analysis of US multinational companies. *Journal of Marketing Communications, 21*(3), 167–184. https://doi.org/10.1080/13527266.2012.740064

Giannarakis, G., Zafeiriou, E., & Sariannidis, N. (2017). The impact of carbon performance on climate change disclosure. *Business Strategy and the Environment, 26*(8), 1078–1094. https://doi.org/10.1002/bse.1962

Gray, R. (2001). Thirty years of social accounting, reporting and auditing: What (if anything) have we learnt? *Business Ethics: A European Review, 10*(1), 9–15. https://doi.org/10.1111/1467-8608.00207

Gray, R., Kouhy, R., & Lavers, S. (1995). Corporate social and environmental reporting: A review of the literature and a longitudinal study of UK disclosure. *Accounting, Auditing & Accountability Journal, 8*(2), 47–77. https://doi.org/10.1108/09513579510146996

Grossman, C. (2017). We Have an Agreement in Paris: So, What's Next for The Private Sector? Retrieved October 18, 2017, from http://www.ifc.org/wps/wcm/connect/news_ext_content/ifc_external_corporate_site/news_and_events/news/we-have-agreement-paris-so-what-s-next-private-sector

Hamann, R., & Acutt, N. (2003). How should civil society (and the government) respond to 'corporate social responsibility'? A critique of business motivations and the potential for partnerships. *Development Southern Africa, 20*(2), 255–270.

Herzig, C., & Godemann, J. (2010). Internet-supported sustainability reporting: developments in Germany. *Management Research Review, 33*(11), 1064–1082. https://doi.org/10.1108/01409171011085903

Herzig, C., & Schaltegger, S. (2011). Corporate sustainability reporting. In *Sustainability communication* (pp. 151–169). Springer.

IPCC. (2013). IPCC Factsheet: What is the IPCC? Retrieved April 1, 2018, from https://www.ipcc.ch/site/assets/uploads/2018/02/FS_what_ipcc.pdf

IPIECA. (2007). *The oil and gas industry and climate change*. IPIECA.

Kent, M. L., & Taylor, M. (1998). Building dialogic relationships through the World Wide Web. *Public Relations Review, 24*(3), 321–334.

Kolk, A. (2004). A decade of sustainability reporting: Developments and significance. *International Journal of Environment and Sustainable Development, 3*(1), 51–64. https://doi.org/10.1504/IJESD.2004.004688

Kraatz, M. S., & Block, E. S. (2008). Organizational implications of institutional pluralism. In R. Greenwood, C. Oliver, R. Suddaby, & K. Sahlin-Andersson (Eds.), *The Sage handbook of organizational institutionalism* (pp. 243–275). Sage.
Laine, M. (2005). Meanings of the term 'sustainable development' in Finnish corporate disclosures. *Accounting Forum, 29*(4), 395–413. https://doi.org/10.1016/j.accfor.2005.04.001
Livesey, S. M. (2002). The discourse of the middle ground: Citizen shell commits to sustainable development. *Management Communication Quarterly, 15*(3), 313–349. https://doi.org/10.1177/0893318902153001
Lounsbury, M., & Glynn, M. A. (2001). Cultural entrepreneurship: Stories, legitimacy, and the acquisition of resources. *Strategic Management Journal, 22*(6-7), 545–564. https://doi.org/10.1002/smj.188
Maignan, I., & Ferrell, O. C. (2001). Corporate citizenship as a marketing instrument—Concepts, evidence and research directions. *European Journal of Marketing, 35*(3/4), 457–484.
Maignan, I., & Ferrell, O. C. (2004). Corporate social responsibility and marketing: An integrative framework. *Journal of the Academy of Marketing Science, 32*(1), 3–19. https://doi.org/10.1177/0092070303258971
Matejek, S., & Gössling, T. (2014). Beyond legitimacy: A case study in BP's "green lashing". *Journal of Business Ethics, 120*, 571–584. https://doi.org/10.1007/s10551-013-2006-6
McQueen, R. (2001). The corporate image-the regulation of annual reports in Australia. *Macquarie Law Journal, 1*(1), 93–128.
Mooney, C. (2007). *The republican war on science*. Basic Books.
Moreno, A., & Capriotti, P. (2009). Communicating CSR, citizenship and sustainability on the web. *Journal of Communication Management, 13*(2), 157–175. https://doi.org/10.1108/13632540910951768
NASA. (2017). Climate change causes: A blanket around the Earth. Retrieved September 10, 2017, from https://climate.nasa.gov/causes/
Newig, J. (2011). Climate change as an element of sustainability communication. In *Sustainability communication* (pp. 119–128). Springer.
Oliver, C. (1990). Determinants of interorganizational relationships: Integration and future directions. *Academy of Management Review, 15*(2), 241–265. https://doi.org/10.5465/AMR.1990.4308156
Oliver, C. (1991). Strategic responses to institutional processes. *Academy of Management Review, 16*(1), 145–179.
Pachauri, R. K., Allen, M. R., Barros, V. R., Broome, J., Cramer, W., Christ, R., Church, J. A., Clarke, L., Dahe, Q., Dasgupta, P., & Dubash, N. K. (2014). *Climate change 2014: Synthesis report. Contribution of Working Groups I, II and III to the fifth assessment report of the Intergovernmental Panel on Climate Change*. IPCC.
Palazzo, G., & Richter, U. (2005). CSR business as usual? The case of the tobacco industry. *Journal of Business Ethics, 61*(4), 387–401. https://doi.org/10.1007/s10551-005-7444-3
Paterson, M. (1996). *Global warming and global politics*. Psychology Press.
Patton, M. Q. (2002). *Qualitative research and evaluation methods*. Sage.
Phillips, N., Lawrence, T. B., & Hardy, C. (2004). Discourse and institutions. *Academy of Management Review, 29*(4), 635–652. https://doi.org/10.2307/20159075
Pulver, S. (2007). Making sense of corporate environmentalism. *Organization & Environment, 20*(1), 44–83. https://doi.org/10.1177/1086026607300246
Qian, W., & Schaltegger, S. (2017). Revisiting carbon disclosure and performance: Legitimacy and management views. *The British Accounting Review, 49*(4), 365–379. https://doi.org/10.1016/j.bar.2017.05.005
Ramus, C. A., & Montiel, I. (2005). When are corporate environmental policies a form of greenwashing? *Business & Society, 44*(4), 377–414. https://doi.org/10.1177/0007650305278120
Souza, M. T. D., Silva, M. D. D., & Carvalho, R. D. (2010). Integrative review: What is it? How to do it? *Einstein (São Paulo), 8*, 102–106.

Spash, C. L. (2016). This changes nothing: The Paris agreement to ignore reality. *Globalizations, 13*(6), 928–933.

Steiner, G. A. (1972). Social policies for business. *California Management Review, 15*(2), 17–24.

Stinchcombe, A. L. (1997). On the virtues of the old institutionalism. *Annual Review of Sociology, 23*(1), 1–18. https://doi.org/10.1146/annurev.soc.23.1.1

Suddaby, R., & Greenwood, R. (2005). Rhetorical strategies of legitimacy. *Administrative Science Quarterly, 50*(1), 35–67. https://doi.org/10.2189/asqu.2005.50.1.35

Taiwan Power Company. (2013). 2012 Annual Report (Rep.). Retrieved December 9, 2017, from http://www.taipower.com.tw/UpFile/AboutUsFile/2012_tpc.pdf

The Royal Society. (2007). Climate change controversies: A simple guide. Retrieved April 1, 2018, from https://royalsociety.org/-/media/Royal_Society_Content/policy/publications/2007/8031.pdf

Torraco, R. J. (2005). Writing integrative literature reviews: Guidelines and examples. *Human Resource Development Review, 4*(3), 356.

United Nations. (1992). United Nations Framework Convention on Climate Change (Rep.). Retrieved October 17, 2017, from https://unfccc.int/resource/docs/convkp/conveng.pdf

United Nations. (2015a). Transforming our world: The 2030 agenda for sustainable development. A/RES/70/1, 21 October. Retrieved October 17, 2017, from https://sustainabledevelopment.un.org/content/documents/21252030%20Agenda%20for%20Sustainable%20Development%20web.pdf

United Nations. (2015b). The Millennium Development Goals Report 2015 (Rep.). Retrieved October 17, 2017, from http://www.un.org/millenniumgoals/2015_MDG_Report/pdf/MDG%202015%20rev%20(July%201).pdf

Vaccaro, A., & Madsen, P. (2009). Corporate dynamic transparency: The new ICT-driven ethics? *Ethics and Information Technology, 11*(2), 113–122. https://doi.org/10.1007/s10676-009-9190-1

Van Halderen, M. D., Bhatt, M., Berens, G. A., Brown, T. J., & Van Riel, C. B. (2016). Managing impressions in the face of rising stakeholder pressures: Examining oil companies' shifting stances in the climate change debate. *Journal of Business Ethics, 133*(3), 567–582. https://doi.org/10.1007/s10551-014-2400-8

Walker, K., & Wan, F. (2012). The harm of symbolic actions and green-washing: Corporate actions and communications on environmental performance and their financial implications. *Journal of Business Ethics, 109*(2), 227–242. https://doi.org/10.1007/s10551-011-1122-4

Weingart, P., Engels, A., & Pansegrau, P. (2000). Risks of communication: Discourses on climate change in science, politics, and the mass media. *Public Understanding of Science, 9*(3), 261–283.

Welbeck, E. E., Owusu, G. M. Y., Bekoe, R. A., & Kusi, J. A. (2017). Determinants of environmental disclosures of listed firms in Ghana. *International Journal of Corporate Social Responsibility, 2*(1), 11.

Whiteman, G., Hope, C., & Wadhams, P. (2013). The vast costs of Arctic change: Methane released by melting ice sheets will have global economic impacts that must be better modelled. *Nature, 499*(7459), 401–403. https://doi.org/10.1038/499401a

Woolfson, C., & Beck, M. (2005). *Corporate responsibility failures in the oil industry.* Baywood Pub.

Yoon, Y., Gürhan-Canli, Z., & Schwarz, N. (2006). The effect of corporate social responsibility (CSR) activities on companies with bad reputations. *Journal of Consumer Psychology, 16*(4), 377–390. https://doi.org/10.1207/s15327663jcp1604_9

Ziek, P. (2009). Making sense of CSR communication. *Corporate Social Responsibility and Environmental Management, 16*(3), 137–145. https://doi.org/10.1002/csr.183

**Muhammad A. F. Allam** has a master's in social sciences in international communication. His wide research interests include the intersection of digital communication and global societal issues such as climate change, diplomacy, inequality, and privacy. He is currently writing his Ph.D. dissertation which explores how socio-economic injustice is manifested in public discourse. He is external Ph.D. candidate at Radboud University, Nijmegen, the Netherlands.

# Sustainable and Smart Living Versus Unworthy Conditions in Portugal

**Sónia Bidarra, Rute Abreu, and Cecília Rosa**

## 1 Introduction

The right to housing is a right fundamental enshrined in many universal/international (United Nations, 1948), European Union (Council of Europe, 1961), and national documents (AR, 1976). Housing can be defined as having a home, a safe place, from which we are not at risk of being evicted and with access to appropriate health and education services, as well as decent employment. It is a pillar of the stability and security of a person or family. Having decent housing is universally considered one of the basic needs of human beings (United Nations, 1976).

In Portugal, the Basic Housing Law, about housing conditions, in paragraph 4 of article 9, promotes environmental sustainability, energy efficiency, fire safety, and the reinforcement of the seismic resilience of buildings and privileges the evolution needs of households and communities (AR, 2019).

Recognizing the importance of what we call "home" as the basis for other fundamental rights, life in society, and physical and mental health itself, the WHO defines healthy housing as one that provides a feeling of home, including a sense of belonging, security, and privacy. Healthy housing also refers to the physical structure of the dwelling, and the extent to which it enables physical health, including by being structurally sound, by providing shelter from the elements and from excess moisture, and by facilitating comfortable temperatures, adequate sanitation and

S. Bidarra · C. Rosa
Instituto Politécnico da Guarda & UDI-IPG, Guarda, Portugal
e-mail: cecirosa@ipg.pt

R. Abreu (✉)
Instituto Politécnico da Guarda & UDI-IPG & CICF-IPCA & CISed-IPV, Guarda, Portugal
e-mail: ra@ipg.pt

 
I. G. Arraiano et al. (eds.), *Corporate Social Responsibility in a Dynamic Global Environment*, CSR, Sustainability, Ethics & Governance,
https://doi.org/10.1007/978-3-031-24647-0_5

illumination, sufficient space, safe fuel or connection to electricity, and protection from pollutants, injury hazards, mold, and pests.

The WHO Housing and Health Guidelines (HHGL) can also have indirect co-benefits for health. Improving thermal insulation, weatherization, and ventilation and installing energy-efficient heating can improve indoor temperatures that support health while also lowering expenditure on energy and reducing carbon emissions (WHO, 2018).

Improving energy efficiency in housing is a great opportunity to promote economic development, environmental stewardship, quality of life, and social equality. Better energy efficiency reduces the pressure of energy use and climate change. Furthermore, improving the energy efficiency of housing constitutes a climate change adaptation measure by better shielding homes from adverse weather conditions.

In accordance with Article 9 of the Treaty on the Functioning of the European Union, Union policies on energy efficiency must be inclusive and, therefore, guarantee the accessibility of energy efficiency measures to consumers affected by energy insecurity. Improvements in the energy efficiency of buildings are expected to benefit vulnerable households, including those affected by precarious energy and, where appropriate, those living in social housing. Member States may already require the parties subject to the obligation to include social objectives in energy saving measures in relation to energy insecurity, and this possibility should now be extended to alternative policy measures and national energy efficiency funds and made an obligation while allowing Member States to maintain full flexibility about their size, scope, and content. If an energy efficiency obligation regime does not allow measures for individual energy consumers, the Member State can take measures to alleviate energy insecurity using alternative political measures (European Parliament and Council of the European Union, 2018a).

In this sense, the European Parliament and the Council of the European Union (2018b) approved the Directive (EU) 2018/2002 on Energy Efficiency; the improvement of energy efficiency along the entire energy chain will bring benefits to the environment, improve the air quality and public health, reduce greenhouse gas emissions, improve energy security by reducing dependence on energy imports from third countries, lower energy costs for families and businesses, help reduce energy insecurity, and will induce increased competitiveness, increased employment, and economic activity in general, thus improving the quality of life of citizens (European Parliament and Council of the European Union, 2018b). Thus, in this research, the authors verify, based on the data from the Energy Certification System, the panorama of the national buildings.

Energy poverty is characterized by families that have access to inadequate levels of essential energy services, due to high energy consumption, low household income, inefficient buildings and appliances, and specific domestic energy needs. It is estimated that more than 50 million households in the European Union live in a situation of energy poverty.

This distinct form of poverty is associated with a series of consequences for the health and well-being of people, with respiratory and cardiac diseases and mental health, aggravated by low temperatures and stress, associated with inaccessible energy bills.

In 2018, about 34 million Europeans were unable to keep their homes warm enough. To maintain a decent standard of living and health, adequate heating, air conditioning, lighting, and energy conditions are essential to power household appliances. Access to energy services is essential for social inclusion. Therefore, the fight against energy poverty can bring several benefits, namely less health expenses, reduction of atmospheric pollution, substituting inadequate heating sources for the purposes for which they are intended, greater comfort and well-being, and improvement of family budgets. Taken together, these benefits would directly boost economic growth and prosperity in the European Union (EC, 2020).

In the data released by Eurostat, in 2019, Portugal is listed as the European Union country where people have less economic conditions to keep their homes properly heated, with 19.2% of the Portuguese people being in a situation of energy poverty. Below Portugal are countries such as Bulgaria (30.1%), Lithuania (26.7%), and Cyprus (2.01%) (Eurostat, 2021). The ability to keep the house warm is one of the nine indicators, defined by Eurostat, structuring the living conditions of individuals and families. It allows us to perceive situations of deficiency in relation to three items, material deprivation rate, or at least four items, severe material deprivation rate. The latter, in 2019, stood at 5.5% in the EU and 5.6% in Portugal.

The rate of severe deprivation of housing conditions, referring to the combination of overcrowded housing and the existence of at least one of four problems: no bathroom (or shower) or toilet installation inside the accommodation, water transfers or rot windows/floor, or insufficient natural light on a sunny day, since 2013 through 2018 has decreased, but still located if slightly above the average UE28 (4.1 and 3%, 9%, respectively). As for the percentage of the total population living in a dwelling with a leaking roof, damp walls, floors or foundation, or rot in window frames or floor, Portugal has 26.9% while the EU28 (13.9%).

## 2 Literature Review

Emerging global challenges, such as climate change, loss of biodiversity, scarcity of resources, migratory movements, demographic changes, pandemics, and rapidly changing economies, have a direct impact at the local level across the Europe, and can worsen disparities in our society.

This assumes the need for sustainable transformation highlighted by the 2030 Agenda for Sustainable Development of 2030, including SDG 3 to ensure healthy lives and promote well-being for all age groups and SDG 11 dedicated to making cities inclusive, safe, resilient, and sustainable, the New Urban Agenda, the Paris Agreement, and the European Commission's Ecological Pact. The Ecological Pact aims to make Europe the first continent in the world with a neutral climate impact and Europe has a strong responsibility as well as the collective capacity to achieve this goal. The New Leipzig Charter provides a political framework for planning and implementing these European and global agreements on an urban scale and also for guaranteeing the availability of sanitation for all at the family level (SDG 6), reaching the goals of renewable energy and energy efficiency (SDG 7), and taking measures to mitigate climate change (SDG 13). The importance of the sectoral approach was recently emphasized in the New Urban Agenda for sustainable urban development established in Habitat III, in 2016. In the same year with the Amsterdam Pact, the EU Urban Agenda dedicated to making cities inclusive, safe, resilient, and sustainable was launched. And in 2019 the European Commission's Ecological Pact appeared, which aims to make Europe the first continent in the world with a neutral impact on the climate and Europe has a strong responsibility as well as the collective capacity to achieve this goal. The New Leipzig Charter provides a political framework for planning and implementing these European and global agreements on an urban scale. Urban transformation is based on the integration of the social, ecological, and economic aspects of sustainable development.

Adequate, safe, and affordable housing and energy supply must meet the needs of different groups in society, including an aging and more diverse population, people with disabilities, young people, and families. At the same time, the number of people with complex housing needs in Europe is increasing, especially in the countries most affected by the financial crisis. In Greece, for example, according to the latest survey by the European Parliament, 36% of households suffer from a lack of energy, while tens of thousands of them do not have access to energy and/or heating.

There is no doubt that the challenge is multifaceted, characterized by houses with low energy efficiency, rising energy prices, and low incomes. Property owners and suppliers across Europe face different challenges, from financing energy-efficient renovations to providing additional services to tenants.

Energy consumption is a concern in many countries, both globally and at EU level; countries have been called upon to take responsibility for strategies to limit global warming. In this regard, the EU set the following targets in 2008: to reduce greenhouse gas emissions by 20% (compared to 1990), increase the share of renewable energy to 20%, and improve energy efficiency by 20% (targets it achieved in 2018).

Energy consumption is a matter of sustainability, since we have a small percentage of the population consuming more than half of all the energy produced worldwide; in addition to countries such as Portugal, they use countries outside the EU to obtain fossil fuels (although in mainland Portugal, according to REN data, more than 60% of the energy consumed comes from renewable energies); there is yet another concern: the weight of the energy bill in the annual total of housing expenses.

The Regulation of Thermal Behaviour Characteristics of Buildings (RCCTE), approved by Decree-Law no. 40/90, of 6 February (MOPTC, 1990), was the first legal instrument that in Portugal imposed requirements on the design of new buildings and major remodeling to safeguard the satisfaction of thermal comfort conditions in these buildings without excessive energy needs, both in winter and in summer.

Member States should apply minimum energy performance requirements for new and existing buildings, ensure certification of the energy performance of buildings, and impose a regular inspection of boilers and air conditioning installations in buildings; in this regard on December 16, 2002, the European Parliament and the Council adopted Directive 2002/91/EC on the energy performance of buildings (European Parliament and the Council of the European Union, 2002).

The abovementioned directive aimed to promote the improvement of the energy performance of buildings in the Community, considering external climatic conditions and local conditions, as well as the requirements for indoor climate and economic profitability. This directive establishes requirements regarding a general framework for a methodology for calculating the integrated energy performance of buildings; application of minimum requirements for the energy performance of new buildings; application of minimum requirements for the energy performance of large existing buildings that are subject to major renovation works; energy certification of buildings; and regular inspection of boilers and air conditioning installations in buildings and, in addition, evaluation of the heating installation when the boilers are over 15 years old.

In Portugal, this directive would be transposed into national law through Decree-Law n° 78/2006, of 4 April (MEI, 2006), which approved the National System for Energy Certification and Indoor Air Quality in Buildings, of Decree-Law n° 79/2006, of 4 April, which approved the Regulation of Energy Systems for Air Conditioning in Buildings, and Decree-Law n° 80/2006 of 4 April, which approved the RCCTE (MOPTC, 2006a, 2006b, 2006c).

According to the Regulation of Thermal Behaviour Characteristics of Buildings (MOPTC, 2006c), the country is divided into three winter climatic zones (I1, I2, and I3) and three summer climatic zones (V1, V2, and V3), Table 1.

In the Region of Beiras e Serra da Estrela, we have eight municipalities in zone I3-V2 characterized by the most demanding winter that they will see. In summer check temperature ranges up daily. The areas of zone I3-V1 are Celorico da Beira, Gouveia, Guarda, and Manteigas, where winter is much more aggressive than summer. The municipality of Belmonte is in zone I2-V2, whose daily thermal amplitudes, in summer, can be considered with average values. In zone I2-V1 area, it is the municipality of Fornos de Algodres, where the summer is mild, and winter more demanding than summer. The municipality of Fundão is in the climatic zone I2-V3, whose summer conditions overlap with those of winter, dry climates, and high thermal amplitudes, with heating needs in winter.

These zones are important for determining the characteristics necessary for the building to play its role in terms of thermal comfort, based on the reference data for winter and summer.

The European Parliament and the Council of the European Union have reformulated Directive 2002/91/EC, through Directive 2010/31/EU of May 19, 2010 (European Parliament and the Council of the European Union, 2010). It establishes the requirements with regard to the common general framework for a methodology for calculating the integrated energy performance of buildings and autonomous fractions; the application of minimum requirements for the energy performance of new buildings and new building fractions; the application of minimum requirements for the energy performance of existing buildings, autonomous

**Table 1** Distribution of municipalities of the Beiras and Serra da Estrela according to climatic zones

| Municipalities | Winter climate zone | Number of degree days (GD) (°C days) | Heating season duration (months) | Summer climate zone | Project external temperature (°C) | Thermal amplitude (°C) |
|---|---|---|---|---|---|---|
| Almeida | I3 | 2540 | 7.7 | V2 | 33 | 16 |
| Belmonte | I2 | 1970 | 7.7 | V2 | 32 | 13 |
| Celorico da Beira | I3 | 2240 | 7.7 | V1 | 30 | 12 |
| Covilhã | I3 | 2250 | 7.3 | V2 | 32 | 13 |
| Figueira Castelo Rodrigo | I3 | 2450 | 8 | V2 | 33 | 16 |
| Fornos de Algodres | I2 | 2060 | 7.7 | V1 | 31 | 13 |
| Fundão | I2 | 1990 | 7 | V3 | 34 | 14 |
| Gouveia | I3 | 2440 | 8 | V1 | 30 | 12 |
| Guarda | I3 | 2500 | 8 | V1 | 31 | 13 |
| Manteigas | I3 | 3000 | 8 | V1 | 30 | 12 |
| Meda | I3 | 2360 | 7.7 | V2 | 33 | 14 |
| Pinhel | I3 | 2390 | 7.7 | V2 | 32 | 15 |
| Sabugal | I3 | 2450 | 7.3 | V2 | 33 | 16 |
| Seia | I3 | 2520 | 7.7 | V2 | 32 | 14 |
| Trancoso | I3 | 2450 | 7.7 | V2 | 32 | 13 |

Source: MOPTC (2006c)

fractions and building components subject to major renovations, constructive elements of the building envelope with a significant impact on the energy performance of the envelope when they are renovated or replaced, and technical building systems when a new system is installed or when the existing system is replaced or upgraded; national plans to increase the number of buildings with almost zero energy needs; the energy certification of buildings or autonomous fractions; regular inspection of heating and air conditioning installations in buildings; and the systems of independent control of energy performance certificates and inspection reports (European Parliament and Council of the European Union, 2010).

The directive was transposed in Portugal by Decree-Law n° 118/2013 of 20 August (MEE, 2013), which also includes the Energy Certification System (ECS), the Regulation of Energy Performance of Residential Buildings (REH), and the Energy Performance Regulation of Commercial and Service Buildings (RECS).

In the residential sector in Portugal, the quality of buildings and their comfort has increased in recent years. The high consumption associated with thermal comfort (heating and cooling) is where the lack of energy efficiency of houses and residential buildings is reflected.

The National Energy Plan and Climate 2021–2030, approved by Resolution of the Council of Ministers n° 53/2020, (PCM, 2020a) includes measures to combat energy poverty and perfect hedging instruments for vulnerable people. To this end, a long-term strategy to combat energy poverty will be developed, with the main objective of obtaining a diagnosis and characterization of the problem, developing monitoring indicators, monitoring strategies, establishing objectives for reducing energy poverty in the medium and long term on national, regional, and local scales, and proposing specific measures to achieve these objectives, b as forms of financing. After the implementation of this strategy in the short term, measures to combat energy poverty will be defined and dimensioned, providing benefits that guarantee the supply of energy to vulnerable citizens or providing support to improve the energy efficiency of homes (PCM, 2020a).

In the context of transposition, Directive (EU) 2018/844 of the European Parliament and of the Council, of May 30, 2018, on the energy performance of buildings and Directive 2012/27/EU on energy efficiency, for the national scope by Decree-Law No. 101-D/2020 (PCM, 2020b).

In 2021, the Long-Term Strategy for the Renovation of Buildings (ELPRE) was approved by the Resolution of the Council of Ministers no. 8-A/2021 (PCM, 2021), which approves the Long-Term Strategy for the Renovation of Buildings (ELPRE). The ELPRE foresees four packages of measures aimed at improving comfort and mitigating energy poverty, increasing energy efficiency, local decarbonization, and gradually increasing comfort. By 2030, the priority will be to improve comfort, mitigate energy poverty, and increase the energy efficiency of 65% of existing residential buildings, built before 1990 and which have the worst energy performance. The measure extends, in 2040, to the remaining residential buildings built until 2016, which will represent 100% of the national park of existing buildings in 2018. In non-residential buildings, with priority for public buildings, the strategy foresees the intervention by 27% by 2030 and 52% by 2040.

## 3 Methodology

Methodologically, this research relies on two different approaches. The first is theoretical analysis based on the literature review of the energy performance of building and the legal regime of the unworthy conditions of the sustainable living of the citizen. The second is an empirical analysis with an exploratory statistical analysis, supported on database of the Agency of Energy (ADENE, 2021).

# 4 Results

An Energy Certificate (CE) of a building or autonomous fraction is the document issued under the National System of Energy Certification and Indoor Air Quality of Buildings (SCE), which reflects the thermal and energy performance of the property, classifying it as energy efficiency and associated energy consumption. The energy class, Fig. 1, can vary on a scale from A+ (more efficient) to F (less efficient). It compares the energy performance of a building or autonomous fraction under current conditions with the performance that it would obtain under minimum conditions, based on reference values or applicable requirements.

In the EC, proposals for improvement measures are presented in terms of energy performance and indoor air quality and comfort, highlighting those of greater economic viability, to reduce the energy consumption of the property and improve comfort.

Under Order n° 15793-J/2013 of the Directorate General for Energy and Geology (MAOTE, 2013) to determine the energy certification of residential buildings, the energy class ratio ($R_{Nt}$) is used:

$$RNt = \frac{Ntc}{Nt}$$

**Fig. 1** Classification of energy certificates by classes. Source: ADENE (2017a)

**Table 2** Distribution of the energy class in residential buildings for housing

| Energy class | Value of $R_{Nt}$ |
|---|---|
| A+ | $R_{Nt} \leq 0.25$ |
| A | $0.26 \leq R_{Nt} \leq 0.50$ |
| B | $0.51 \leq R_{Nt} \leq 0.75$ |
| B− | $0.76 \leq R_{Nt} \leq 1.00$ |
| C | $1.01 \leq R_{Nt} \leq 1.50$ |
| D | $1.51 \leq R_{Nt} \leq 2.00$ |
| E | $2.01 \leq R_{Nt} \leq 2.50$ |
| F | $R_{Nt} \geq 2.51$ |

Source: MAOTE (2013)

where $N_{tc}$ corresponds to the value of the annual nominal needs for primary energy [$kWh_{EP}/(m^2$ year)] and Nt corresponds to the regulatory limit value for the annual nominal needs of primary energy [$kWh_{EP}/(m^2$ year)]. The annual requirement values are calculated in accordance with the Energy Performance Regulation of Residential Buildings (REH). To sort the buildings or separate fractions, a scale where each range of values $R_{Nt}$ is used corresponding to a power class presented in Table 2.

According to article 3 of Decree-Law n° 28/2016, the Energy Certification System (ECS) covers buildings or fractions, new or subject to major intervention, under the terms of the REH and RECS. When, however, a fraction already built is not constituted as an autonomous fraction according to a constitutive title of Horizontal Property (HP), it is only covered by the SCE from the moment it is given on lease. Also covered by the SCE are all buildings or fractions that exist from the moment of their sale, donation in fulfillment, or lease after the entry into force of this law (ME, 2016). Since July 2008, all new buildings must have a valid certificate. In the case of existing buildings, since 2009 they must have a valid certificate, at the stage of entering the respective sale, rental, or lease contract. Of the energy certificates issued, the largest percentage is for residential buildings 88.19% and 88.83%, in Portugal and in the *Beiras and Serra da Estrela* Region, respectively.

Table 3 shows that the number of energy certificates issued by ECS to residential buildings for housing, until March 23, 2021, in Portugal, was 1234.062, with the majority (23.7%) relating to energy efficiency class D, the sixth on a scale of eight, since the energy class ratio is between 1.51 and 2.00 $kWh_{EP}/(m^2$. year). And A + class is more efficient and F less efficient. A fraction that complies with the minimum required by the current regulations will be classified in energy class B-.

Until that same date, 20,551 energy certificates for residential buildings had been issued for the Beiras and Serra da Estrela region, with the majority (25.2%) corresponding to the energy efficiency class F. Certificates with this classification are higher in number in the municipalities of Figueira Castelo Rodrigo (159), Fornos de Algodres (120), and Manteigas (115), which corresponds to 46.4%, 43.2%, and 41.5%, respectively. Of the municipalities of Beiras and Serra da Estrela, those with the highest percentage of F-rated residential buildings are Figueira de Castelo Rodrigo (46%), Fornos de Algodres (43%), and Manteigas (42%). In the case of residential buildings, the highest percentage of energy certificates issued

**Table 3** Distribution of energy certificates for residential buildings for housing, 2021

| | A+ | A | B | B− | C | D | E | F | Total |
|---|---|---|---|---|---|---|---|---|---|
| Portugal | 28.937 | 127.435 | 128.301 | 95.398 | 277.400 | 292.930 | 180.189 | 103.472 | 1234.062 |
| Beiras e Serra da Estrela | 1.144 | 1.437 | 2.201 | 1.545 | 2.209 | 3216 | 3.617 | 5182 | 20.551 |
| Almeida | 28 | 18 | 43 | 27 | 21 | 23 | 55 | 129 | 344 |
| Belmonte | 16 | 25 | 39 | 50 | 54 | 74 | 113 | 143 | 514 |
| Celorico da Beira | 29 | 31 | 56 | 33 | 31 | 63 | 82 | 217 | 542 |
| Covilhã | 219 | 515 | 692 | 457 | 870 | 1.170 | 1.077 | 1.156 | 6.156 |
| Figueira Castelo Rodrigo | 34 | 17 | 18 | 21 | 13 | 30 | 51 | 159 | 343 |
| Fornos de Algodres | 12 | 20 | 43 | 22 | 9 | 18 | 34 | 120 | 278 |
| Fundão | 125 | 177 | 377 | 242 | 279 | 473 | 532 | 506 | 2.711 |
| Gouveia | 106 | 42 | 79 | 73 | 65 | 106 | 177 | 381 | 1.029 |
| Guarda | 192 | 297 | 326 | 295 | 530 | 673 | 642 | 777 | 3.732 |
| Manteigas | 15 | 7 | 24 | 12 | 21 | 35 | 48 | 115 | 277 |
| Meda | 17 | 23 | 59 | 32 | 15 | 27 | 59 | 135 | 367 |
| Pinhel | 41 | 38 | 93 | 57 | 20 | 42 | 51 | 160 | 502 |
| Sabugal | 61 | 41 | 97 | 67 | 50 | 77 | 138 | 314 | 845 |
| Seia | 179 | 149 | 162 | 101 | 211 | 363 | 490 | 675 | 2.330 |
| Trancoso | 70 | 37 | 93 | 56 | 20 | 42 | 68 | 195 | 581 |

Source: ADENE (2021)

corresponds to buildings that already exist: 77.45% for Portugal and 78.52% for Beiras and Serra da Estrela.

As previously mentioned for residential buildings, they are covered by the SCE when new, when subject to major intervention, or in the case of sale or lease after December 1, 2013, of the same or fraction thereof. In Table 4, the authors analyze the number of certificates issued for existing housing buildings, from 2014 to 2016, and it has decreased and slightly increased in 2017, followed by a decrease until 2020. In the Beiras and Serra da Estrela region, it decreased in 2015; compared to the previous year, it increased until 2017 and until 2020 it decreased. This fact can have its justification in the validity of 5 years of the CE for the case of housing buildings that do not undergo major interventions or in the case of leases if the tenant continues 5 years later and has not changed.

The certificates issued for existing housing buildings are in greater numbers in the municipalities of Covilhã, Guarda, Fundão, and Seia, with 4625, 3071, 2011, and 1955 certificates, respectively, a fact that can be justified by the growing demand for housing and since this certificate becomes mandatory (Decree-Law n° 118/2013) (MEE, 2013) in case of sale or lease of the property.

The Regulation of Thermal Behaviour Characteristics of Buildings that establishes the rules to be observed in the design of all residential buildings and the air conditioning systems without service buildings centralized so that the requirements of thermal comfort, be it for heating or cooling, ventilation to guarantee air quality inside buildings, or domestic hot water needs, can be met without excessive energy expenditure. The pathological situations in the building elements caused by the occurrence of surface or internal condensations are also minimized, with a potential negative impact on the durability of the building elements and the quality of the indoor air.

The Energy Pre-Certificate (PCE) is the SCE certificate for new buildings or fractions in new buildings, as well as for buildings or fractions subject to major interventions, issued in the design phase before the start of construction or major intervention.

A concern of housing policies is undoubtedly the need to rehabilitate the aging housing stock. In general terms, the certificates issued for construction projects are greater in number than those issued for rehabilitation projects.

In the region of Beiras and Serra da Estrela, from 2014 to March 23, 2021, a total of 834 certificates were issued for projects for the rehabilitation of buildings or fractions for housing purposes (Table 5). The municipalities with the highest number of these certificates were Covilhã (200), Guarda (154), and Seia (98).

During the licensing procedure of a building or permit building, it is not issued an energy certificate, but a PCE which confirms that the project meets the requirements imposed by the regulations. In practice, it corresponds to a "project certificate" since it has an identical format and the same type of information as a certificate.

The certificates issued for construction projects (Table 6), in the region of Beiras and Serra da Estrela, from 2014 to March 23, 2021, were 2607.

The municipalities with the highest number of PCEs issued for the construction of buildings intended for housing were Covilhã (903), Fundão (428), and Guarda (364).

**Table 4** Distribution of energy certificates for existing buildings for housing, 2014–2021

| | Existing buildings for housing | | | | | | | | |
|---|---|---|---|---|---|---|---|---|---|
| | 2014 | 2015 | 2016 | 2017 | 2018 | 2019 | 2020 | 2021 | Total |
| Portugal | 147.690 | 137.733 | 131.384 | 140.630 | 137.011 | 127.693 | 111.452 | 22.216 | 955.809 |
| Beiras e Serra da Estrela | 2.302 | 1.991 | 2.209 | 2.552 | 2.385 | 2.300 | 2.007 | 390 | 16.136 |
| Almeida | 31 | 37 | 39 | 38 | 38 | 41 | 27 | 13 | 264 |
| Belmonte | 87 | 45 | 59 | 59 | 70 | 51 | 37 | 12 | 420 |
| Celorico da Beira | 79 | 58 | 78 | 57 | 60 | 50 | 39 | 10 | 431 |
| Covilhã | 535 | 462 | 565 | 927 | 706 | 667 | 645 | 118 | 4.625 |
| Figueira Castelo Rodrigo | 56 | 26 | 36 | 38 | 37 | 39 | 36 | 2 | 270 |
| Fornos de Algodres | 23 | 25 | 24 | 37 | 40 | 26 | 23 | 5 | 203 |
| Fundão | 280 | 193 | 248 | 281 | 335 | 340 | 278 | 56 | 2.011 |
| Gouveia | 140 | 79 | 141 | 139 | 118 | 104 | 86 | 21 | 828 |
| Guarda | 498 | 495 | 457 | 420 | 432 | 378 | 343 | 48 | 3.071 |
| Manteigas | 23 | 16 | 32 | 42 | 33 | 48 | 38 | 4 | 236 |
| Meda | 37 | 34 | 42 | 34 | 42 | 36 | 41 | 7 | 273 |
| Pinhel | 43 | 79 | 66 | 73 | 64 | 82 | 74 | 21 | 502 |
| Sabugal | 83 | 112 | 104 | 72 | 105 | 81 | 70 | 10 | 637 |
| Seia | 347 | 278 | 255 | 265 | 251 | 284 | 227 | 48 | 1.955 |
| Trancoso | 40 | 52 | 63 | 70 | 54 | 73 | 43 | 15 | 410 |

Source: ADENE (2021)

**Table 5** Distribution of energy certificates for rehabilitated buildings for housing project, 2014–2021

| | Rehabilitated buildings for housing project | | | | | | | | Total |
|---|---|---|---|---|---|---|---|---|---|
| | 2014 | 2015 | 2016 | 2017 | 2018 | 2019 | 2020 | 2021 | |
| Portugal | 225 | 435 | 728 | 1.013 | 1.065 | 1.102 | 1.141 | 255 | 5.964 |
| Beiras e Serra da Estrela | 47 | 42 | 87 | 98 | 159 | 187 | 177 | 37 | 834 |
| Almeida | – | 1 | 6 | 2 | 5 | 2 | 3 | – | 19 |
| Belmonte | – | 1 | 1 | 2 | 2 | 3 | – | – | 9 |
| Celorico da Beira | 7 | – | 5 | 2 | 4 | 8 | 3 | 1 | 30 |
| Covilhã | – | 8 | 8 | 29 | 33 | 74 | 33 | 15 | 200 |
| Figueira Castelo Rodrigo | – | 2 | 2 | 1 | 1 | – | 4 | – | 10 |
| Fornos de Algodres | 3 | – | 1 | – | – | 4 | 2 | – | 10 |
| Fundão | 2 | 5 | 14 | 6 | 36 | 9 | 10 | – | 82 |
| Gouveia | 3 | 8 | 7 | 8 | 20 | 13 | 9 | – | 68 |
| Guarda | 17 | 4 | 19 | 16 | 17 | 25 | 47 | 9 | 154 |
| Manteigas | 2 | – | – | – | 1 | – | 9 | – | 12 |
| Meda | 1 | 1 | 3 | 2 | 1 | 3 | 1 | 2 | 14 |
| Pinhel | 1 | 5 | 3 | 6 | 5 | 17 | 16 | 3 | 56 |
| Sabugal | 3 | 1 | 7 | 12 | 9 | 9 | 10 | – | 51 |
| Seia | 8 | 5 | 11 | 10 | 19 | 16 | 23 | 6 | 98 |
| Trancoso | – | 1 | – | 2 | 6 | 4 | 7 | 1 | 21 |

Source: ADENE (2021)

The municipalities of Covilhã, Fundão, Guarda, and Trancoso are the ones with the highest number of certificates issued for construction projects to the detriment of the certificates issued for rehabilitation projects.

## 5 Discussion

Households in the European Union face the difficulty of having their homes heated and paying their energy bills. Recognizing the need to combat the risk of energy poverty faced by households that do not have the means to enjoy essential energy services guarantees an adequate standard of living. Effective programs, such as financing schemes that allow families to renovate their homes, can reduce the amount of energy bills and help the environment. In 2020, the Commission will develop guidelines to help Member States to address the issue of energy poverty (European Commission, 2019).

In a study developed by ADENE, in July 2017, within the scope of the awareness campaign on energy efficiency in private housing, it indicated that consumers spend on average about €112.00 in the consumption of energy and water in the home. Electricity is the source of energy that implies the highest expenses, followed by gas

**Table 6** Distribution of energy certificates for new buildings for housing project, 2014–2021

| | New buildings for housing project | | | | | | | | |
|---|---|---|---|---|---|---|---|---|---|
| | 2014 | 2015 | 2016 | 2017 | 2018 | 2019 | 2020 | 2021 | Total |
| Portugal | 669 | 949 | 1.302 | 1.532 | 2.057 | 2.057 | 2.157 | 451 | 11.174 |
| Beiras e Serra da Estrela | 238 | 384 | 289 | 446 | 357 | 436 | 403 | 54 | 2.607 |
| Almeida | 7 | 4 | 6 | 7 | 8 | 5 | 2 | 2 | 41 |
| Belmonte | 2 | 5 | 29 | 6 | 8 | 11 | 2 | 3 | 66 |
| Celorico da Beira | 4 | 9 | 11 | 9 | 3 | 7 | 4 | 1 | 48 |
| Covilhã | 77 | 183 | 34 | 188 | 90 | 161 | 158 | 12 | 903 |
| Figueira Castelo Rodrigo | 9 | 7 | 5 | 5 | 10 | 2 | 5 | – | 43 |
| Fornos de Algodres | 2 | 4 | 7 | 10 | 10 | 5 | 10 | – | 48 |
| Fundão | 30 | 49 | 59 | 85 | 52 | 72 | 71 | 10 | 428 |
| Gouveia | 10 | 17 | 11 | 12 | 14 | 10 | 8 | 1 | 83 |
| Guarda | 21 | 28 | 49 | 39 | 63 | 99 | 60 | 5 | 364 |
| Manteigas | 1 | 4 | – | 3 | 5 | 4 | 4 | – | 21 |
| Meda | 9 | 5 | 11 | 8 | 18 | 10 | 4 | – | 65 |
| Pinhel | 16 | 10 | 11 | 9 | 10 | 8 | 10 | 9 | 83 |
| Sabugal | 14 | 18 | 20 | 20 | 23 | 17 | 9 | 5 | 126 |
| Seia | 24 | 27 | 23 | 27 | 29 | 17 | 33 | 2 | 182 |
| Trancoso | 12 | 14 | 13 | 18 | 14 | 8 | 23 | 4 | 106 |

Source: ADENE (2021)

and water. Many of the interventions carried out in the buildings are aimed at increasing energy efficiency in housing, which translates into a positive impact, namely in the relief of the monthly budget through the reduction of energy expenditure. As a result of the interventions, it is also worth mentioning the importance of increasing thermal comfort in improving people's health conditions (ADENE, 2017b).

In Portugal, with an extremely old housing stock, the lack of sustainability of the buildings means that people must use more energy to satisfy their needs, whether for heating, cooling, ventilation, dehumidification, or natural light. This is reflected in their expenses, but which for many become unaffordable.

According to the general statistics of the energy certification of buildings available on the website of the SCE, available by ADENE, the number of energy certificates issued for day March 23, 2021, in Portugal was 1,234,062, with the majority (23.7%) relating to energy efficiency class D, the sixth on a scale of eight, which does not meet the minimum requirements required by current regulations.

This class arises from the classification of the thermal behavior of the building's housing elements, such as the roof, walls, windows, and floors. These will contribute to energy consumption associated with air conditioning and thermal comfort in the home. To achieve this comfort, you must consider the gains (in the summer) and heat losses (in the winter) of the house.

Based on climatic zones, in the region of Beiras and Serra da Estrela, the 15 municipalities are divided between zones I2 and I3 in winter and V1, V2, and V3 in summer. Hence, this region is dominated by the severity of the winter because the 12 municipalities belong to I3 zone. Eight of these are characterized by high thermal amplitudes (I3-V2); only in one of the municipalities is summer more pronounced than winter (I2-V3). For the region of Beiras and Serra da Estrela, 20.551 ECS had been issued for residential buildings, of which 5.182 correspond to the class of energy efficiency F, characterized by the worst energy performance.

## 6 Conclusion

This research aims to bring the issue of unsustainability of housing to the discussion. As part of the research developed in themes of sustainability and housing in the framework of the master's thesis in Integrated Systems of Management (Environment, Quality, Safety, and Social Responsibility). The subject of research about housing is of great importance to the quality of life of the citizen that demands a place to live that he or she calls home. It can have influence on our health, our education, and our role in society.

The region of Beiras and Serra da Estrela has territories of different characteristics, with higher altitude varying between 845 m in Almeida municipality and 1993 m in the Covilhã, Manteigas, and Seia municipality. This region corresponds to about 7% of the territory of mainland of Portugal, with the smallest area being Belmonte municipality with 118.76 km$^2$ and the largest area Sabugal municipality with 822.70 km$^2$. Another important aspect is the age of building because the preponderance of buildings aged, with the worst energy performance, in a region mainly characterized by the severity of the winter.

If on one side the lack of quality of the air inside, or the absence of ventilation, compromises the health, the lack of natural lighting affects psychologically. If there was something that this pandemic brought, it was the change of our idea of home, because of the time we have been forced to stay in it and how it was or not adequate to our needs. It represented one challenge for everyone, but for those who have housing needs, it was even more painful, such as in the case of most seniors who were isolated, many of them in old houses and with little heat.

With a older housing stock and consequent low performance energy, Portugal should be worth the policies of rehabilitation and improvement of efficiency energy.

The limitation of this research points out the challenge of the theme of certification energy, with its terms and requirements without help. This scenario of difficulties is compounded by the lack of economic capacity of households, especially of elderly citizens. A household to be classified as a "financially deprived situation," so its average monthly income (RMM) must be less than four IAS, so the value is € 1715.60. In 2017, the RMM per household, in Portugal, is lower (€ 1474.17) than the value of the household classified in "situation of financial need" (€ 1715.60). It can also be seen that all municipalities in the region of Beiras and Serra da Estrela also have lower values, with the municipality with the lowest RMM being Mêda (€ 1071.58) and the highest RMM being Guarda (€ 1518.58).

The first challenge of this research is demanding the awareness of better housing and promoting the participation mechanisms, specifically emphasis on the (isolated) elderly, which need to be strengthened on rural and mountain areas. The second challenge of this research is in terms of demographics characterized by a low-density population 33.6 persons per $km^2$, in 2019, with the most worrying being Figueira de Castelo Rodrigo (11.0) and Almeida (11.4) and Covilhã the municipality with a greater number 84.2 people per $km^2$.

This research is a second step of research project that aims to help the European Union to encourage the Member States to use European funds and to ensure transition between unworthy conditions and social inclusion. Indeed, this will help to improve the situation of housing for each citizen, but it is a long road to the main well-being of the citizen.

**Acknowledgment** The authors wish to thank Professor David Crowther, Jose Angel Perez Lopez, and two anonymous referees for their helpful comments on earlier versions of this paper. Also, the organizing committee (Professor Irene Arraiano, in particular) and scientific committee (Samuel Olu Idowu, Réne Schmidpeter and Nicholas Capaldi of Global Corporate Governance Institute). The second author is financed by national funds through FCT - Foundation for Science and Technology, I.P., within the scope of multi-annual funding UIDB/04043/2020.

## References

Agência para a Energia (ADENE, 2017a). *Manual da Etiqueta Energética*. fevereiro de 2017. Lisboa: ADENE.

Agência para a Energia & Consulmark (ADENE). (2017b). *Estudo de mercado no âmbito das campanhas de sensibilização e de promoção da eficiência energética na Habitação Particular-Relatório Final*. 7 de julho de 2017. Lisboa.

Agência para a Energia (ADENE, 2021). *Certificação Energética de Edifícios*. Available: https://www.sce.pt/estatisticas/. Lisboa: ADENE.

Assembleia da República (AR). (1976). *Constituição da República Portuguesa*. Diário da República n° 86/1976, Série I de 10 de abril de 1976. Lisboa.

Assembleia da República (AR). (2019). *Lei n° 83/2019 - Lei de bases da habitação*. Diário da República n° 168/2019, Série I de 3 de setembro de 2019. Lisboa.

Council of Europe. (1961). *European social charter*. Council of Europe.

European Commission (EC). (2019). *Comunicação da Comissão ao Parlamento Europeu, ao Conselho Europeu, ao Conselho, ao Comité Económico e Social Europeu e ao Comité das Regiões - The European Green Deal*, de 11 de dezembro de 2019.

European Commission (EC). (2020). *Recomendação (UE) 2020/1563 da Comissão de 14 de outubro de 2020, sobre a pobreza energética*. Jornal Oficial da União Europeia de 27 de outubro de 2020.

European Parliament and Council of the European Union. (2012). *Diretiva 2012/27/UE do Parlamento Europeu e do Conselho de 25 de outubro de 2012 relativa à eficiência energética, que altera as Diretivas 2009/125/CE e 2010/30/UE e revoga as Diretivas 2004/8/CE e 2006/32/CE*. Jornal Oficial da União Europeia de 14 de novembro de 2012.

European Parliament and Council of the European Union. (2002). *Diretiva 2002/91/CE do Parlamento Europeu e do Conselho de 16 de dezembro de 2002 relativa ao desempenho energético dos edifícios*. Jornal Oficial das Comunidades Europeias de 04 de janeiro de 2003.

European Parliament and Council of the European Union. (2010). *Diretiva 2010/31/UE do Parlamento Europeu e do Conselho de 19 maio de 2010 relativa ao desempenho energético dos edifícios*. Jornal Oficial da União Europeia de 18 de junho de 2010.

European Parliament and Council of the European Union. (2018a). *Diretiva (UE) 2018/844 do Parlamento Europeu e do Conselho de 30 de maio de 2018 que altera a Diretiva 2010/31/UE relativa ao desempenho energético dos edifícios e a Diretiva 2012/27/UE sobre a eficiência energética*. Jornal Oficial da União Europeia de 19 de junho de 2018.

European Parliament and Council of the European Union. (2018b). *Diretiva (UE) 2018/2002 do Parlamento Europeu e do Conselho de 11 de dezembro de 2018 que altera a Diretiva 2012/27/UE relativa à eficiência energética*. Jornal Oficial da União Europeia de 21 de dezembro de 2018.

Eurostat. (2021). *Inability to keep home adequately warm- Income and living conditions EU-SILC survey*. Obtido de Eurostat, o serviço de estatística da União Europeia. Availability: https://ec.europa.eu/eurostat/web/products-eurostat-news/-/DDN-20200106-1

Ministério da Economia (ME). (2016). *Decreto-Lei n° 28/2016*. Diário da República n° 119/2016, Série I de 26 de junho de 2016. Lisboa.

Ministério da Economia e da Inovação (MEI). (2006). *Decreto-Lei n° 78/2006*. Diário da República n° 67/2006, Série I-A de 4 de abril de 2006. Lisboa.

Ministério da Economia e do Emprego (MEE). (2013). *Decreto-Lei n° 118/2013*. Diário da República n° 159/2013, Série I de 20 de agosto de 2013. Lisboa.

Ministério das Obras Públicas, Transportes e Comunicações (MOPTC). (1990). *Decreto-Lei n° 40/90*. Diário da República n° 31/1990, Série I-A de 6 de fevereiro de 1990. Lisboa.

Ministério das Obras Públicas, Transportes e Comunicações (MOPTC). (2006a). *Decreto-Lei n° 78/2006*. Diário da República n° 67/2006, Série I-A de 4 de abril de 2006. Lisboa.

Ministério das Obras Públicas, Transportes e Comunicações (MOPTC). (2006b). *Decreto-Lei n° 79/2006*. Diário da República n° 67/2006, Série I-A de 4 de abril de 2006. Lisboa.

Ministério das Obras Públicas, Transportes e Comunicações (MOPTC). (2006c). *Decreto-Lei n° 80/2006*. Diário da República n° 67/2006, Série I-A de 4 de abril de 2006. Lisboa.

Ministério do Ambiente, Ordenamento do Território e Energia (MAOTE). (2013). *Despacho (extrato) n° 15793-J/2013*. Diário da República n° 234/2013, 3° Suplemento, Série II de 3 de dezembro de 2013. Lisboa.

Presidência Conselho de Ministros (PCM). (2018). *Decreto-Lei n.° 37/2018*. Diário da República n.° 106/2018, Série I de 04 de junho de 2018. Lisboa.

Presidência do Conselho de Ministros (PCM). (2020a). *Resolução do Conselho de Ministros n° 53/2020 - Aprova o Plano Nacional Energia e Clima 2030 (PNEC 2030)*. Diário da República n° 133/2020, Série I de 10 de julho de 2020. Lisboa.

Presidência do Conselho de Ministros (PCM). (2020b). *Decreto-Lei n.° 101-D/2020*. Diário da República n.° 237/2020, 1° Suplemento, Série I de 07 de dezembro de 2020. Lisboa.

Presidência do Conselho de Ministros (PCM). (2021). *Resolução do Conselho de Ministros n° 8-A/2021 - Aprova a Estratégia de Longo Prazo para a Renovação dos Edifícios (ELPRE)*. Diário da República n° 23/2021, 1° Suplemento, Série I de 3 de fevereiro de 2021. Lisboa.

United Nations (UN). (1948). *Universal declaration of human rights*. UN.

United Nations (UN). (1976). *Vancouver declaration on human settlements*. Report of habitat: United Nations conference on human settlements. UN.

World Health Organization (WHO). (2018). *WHO housing and health guidelines*. WHO.

**Sónia Bidarra** is a student of the master's degree in Integrated Management Systems (Environment, Quality, Safety, and Social Responsibility) from the Polytechnic Institute of Guarda. Currently, Sónia is preparing her dissertation to obtain a master's degree. For the 2 years since May 2020, she has been working at UDI of the Polytechnic Institute of Guarda. She has Degree in Management of the Polytechnic Institute of Guarda. Sónia presented the article "A Gender Equality perspective in Polytechnic Institute of Guarda" and the article "Risk Management in Local Housing Strategy" in ICAFI 2020. She regularly develops scientific activities related to the participation and organization, conferences, and meetings. Furthermore, Sónia promotes research projects related to local housing strategies. Sónia presented the Project "ANIMARK—Intelligent Monitoring System for Animals," in the Immersion Program BfK Ideas 2020; collaborated in the Project "UNDERGUARDA" in the "17th Edition of Poliempreende—Projects of Entrepreneurial Vocation"; and presented the Business Idea "Rumex Induratus—The discovery for New Village,", in the Immersion Program BfK Ideas 2019

**Rute Abreu** is Professor since 1990 in Accounting and Finance at the Polytechnic Institute of Guarda, Portugal. Rute teaches financial accounting, auditing, quality, environment and safety standards, sustainability, and social responsibility management. Rute has a master's degree in industrial engineering from the Universidade Nova de Lisboa, Portugal, and a Ph.D. in Accounting and Finance from the Universidad de Salamanca, Spain. Actually, she is Accounting Degree Director and Director of master's degree in Integrated Management System. Usually, she develops activities with Social Responsibility Research Network and Global Corporate Governance Institute. Rute researches at the Center for Research in Accounting and Taxation of the Polytechnic Institute of Cávado and Ave (CICF-IPCA), Portugal, on CSR, corporate governance, accounting, and firm valuation, and it has been published in several journals. Also, she participates in the blind review process for several journals and helps as Editor on *The Social Responsibility Journal* and *Accounting and Management Review of Portuguese Chartered Accountants*. Regularly, she develops scientific activities related to participation and organization, all over the world, conferences, and meetings. Additionally, Rute promotes research projects related to accounting education, tax citizenship, management quality systems in social economy, and performance analysis at beauty firms that transfer knowledge to society.

**Cecília Rosa** is Professor at the Department of Mathematics of Escola Superior de Tecnologia e Gestão at the Polytechnic Institute of Guarda, Portugal. She holds a Ph.D. in Mathematics from the University of Beira Interior, Portugal; Applied Mathematics Master from the University of Porto, Portugal; and a Degree in Mathematics from the University of Coimbra, Portugal. She is a collaborating member of the Research Group in Applied Mathematics of the Research Center Instituto de Telecomunicações (IT) and a member of the Research Center Unit for the Development of the Interior (UDI) of the IPG. Her main research is in applied mathematics, namely numerical algorithms where we study metapopulation dynamics with application to demographics; application of optimal control techniques to compartmental problems of epidemiology; mathematics applied to management; and data analysis and artificial intelligence. Additionally, Cecília is a member of the research project: Local Housing Strategy in the municipality of Figueira de Castelo Rodrigo, Portugal.

# Institutional Roles in Promoting Corporate Social Responsibility in Tanzania

**Ibrahim Ramadhani Kikwiye**

## 1 Introduction

Corporate social responsibility (CSR) activities are confronted by various institutions in the country in which both multinational enterprises and their local counterparts operate (Demir & Budur, 2019; Khan et al., 2021; Kim et al., 2018; Kumi et al., 2020; Long et al., 2020). This chapter seeks to determine the roles played by institutions in promoting or hindering the CSR activities. Selected company managers are used to justify the contentions of institutions as playing roles in making CSR practices. This study is performed in a developing country of Tanzania by examining few and selected companies. Companies both multinationals and their local counterparts operate to serve the purpose for which they were established and thus contribute to the well-being of the stakeholders in various forms including CSR (Detomasi, 2008). While companies do so, they interact in many ways with institutions. These institutions impacting on the scale of operations, including those related to CSR, are composed of the national government regulatory domain, non-governmental organisations (NGOs), and community or beneficiary groups (Dentchev et al., 2015). The interactions with institutions may in one way or another shape the form and conduct of companies' participation in CSR (Steurer, 2010). These institutions are expected to create some form of roles in order to encourage and provide a good CSR environment.

Governments do have powers and are an authority in providing polices and strategies targeting socio-economic development (Wu & Davidson, 2011). The governments are proponents of CSR because they derive from it benefits which can revamp the welfare of their citizens in their country, and particularly it is

I. R. Kikwiye (✉)
Department of Accounting and Finance of the Institute of Finance Management (IFM), Dar es Salaam, Tanzania



I. G. Arraiano et al. (eds.), *Corporate Social Responsibility in a Dynamic Global Environment*, CSR, Sustainability, Ethics & Governance,
https://doi.org/10.1007/978-3-031-24647-0_6

important to the developing countries (Steurer, 2010). According to Albareda et al. (2008), governments especially from the developing country contexts experience development challenges which cannot be handled by themselves without the contributions from companies in form of CSR. As such, governments play four roles of mandating, facilitating, partnering, and endorsing in ensuring that CSR is practised and companies become legitimate operators (Fox et al., 2002). Likewise, government can play roles of regulating, brokering, warranting, and creating enabling environment for companies to flourish and then practise CSR (Albareda et al., 2008).

Similarly, government agencies have a role to play in making CSR a reality by companies in developing countries. Government agencies provide checks and balances on operations, inspection, and setting and administering regulations which are intended to deter any illegitimate practices of companies for stakeholders and communities in general (Greenwood et al., 2002). Furthermore, community members as a subset of institutions also play a role in making companies indulge in CSR. According to Wu and Davidson (2011: 114), community groups "serve as an unofficial watchdog and organisational whistleblower, ready to raise red-flag of criticisms if the corporation strays onto a path or policy deemed inappropriate by the group". The community can as well use lobby groups and public outcry using violent or non-violent means to make companies practise CSR (Yang & Rivers, 2009). And, finally, NGOs as a third sector exist to serve the interests of people establishing it or the interest of the wider community (Michael, 2003). By working on behalf of the society, NGOs can voice against any unwanted action(s) of the companies and have the ability to put into effect the practice of CSR.

The institutional influences have become the cornerstone of CSR practices in developed countries, while the same is not much documented in the context of developing countries including Tanzania (Amalados & Manohar, 2013). There exists a gap in the literature due to studies conducted previously not documenting the roles of institutions in a fashion suggesting government, NGOs, and beneficiary communities. Past studies have delved much on the institutions on a single basis (Kumi et al., 2020). The study of Long et al. (2020) was oriented towards establishing the role of government as the owner of some companies and its effect in the relationship between CSR and firm financial performance in the Chinese context. Demir and Budur (2019) studied on the roles of leadership ways and how they interact with NGOs in CSR activities and findings showed that companies which are oriented towards good ethical leadership experience institutional CSR with NGOs. Similarly, Kim et al. (2018) studied on how beneficiary community influences the CSR practices and found that they affect the practices of CSR symbolically as well as practically. However, few known studies such as that of Khan et al. (2021) in Pakistan, touched on the institutions in an integrated fashion. In this study, Khan et al. (2021) found that peer pressure contributed greatly to enforcing companies practising CSR. Despite the few exceptions, the concentration of most studies on single institutional influencing mechanisms creates a vacuum in understanding other institutional frameworks such as government, NGOs, and community members in relation to practising CSR (Kumi et al., 2020). The chapter aims at pointing out the roles played by the institutions in shaping the practice of

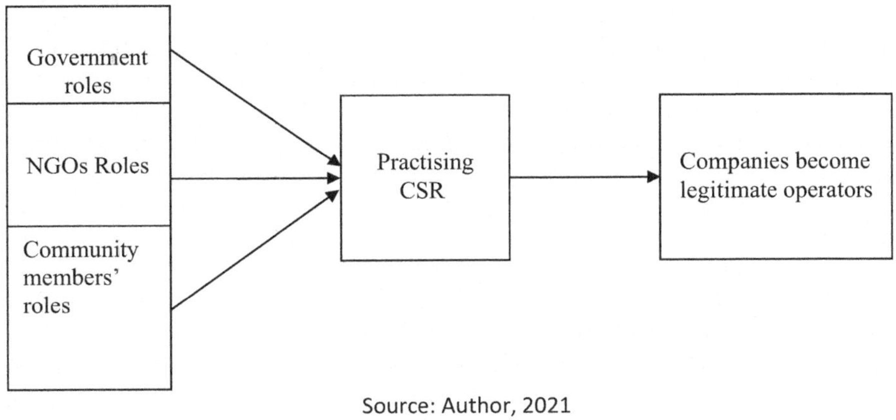


**Fig. 1** Conceptual framework of institutional roles, CSR practices, and legitimacy theory. Source: Author, 2021

CSR in the context of Tanzania and fills the gap in the literature. Specifically, the chapter seeks to address how the government organisations, NGOs, and beneficiary groups play a role in ensuring CSR practice as well as understanding the repercussions for companies which do not engage in CSR. In research questions one and two, selected managers of companies are involved in justifying the roles of institutions in practising CSR. Research questions three, To address the objectives, the chapter raises three research questions as follows:

*Research Question 1.* How the government plays roles in promoting companies to practise CSR in Tanzania?
*Research Question 2.* How do NGOs and beneficiary groups play roles in promoting CSR practices in Tanzania?
*Research Question 3.* What happens to non-complying companies with the institutions of government, NGOs, and community members?

The research questions are answered through the use of semi-structured interviews with government officers, company and NGO managers, and community members as well. The information through which the research questions are answered was undertaken in the place of work and in the field for beneficiary members.

The research questions feature the conceptual framework and legitimacy theory in addressing the institutional pushes to practise CSR and become legitimate. The conceptual framework is presented in Fig. 1.

The conceptual framework above exhibits a relationship between institutions and companies in realising the practising of CSR. Without such interactions, the roles of institutions become unavailable. Government play roles through regulations, rules, and laws, while NGOs are the watchdogs through which communities at both local and international levels can raise voice if the company is misbehaving and not practising CSR (Wu & Davidson, 2011). Community members raise concerns over the relationship which repels their recognition through CSR. Community members can raise the voice and boycott the commodities or operations of the multinational and/or local companies. While each player has its role to play, it is of ultimate usefulness to have all actors active in ensuring that the multinational companies and their local counterpart feel the pressure from all stakeholders and as such they will be able to participate in CSR. The framework presents the players (government, managers, NGOs, and community members) and how their integrated roles provide more concerted repercussions. While companies surrender to institutional roles, they ultimately become legitimate operators.

The rest of the chapter is structured as follows. The next section is about theoretical approach, while Sect. 3 contains literature review on the role of institutions in CSR. Section 4 highlights the methodology used, and Sect. 5 contains data sources and data analysis. Finally, Sects. 6 and 7 contain findings and discussions and conclusion and recommendations, respectively.

## 2 Theorical Approach: Legitimacy Theory

The study was undertaken in the light of legitimacy theory which is defined as "generalized perception or assumption that the actions of an entity are desirable, proper, or appropriate within some socially constructed system of norms, values, beliefs, and definitions" (Suchman, 1995: 574). The legitimacy theory focuses on the government, agencies, NGOs, and community members as institutions enforcing companies to be CSR responsive through peer pressure which ultimately targets achieving of legitimacy (Khan et al., 2021). Companies exist to operate in line with institutions' expectations, hence achieving legitimacy and a licence to operate (Hah & Freeman, 2014).

Similarly, legitimacy theory addresses issues pertinent to organisations, stakeholders, and general society as they depend on one another (Patriotta et al., 2011). The relationship that exist between companies' operations on the one hand and the regulatory agencies, NGOs, and community members on the other hand can be used to determine the relationship that exists between them regarding the CSR (Moffat et al., 2016). Companies are being legitimate because of the rights conferred to them by society, when the policies and strategies of such company are congruent with that of the government, agencies, NGOs, and the society in which they operate (Magness, 2006).

According to Suchman (1995: 575), "organizations seek legitimacy for many reasons, and conclusions about the importance, difficulty, and effectiveness of legitimating efforts may depend on the objectives against which these efforts are measured" by oversight bodies. That implies that application of the legitimacy theory will depend on the condition facing the organisation such as that of regulatory agencies, NGOs, and community members, and thus, it differs from place and place and time to time.

Moreover, legitimacy theory acts as a mediator when management faces problems with regulatory agencies, NGOs, and surrounding communities to make corrections of their past mistakes by adopting legitimate behaviours (Magness, 2006). According to Guthrie and Parker (1989: 343) although legitimacy theory is reactive in nature, "it agrees that business operates in society via a social contract where it agrees to perform various socially desired actions in return for approval of its objectives, other rewards and its ultimate survival. It therefore needs to disclose enough information for society to assess whether it is a good corporate citizen."

## 3 Literature Review on the Role of Institutions in Corporate Social Responsibility

### *3.1 Role of Government in CSR*

Governments possess ultimate power and authority in terms of policies and strategies targeting social and economic developments (Wu & Davidson, 2011). Consequently, governments provide a good regulatory framework for companies to earn profits and set guidelines that encourage more voluntary CSR (Steurer, 2010). Governments, especially in the developing country context, such as Tanzania are huge supporters of the CSR agenda as they derive benefits which can improve the welfare of the people in the country (Steurer, 2010). Despite the fact that CSR activities in most developing countries are voluntary in nature, regulations are used as a mechanism to oversee and control them (Bartle & Vaas, 2007). As Albareda et al. (2008) suggest, the challenges faced by governments in meeting community demands cannot be handled by the state itself. It is important to acknowledge government partnering with business organisations with the common goal of improving CSR (Li & Miao, 2020).

Fox et al. (2002) suggested four roles of governments that are used to accomplish CSR within the business community as mandating, facilitating, partnering, and endorsing. Other roles according to Albareda et al. (2008) include regulating, brokering, warranting, and creating an enabling environment for the companies to realise their objectives as well as becoming legitimate and obtain a licence to operate. When companies obtain legitimacy, they are confident to be supplied with resources and achieve support of institutions (Pfeffer & Salancik, 1978). The mandating by government refers to observing the minimum standards that businesses should abide by following laws and codes that are established by national governmental systems (Dentchev et al., 2015). Facilitating means that the government offers incentives to companies so that they become active in CSR matters. The most common incentives that are offered to companies are tax incentives although developing countries still face challenges to amend tax laws so as to take into account charities as a reason for incentives (Saleh, 2019). In their facilitating role, governments provide various encouragements to attract or motivate businesses to become more compliant with practising CSR (Albareda et al., 2007).

Similarly, the government can collaborate with companies through partnering by integrating the skills and input of public and private sectors. Albareda et al. (2007) and Steurer (2010) stress that partnering is used by governments to make collaborations and agreements with companies and other stakeholders to undertake joint programmes in order to improve social and environmental conditions. Even companies which fear being sidelined by not participating in CSR prefer to partner with the government and other stakeholders (Blowfield & Murray, 2011). For instance, governments have partnered in the PPP between state and private companies, mostly involved in road infrastructure in Brazil (Griesse, 2007).

## 3.2 Roles of Other Institutions (Community Members, NGOs, and Government Agencies)

As a subset of the civil society sector, community voice is another group that influences multinational and local companies to practise CSR. According to Lin et al. (2019), stakeholders such as beneficiary or community members are one of the most important drivers of companies to indulge in practising CSR. The community can voice its demands and play an essential role in the business decisions involving CSR. The community might form coalitions of unemployed and redundant, consumer groups, farmers and livestock keepers, environmentalists, human rights activists, health professionals, women and children, disabled people, and other minorities formed by like-minded individuals to advance their claims towards a company (Nazir & Islam, 2020). According to Wu and Davidson (2011: 114), community groups "serve as an unofficial watchdog and organizational whistle-blower, ready to raise a red-flag of criticisms if the corporation strays onto a path or policy deemed inappropriate by the group". According to Yang and Rivers (2009), communities may also use various lobby groups in the company and sideline those not paying attention to CSR in their surrounding areas. They view that a community's voice plays a significant role in influencing CSR practices by companies through public outcry against them using violent or non-violent means (Melubo et al., 2019).

As an influencing mechanism for companies to adopt CSR practices; NGOs exist to serve the interest of the few people establishing them or the interest of the wider community (Michael, 2003). NGOs work on behalf of the society surrounding them; they aim to influence policy-makers' decisions to be included or represented without becoming part of the management. In addition, NGOs have educational roles for a wider number of stakeholders with whom they cooperate in areas of their specialisation (Bebbington, 2004).

According to Bebbington and Farrington (1993), NGOs take part in some public sector activities in order to highlight the inefficiencies and bureaucracy and to encourage participatory development and enhance transparency. This calls for immediate measures to address such conditions by providing goods and services and CSR practices that see different organisations, businesses, and governments forging new relationships (Griesse, 2007). Where this applies, some functions related to it are taken over by NGOs to improve social standards (Beschorner & Muller, 2007). It is important for the third sector to align CSR objectives rather than relying on governmental efforts which are sometimes considered unsuccessful (Michael, 2003). Additionally, Collins (2012: 383) states, "governments; exercise control by developing and enforcing rules and regulations, businesses provide goods and services, and NGOs provide goods and services not met by governments and businesses."

Studies relating to institutional roles on CSR have increased in developing countries generally. Long et al. (2020) studied the role of government intervention in Chinese context and found that the involvement of government in an ownership of companies decreases the relationship between CSR and performance. Khan et al. (2009) in Bangladesh found that the government pressurises banks to practise CSR. Another study by Shauki (2011) in Indonesia found that the government's use of incentives to induce companies to practise CSR was an essential element. Sobhani et al. (2009) in Bangladesh found the government took major steps towards an institutional framework by observing corporate governance and the stock exchange markets. Xu and Liu (2020) as well studied the role of government promoting CSR through incentives in China and the findings indicated that the politically connected companies demonstrate better CSR than those without connections. Koku and Gustafsson (2003) in Ghana found that the institutional framework of district authorities, NGOs, and the local community was key to accomplishing environmental protection strategies. Matten and Moon (2008) studied a conceptual framework for a comparative understanding of CSR and found that CSR was more induced by use of incentives than regulations in the USA as opposed to companies in Europe. Ararat (2008) in Turkey found that the government promoted CSR by mandating a capital market board to establish inclusive guidelines that were then followed by all companies listed on the Istanbul exchange market.

A study conducted in Australia and Hong Kong by Lin et al. (2019) regarding the influence of stakeholders revealed that they used both cooperated and aggressive mechanisms to influence companies to take part in CSR. Moreover, in the Chinese context Li and Miao (2020) studied the role of collaboration of the government and companies in the pursuit of CSR and found that both direct and indirect collaborations enhance CSR practice. In the DRC, Maria and Devuyst (2011) found that mining companies experienced problems with both central and local governments that were not participating in environment protection and CSR. Additionally, Okoye (2012) studied the role of government in relation to CSR in Nigeria and found that there was less partnering with companies to protect the environment. In Kazakhstan, findings show that the government was more involved in CSR on economic and legal issues, while NGOs require businesses to focus on ethical and discretionary issues (Mahmood & Humphrey, 2013). Furthermore, Tan (2009) studied institutional structures in China and found that the government is not strong in enforcing MNEs, while NGOs are very strong and pressurise the MNEs to adopt ethical standards. Baughn et al. (2007) studied CSR in relation to environmental responsibility in Asian countries and found that the level of awareness is high and determined by economic development, political factors, and government influences.

# 4 Methodology

This study is guided by multi-case study design through which interactions with participants to gather information using semi-structured interviews were undertaken (Yin, 2012). The multi-case study is "an in-depth exploration from multiple perspectives of the complexity and uniqueness of a particular project, policy, institution, programme or system in a real life context" (Simons, 2009: 21). According to Yang (2010), a case can take various dimensions such as individual persons, groups, households, organisations, community, and/or nations. These multiple sources oriented from different organisations of government, companies and NGOs managers, and the beneficiary group.

The multi-case study design investigates the phenomenon by comparing the chosen cases leading to an in-depth understanding of the subjective meanings obtained from participants (Punch, 2014). This understanding is derived from the interaction with participants from government officers, companies, NGOs, and the local communities on how they play the roles in influencing the CSR practices. Each particular organisation (government, company, NGOs, and beneficiary groups) was treated as a single case and when combined they made up multi-cases (Bryman, 2016).

# 5 Data Sources and Data Analysis

## *5.1 Data Sources*

The data for this study were obtained from participants from government officers, companies and NGOs' managers, and community members. Government officers were selected based on their participation in decision making regarding the roles in fostering CSR practices. Similarly, officers with responsibility of sanctioning the conduct of companies against any social and environmental issues, which are core activities performed in the name of CSR, were also involved making a total of four interviewees from the government institutions. Similarly, NGOs were also selected on the basis of participation in CSR activities through being watchdogs against any social and environmental violations of companies (Fiaschi et al., 2017). Other NGOs were involved in the data collection process as they are advisors regarding the practices of CSR. This category made up a total of three NGOs interviewed. Finally, community members or beneficiary groups living around companies' areas of operations such as mining sites and oil and gas extraction areas were also used.

The participants in this category were composed of four focus group discussions from communities in mining and oil and gas production areas. However, their contributions were more or less and therefore four best selected statements from participants were used as evidence to support findings. The study selected purposively five managers from manufacturing, production, and bank and one director of operations in oil and gas company, and they were interviewed so as to provide supporting and/or opposing views of the institutions in promoting CSR. Four of the six companies were multinationals, while two of them were local counterparts. The evidence from company managers provides a basis through which the findings of the roles of institutions in CSR in Tanzania would be robust.

The data collection was done over a period of 2 years from 2018 to mid-2020 using semi-structured interviews with government officers, companies, and NGOs, while beneficiary communities interacted using focus group discussion (Merchant & Van de Stede, 2006). The interviews took place in the offices of the participants after seeking consent to do so. The number of participants from government officers, managers, and NGOs totalled 15, while three focus group discussions composed of six to ten members (women, men, and youth) were interviewed to determine how they play a role in influencing companies to practise in CSR. The composition of focus group discussions was gender inclusive and age sensitive (Fawcett & Pockett, 2015).

The data were collected sequentially in participants' offices and villages of abode beginning with all interviews and thereafter focus group discussions followed. All participants were given information in advance, and their consent was obtained to undertake the research through official letters from the government (Gray, 2009). The semi-structured interviews were in English, while the focus group discussions were in the local language which was then translated into English by a researcher. The two methods used interview guide, and all categories used similar questions to allow comparisons from one interviewee to another (Flick, 2014). The time taken for a semi-structured interview averaged 1 h and 10 min, while the time for focus group discussions was more than 2 h and 30 min.

## 5.2 Data Analysis

All semi-structured interviews and focus group discussions were audio recorded word for word and then transcribed using qualitative coding to determine the ensuing themes (Paulus et al., 2014). Thematic analysis was adopted to determine themes that were obtained from transcripts. Unwanted information was sorted and removed to allow the obtained themes to be interpreted and classified according to their similarities for generalisation purposes (Flick, 2014). The thematic analysis resulted in many themes including incentives, dialogue, and awareness during registration, directives, education, meetings, and request through letters, roles, punishments, and directives. Other themes were fines and penalties, closure of business, court cases, and fines. The themes that reflected the topic under investigation from SSIs and FGDs were combined and put together for the presentation of the results.

# 6 Findings and Discussions

The findings of this study are based on the objective of determining the roles of institutions and beneficiary communities in promoting CSR practices in Tanzania. The institutions are government and NGOs. The study determined the roles of institutions in Tanzania in stimulating companies to be more willing to participate in CSR practices. This section details how the government plays roles in ensuring that the maximum benefits of CSR are achieved. The findings are presented below.

*Research Question 1.* How the government plays roles in promoting companies to practise CSR in Tanzania?

## *6.1 The Roles Played by the Government*

The roles of government are discussed in terms of incentives to encourage companies, rewards, and punishments for abiding by and violating companies as well as the awareness of CSR during their registration process. Other roles of government include the directives, dialogue with companies, and the awareness during registration and operations.

### 6.1.1 The Role of Providing Incentives to Encourage Companies to Be More Compliant

The government discourages giving incentives to companies with the intention of practising CSR. The officers feel that the incentives issue is dependent on policy formulation. As long as there are no such policies guiding the conduct of CSR, the government cannot indulge in incentives because companies are expected to participate in CSR of their own accord. Similarly, another participant explained that decisions related to incentives have to be made by the minister of energy and minerals. The government office cannot be involved with giving incentives as that is outside their domain of operations. The agency provides an advisory role only to the ministry. To support the arguments, the principal officer from the ministry stated:

> The government doesn't give incentives, and we actually discourage incentives, but CSR helps companies to get the social licence to operate. And the social licence is something the company has to get from the community and not being imposed to them. It is the companies' responsibility to ensure their presence are felt and accepted by the community. Government Officer 1

> Giving incentives to the companies are a policy issue. The policy issues rely upon the Minister to make decisions. Our office has advisory roles to the higher authorities regarding matters connected with auditing which has something to do compliance and issue of advice thereafter. Government Officer 2

The findings concerning government taking the role of providing incentives to companies to be responsive to CSR reflect several prior studies such as those of Matten and Moon (2008) and Xu and Liu (2020). Matten and Moon (2008) found that the incentives are useful in attracting companies to participate in CSR instead of using regulations in US-based companies. According to Xu and Liu (2020), CSR is well practised by companies which are offered incentives in China. Regulations alone cannot work out as most of developing countries including Tanzania still experience voluntary CSR (Griseri & Seppala, 2010).

On the other hand, the only form of incentive that encourages companies to participate in CSR is that they get tax exemption (Galaskiewicz, 1997). The government office responsible for registration and promotion investment in Tanzania viewed that companies get tax exemptions on capital intensive expenditure when they invest in Tanzania. This finding is consistent with Saleh (2019) who found tax incentives as a catalyst for promoting CSR. The officer dealing with investment matters stated:

> We have no laws, and CSR is not mandatory. Companies do it on their own willingness. Generally, we offer incentives to investors in terms of tax exemptions on capital intensive equipment and machinery in the first years of their operations... They are tax exempt to encourage their fully operations. Government officer 3.

### 6.1.2 Directives from the Ministries and Agencies

The directives from the government are connected to the mining and oil and gas guidelines developed to give guidance on the need for companies to consider communities through CSR in their operations (Babiak & Trendafilova, 2011). The guidelines' emphasis is on caring about social, economic, and environmental matters for the communities. The guidelines give the directions on what companies are expected to do without conditions as CSR is not mandated. The regulatory agencies also share the guidelines with companies for the mutual benefit of the parties—a condition at parity with legitimacy theory requiring appropriate role in society (Palazzo & Scherer, 2006). Upon agreeing with guidelines, they become directives, subject to implementation by companies (Campbell, 2006). From time to time, the government follows up on whether they are being implemented or not. The government officer stated that:

> Yeah, at the Ministry we have the natural gas policy 2014 which gives policy guidance on how CSR should be undertaken in the areas concerned. Government Officer 1

Similar views were provided by the government agency's officer for having directives given to the companies in mining and oil and gas sectors to guide them in their operations. However, the agency ensures that there are consensus and agreement made about the contents which then become binding during their operations. The officer stated:

> Yes, there are many directives given by this institution. The directives are then shared between government agency and companies. On agreeing the contents, the directives become legally binding, although we know CSR is not mandatory. We give them certificates of compliance and if the company indicates that it will participate in CSR, then it will be subjected to audit. Government officer 4.

In justifying the government's agency regarding setting and giving directives which are followed by companies, the companies provided supportive arguments from a company in the oil and gas sector, which stated:

> You must have an environment impact assessment (EIA) on hand before and during operations. To ensure directives are followed, they hire consultants who go to the field and see if operations are compliant with EIAs: CSR Manager manufacturing Company.

The findings on directive as an institutional role to participate in CSR are consistent with that of Carini et al. (2018) who found that directives as a role must be applied with care. The discussions and agreements between regulatory bodies and companies in Tanzania are apt to the issues that may happen if directives are given without consideration.

However, the guidelines have not been formulated to consider other sectors yet. Once all sectors are aware of the guidelines, this will establish equal opportunities for companies' participation in CSR, albeit on different levels. The guidelines would also consider wider stakeholder participation and engagement. The government officer stated:

> We expect to have guidelines and all companies will follow them when investing in the country. The guidelines are formulated in consultation with local authority leaders and the communities where discussion and agreement on CSR activities is made. Government officer 3.

### 6.1.3 Dialogue with Companies

Dialogue is also useful in influencing companies to participate in CSR, especially where laws regarding CSR are still lacking (Campbell, 2006). The participants suggested that dialogue is provided on a limited scale since such a thing is new and probably will become the norm in the near future. Dialogue takes place in the course of investing in the oil and gas sector no matter how long the investments last. The government officer argued:

> When the companies come to invest in Tanzania in oil and gas; they are expected to participate in CSR. Even if they are doing a project of three months they have to leave something behind in form of CSR. Government Officer 1.

The government officer responsible for oil and gas in the ministry was probed further to explain such participation in the dialogue, while the sector is still immature. His explanation indicated that the gas sector has been in operation for a decade now compared to the oil sector, which is still at the initial stages of exploration. He stated:

> The oil and gas sector is not mature yet, but we had the activities since 2004 when we started producing gas. The company operating in Lindi region offers scholarships to secondary school students in that region. Government Officer 1.

### 6.1.4 Awareness During Registration and Operations

Some form of awareness is raised about CSR during the registration process of both MNEs and local companies. The government office responsible for environmental matters is tasked with the duty of issuance of procedures regarding the need for companies to protect the environment. The influence becomes apparent when the government office issues an environment permit as a go-ahead for operations, whereas non-compliance becomes punishable (Christmann, 2004). The government officer stated:

> We give them guidelines on environmental matters during registration. But they should get the environmental permits as a clearance of operations. Government officer 3.

However, normative influences are evidenced from mining companies using the mining development agreements (MDAs) which specifies the role of companies through CSR. With these agreements companies are made aware that CSR is part of what they have to do (Campbell, 2006). The government officer concerned with mining activities explained that the awareness is raised during discussions and agreement on implementing CSR activities in their operations. He stated that:

> The mining companies have (MDAs) which contain clauses on CSR. We agree with companies to follow those clauses on what companies' will do. We then enforce on the implementation of the contents of the clauses. Government officer 1.

The companies provided justifiable evidence regarding their participation with government and other institutions to deliberate on issues concerned with CSR. Through these awareness campaigns, forums and consultations are conducted and agreement reached out on what CSR should be given priority by recipients themselves instead of companies taking on initiatives on their own. The marketing manager stated:

> We have the open door policy for the community come to us whenever they have an issue. Marketing Manager—Bank AB.

> We are issued with guidelines. If you are talking about social investment; these are issues in CSR which have an impact assessment. Manager—Manufacturing company

## 6.2 The Roles of NGOs and Community Members

*Research Question* 2. How do NGOs and beneficiary groups play roles in promoting CSR practices in Tanzania?

This question details the roles of NGOs and community members in ensuring that companies practise CSR. The findings are divided into two parts. Part one details findings about the roles of NGOs, while part two contains the findings about the roles of beneficiary communities in CSR.

### 6.2.1 The Roles of NGOs

NGOs as institutions and watchdogs for communities play a role in influencing multinational and local companies in practising CSR for the well-being of the society surrounding their operations (Arenas et al., 2009). These roles are described in the form of incentives to encourage companies, the dialogue, the actions, and whether or not the actions are helpful to deter violating companies from their previous bad practices.

#### 6.2.1.1 Incentives to Encourage Companies to Participate in CSR

NGOs offer incentives through involving other stakeholders to allow other organisations in the similar industry or services to copy their practices (Guay et al., 2004) to participate in CSR contributions, especially those directed at health services. The NGO in the health sector developed a system which has been sub-contracted and adopted by a public institution. The NGO manager supported the statements by stating that:

> We involve other stakeholders so that they can appreciate what we do. We participate in charities, eyesight, health scams, test HIV/AIDS, diabetes, and blood pressure for free. The National Social Security Fund (NSSF) adopted our system and implemented it in its programmes. Project Manager: NGO 1.

Similarly, copying influence occurs due to NGOs' involvement in rewarding mechanisms to inspire and inform others to get involved in joint efforts and become responsible. The initiative to provide rewards to companies that do well in CSR is a big incentive leading to an increase in the scope of participation by many companies. The NGO community relations manager stated:

> We reward the company which does well in CSR. We have corporate award, which goes to the best overall performing company in environmental reporting. Unfortunately, three years ago no company won it based on our strict assessment criteria. We want companies to preserve, protect and observe human and labour rights, fight corruption, and stop land grabbing. Community Relationship Manager: NGO 2.

#### 6.2.1.2 Dialogue with Companies on CSR Matters

Stakeholder dialogue with companies is inevitable to inform what is going on in the environment in which they operate and bring a consensus between antagonistic parties (Campbell, 2006). As stakeholders, NGOs convene advocacy campaigns in the areas of their specialisation (Radaelli, 2000) to discuss issues with community members. The manager from the human rights department explained that companies interact with NGOs through the sharing of research findings on the environment, human rights, labour rights, and ethics in general. The manager stated:

> We call upon the companies to discuss the research findings with them. We actually plan to have a roundtable discussion with companies. We are in discussion with the Tanzania Private Sector Foundation (TPSF) and corporations for wider participation. We use research results to guide companies and if they are not in the meeting we expose the findings in the media. Community Relationship Manager: NGO 3.

The contention of the managers regarding dialogue with companies was vindicated by the production company manager that they do involve with the NGOs in the pursuit of CSR. The manager stated that:

> Most of the time the NGOs dealing with education and health services have been coming to us asking for our involvement in CSR whenever they have projects for the communities. Manager—Production company

This finding is in agreement with the connotation that there exists a barrier of communication between companies and the communities to give room for them to discuss matters affecting their lives as a result of companies' operations. This is supported by the study of Liao et al. (2017) who found a disparity between European and Chinese contexts where the latter experienced low level of communication compared to companies based in the former context.

### 6.2.2 The Roles Played by Beneficiary Community

The second part of the question illuminated the roles of beneficiary community members in guiding companies in participating in CSR.

The beneficiary community brings another perspective of CSR which cannot be obtained from the perspective of government and NGOs. Essentially, full participation of communities implies that decision-makers take concerns for their well-being into consideration through practising CSR. Some communities have a limited series of forums to air their grievances against and/or provide useful suggestions to companies. Convening meetings with community members and other stakeholders, such as companies, government, and NGOs, are central to the welfare of all parties. When such gatherings are undertaken, their resolutions become a backbone in accomplishing CSR practices (Brammer et al., 2012). Through this, a consensus is reached to reduce any misunderstandings. The following are the roles through which beneficiary community influences the companies to practise CSR.

#### 6.2.2.1 Stakeholder Meetings with Companies, Government, and Communities on CSR

Stakeholder meetings and written messages rarely take place for members of communities, though they are used to coerce companies to practise CSR (Høvring et al., 2018). There is no agreement on how to utilise CSR, and community members blame the government and companies for failure to compromise on priorities. LK (female), aged 45 years, stated:

> The meetings are confined within the council's premises involving the company and the community members. No emphasis on ensuring contributions goes directly to the intended community. The money that goes directly to councils has no direct benefit to this ward where the gold mine is located. We should have benefited before other wards and villages. (Focus group 1: Mining).

In addition, the promised CSR contributions take too long to be realised by needy communities. Accordingly, the village chairman, a male aged 41 years, stated:

> They come and we discuss issues in the meeting and we tell them our demands. They ask us write to write request and then obtain promises. If this happened in February, you will hear it again in February next year. Recently we requested roofing material for school buildings. They promised to contribute 150 iron sheets. (Focus group 3: Mining).

Another participant stated that the administrative hierarchy denies them the right to take part in the forums. For effective communication with people at grassroots level, the government should start meeting with stakeholders in those areas and move upwards to the district, regional, and national levels. He also insisted the local government call annual general meetings to discuss development issues, especially those linked to important services complemented by CSR. RM, a young male, stated:

> Based on the government administrative structure; we have District Commissioner (DC), District Executive Director (DED), Ward Councillor (WC), Ward Executive Officer (WEO), Village executive officer (VEO) and street administrators. For the government to be aware of what people are in need should be initiated from the lower level. We need general meetings with the gold mine so that we can sort out social and development issues facing us. (Response of group 2: gold mining).

Efforts to interact with many stakeholders in order to hold dialogue fail due to leadership challenges at all administrative levels of the government, including the Parliament of Tanzania. When communities are ready and willing to provide their opinions on how best to handle CSR, the government fails to play its part. A participating member had personally visited central and local government at the district level to ask them to allow community members to meet the inspection team from the parliament of Tanzania. RM, a young male, stated:

> Last year the Parliament committee for environment visited this locality. Instead of making environment inspection and report to the Speaker of the Parliament; they ended up inspecting the headquarters and machineries of the mining company only. Community members were not involved because the process was manoeuvred by the DC. I personally requested without success the DC and DED to ask the inspection team to come and see what is actually happening on the ground. Yet they go to press and report how the investor takes

> care of the environment. So you see, if the inspection committee had come to people they would have been well informed. (Focus group 2: gold mining).

Similar opinions were provided by members of the mining and oil and gas sectors. The participants explained that local leaders do not attend the meetings with community members in the local areas where companies operate. Even on occasions when top officers from the government have arranged to meet people and leaders, they simply do not turn up. The local community chairman, an adult male, stated:

> One day the resident commissioner for minerals informed us that he will come to meet people of this locality. As leaders we invited the people and they attended. We waited there for many hours. Unfortunately he went to the gold mine offices and later on left without having met the people as promised before. Nowadays government officers from the ministry visit the mine and go back without meeting people to see how we are treated by our neighbour; the gold mine. (Focus group 1: gold mining).

A member of the oil and gas sector stated that they have had no meetings with investors and/or government with the intention of forging relationships for the sake of stakeholders' well-being. Mr. MSW, an adult, stated:

> We haven't convened any meetings. We have many problems, as if the gas and other resources are not coming from our region. (Focus group 1: Oil and Gas).

Apart from a lack of meetings with companies, government officers, and community members, the communities in which mining and oil and gas operate face communication setbacks. This is due to a lack of response in regard to applications they send to companies requesting some contributions towards the society. More often than not, most of the applications are unsuccessful. According to the ward executive officers, many applications have been sent and continue to be sent, but few of them have been successful. The views from another FGD is provided by OM, male and adult. He was concerned with a communication breakdown between companies' employees and the people around the company. He warned that having no interaction creates a sort of hatred among the stakeholders. The following are the responses from two focus groups. Mr. OM, an adult, stated:

> In general, there is no meeting. We have tried to organize meetings and have ended in vain. I talked to community officer to ask the company assist us in sports and he told me to discuss the issue with the District Commissioner. Channelling all applications to District Commissioner is also not appropriate because he is a human with sentiments and reprisals. We would like our officers make everything transparently for the benefit of communities around. If this does not change the level of hate between communities and two mining companies will continue to rise. (Focus group 3: gold mining).

Mr. M, Ward officer and adult, explained:

> It is true; we experience difficulties when we approach managers for assistances. You can write an application letter stating your society needs, but at the end of the day no feedback is provided. Recently, we had many projects for schools and laboratory. We requested stakeholders for assistances by sending letters to them to complement citizens' contributions. As well, we made application through the District Commissioner, and one company has assisted us building two classrooms and a staff room. (Focus group 2: Oil and Gas).

The analysis of data shows that even if there is a communication breakdown between companies and community members, there are few promises from companies which are either delayed or not accomplished at all despite frequent follow-ups. The government officers prevent the communities from discussing CSR projects in their areas. Communities surrounding mining companies plan and forward their requests to the local government for funding. Due to lack of funds, the projects remain uncompleted. It had been their expectation that the gold mine would provide the funds for the implementation of these projects or construct the projects and hand them over to the community members as CSR in the local areas. CN, a male, aged 36 years, argued that:

> At the village office, we plan projects and write to the council for implementation each and every year, but nothing comes out. (Focus group 1: gold mining).

There was another agreed project, which to date had not been fulfilled; according to a group member the project is about youth empowerment through loans. The participant blamed local government leadership for failure to initiate the project even though the funds were issued by the mining company. The community requests are followed up by reminders, which are not responded to in time. The following were responses from focus group participants.

CN, male aged 36 years, argued:

> I am a chairman of the youth group in the village. One day we went to a seminar which was organized by the community relations officer. The seminar concluded that youth will be empowered through loans. It is now three years since the seminar and we have not been empowered. The fund was released but did not reach the beneficiaries. (Focus group 2: gold mining).

RM, a young male, claimed:

> They haven't responded to any of the letters. Every time we put our requests in writing and deliver it to the company, but they don't fulfil their promises. The company brought the suggestion box for people to write complaints about negative impacts of mining operations. We have complained a lot, but they don't collect the complaints. (Response of group 3: gold mining).

Supporting evidence from companies regarding interaction with communities show that managers who indulge in CSR have started contacting them through their representatives and allow them to be part of the decisions of the most important needs. The sales and marketing and business development manager and the director of operations stated, respectively:

> All supports we provide are normally initiated by the community members themselves. Sales, Marketing and Business Development Manager.

> Beneficiary influence us. When they raise the need of support they conduct meetings, and their meeting outputs are sent to us. We are normally invited in the meetings they raise issues and we tell them the ways that we have to support them. Director: Oil and gas

The findings obtained from communities reflect the fact that there are problems associated with communication between companies and stakeholders regarding dialogues, convening meetings, and request letters. There is a barrier of communication, a situation that welcomes poor relationships. Companies have always not acted promptly due to logistic and financial arrangements. Delayed responses signal for the possibility of facing legitimacy concerns. And where companies backtrack, many efforts may be needed to repair reputation at stake. The results for community members are comparable to Mutti et al. (2012) in Argentina, where CSR communication, dialogue, and involvement of communities surrounding the companies' areas of operations were limited.

## 6.3 *What Happens to Non-Complying Companies*

*Research Question 3.* What happens to non-complying companies with the institutions of government, NGOs, and community members?

The question seeks to determine the implications of companies which are not compliant to the roles played by institutions and the beneficiary community. The findings are based on the actions taken against companies which do not comply.

### 6.3.1 Actions Taken Against Violating Companies

The question was intended to establish whether or not the government institutions take action against violating companies (Frumkin & Galaskiewicz, 2004). Sometimes companies are found guilty of violating social and environmental issues intentionally. The government, through its institutions, abides with imposing fines and penalties and sometimes halts the operations of the companies in question. The agency's officer stated:

> From the environment act 2004, there is a provision which requires a person polluting the environment to pay fees and spend money to rejuvenate the environment (technically called polluter based principle). The company should rejuvenate the environment so that it will not cause any side effects. Government officer 4.

A probe question was raised to the principal environmental officer to give examples of fines, closing the operations, or the punishments imposed upon companies. The officer responded that there are various actions where the intention is to deter companies from those violations (Frumkin & Galaskiewicz, 2004). He stated:

> Some companies have been ordered to close their operations because they are polluting the environment in which community resides. Recently, a company polluted the environment by producing excessive dust and its production has been suspended. Yesterday, another company in gold mining was stopped to produce because of polluting the environment. Two other industries have faced the same sanctions in Morogoro region. Government officer 3.

A probe question was raised to understand whether the halting of operations is temporary or indefinite. The officer explained that companies involved in the violations have all been compliant and pleaded guilty. The companies' mistakes are not beyond repair and halting operations is normally temporary and no instances have led to permanent closure. The officer stated:

> Normally for production industries we halt the operations until when the anomaly is rectified. There are no instances where the closure has been indefinite. Closing the industry implies that it has problems that cannot be corrected, something which is difficult to experience. Government officer 1.

However, if the violations are repeatedly carried out there is a chance that a company will be closed indefinitely. Closing the company is authorised by the minister, but this is preceded by revoking the licence to operate. The punishments imposed have been used as a deterrent to other companies in the industry or across industries. The planning and development manager stated:

> Yes, if the fault is done twice or more, the company will be closed. The policy provides that if you are fined first, second, the third time will call for stopping or revoking the license. Thus the company is ordered to pack things and go. The fines are not a good thing for a company, and it gives a signal to other players in the market to take cautions not to misbehave. Government officer 4.

NGOs, like the government, also take action against violating companies through their advocacy campaigns to resolve misunderstandings between stakeholders (Arenas et al., 2009). Legal and human rights NGOs file cases in the courts against companies' misuse of the environment and abuse of human, labour, and compensation rights. NGOs have also been defending for free cases in the courts on behalf of community members whose rights are denied by the companies. In supporting the disputes, the manager validated:

> One gold company was involved in many violations to surrounding communities and we filed a case in the local courts but it didn't work. We transferred the case to London and the court found it guilty and was fined. Still, we have many cases filed against companies' misbehaviours, environment pollution and land grabbing. There is a Nyamuma case in which peoples' land was taken out for a game reserve without compensations. We won the case and the land was retained. Recently, is Magamba case where investor grabbed land from the small scale miners; we defended them and won. Manager NGO 2.

NGOs are also involved in resolving conflicts that ensue between governments and companies, and companies and community members. NGOs normally work in consultation with the government on issues related to private health sectors in general as an action to end conflicts. She stated that:

> There are many instances; we stand as an umbrella organization for the private hospitals. We advocate for the private health sector on quality and inspection and have advised the government not to close private hospitals once they find anomalies. Nowadays we collaborate with government in inspecting the private health facilities. If there is a problem we discuss and agree to allow the hospital to operate while amending the shortcomings. Manager: NGO 1.

However, depending on the nature of the NGO and the sector in which it is located, it is unusual to see other NGOs participate in taking actions against violating companies. The manager of an NGO stated that they play a part through educating community members on various rights and CSR generally. He stated:

> Unfortunately, we don't have the power to sue them. But we have a power to educate them on their rights and make them speak on their own behalf. We call upon companies violating the rights of citizens to stop operating in those parameters. Manager: NGO 3.

Finally, the study determined if the actions taken against violators are useful or not; the NGOs alluded that the actions taken against violating companies are correct in shaping the conduct of the companies and deter them from previous behaviours. The punishments, including court rulings and fines, are incentives for good behaviour. The manager stated:

> I think they behave; because they have given out huge amount of money. There are still challenges in Tanzania because the court system is slow and companies have a lot of money. When you take a case to court of law, it is really difficult to get through. Some companies are being charged meagre amounts and they don't bother as long as they have taken hundreds of hectares of land. Community Relationship Manager: NGO 2.

Companies that support non-complying ones face serious actions against their bad practices. The companies follow the directives given by the government and government agencies, especially those in connection with environmental matters. The government agency is determined to ensure there is environmental protection. The manager stated that:

> They are [government agencies] effective these days and if they give you directives they allow a period of time to accomplish them and later you will be audited. If you are non-compliant, you will be fined and can even close your operations. Manager—manufacturing company

The findings regarding what happens to the non-complying companies do correspond to the study of Arenas et al. (2009) who found that legal actions against non-complying companies are taken along with fines and penalties being imposed on companies who fail in the court cases. Being non-compliant in CSR means that the companies involved would lose legitimacy as operators and in the end escalated conflicts with surrounding communities may occur and jeopardise the scale of operations and such losses may follow (Idemudia & Ite, 2006).

# 7 Conclusions and Recommendations

## *7.1 Conclusions*

The study was undertaken to determine the roles of institutions in shaping the conduct of CSR activities by companies in Tanzania. Based on the findings elaborated above, it is concluded that the roles of institutions are a catalyst for companies practising CSR. The government provides the roles in various forms such as giving incentives, dialogues with companies to participate in CSR, awareness during registration, and directives. Similarly, NGOs roles reflect those of government as incentives and dialogues are also used by them in persuading companies to indulge in CSR. Furthermore, beneficiary communities' roles include use of meetings and request letters to companies as a way to make companies undertake CSR activities.

Regarding the companies which do not heed to practise CSR, they are confronted in one way or another with stringent actions of government. Non-compliant companies face punishments such as payment of fines and fees and closure of business operations. NGOs deal with non-complying ones through court cases, fines, and educating companies.

## 7.2 Recommendation for Policy and Future Research

### 7.2.1 Recommendation for Policy-Makers

To a large extent, the institutions involved in ensuring companies practise CSR seem not to understand well the concept of CSR. Most of the community members cannot differentiate between CSR and the legal obligations of companies to the extent that they consider CSR contributions as compulsory and legally binding. They have no or little knowledge that CSR is still practised on a voluntary basis in Tanzania as in most of the developing countries and that companies have discretion to undertake it or not. In fact, companies provide CSR just to be good citizens and become legitimate operators, but when they do not do so, they are blamed by the communities and NGOs. There is a need, therefore, to provide training and awareness initiatives to government officers, communities surrounding companies' areas of operations, and NGOs to understand what is CSR, its roles in meeting development challenges, and how it catalyses good stakeholders' relationship in line with the bounds of legitimacy theory. The training and awareness campaigns could be organised by government, companies, and NGOs on an individual or collective basis.

Equally, even if companies provided evidence supporting their interaction with community members through meetings and other forms of communications, it was to a large extent refuted by most of the community members in the focus groups. This situation is an indicator that all is not well with companies and stakeholders surrounding their areas of operations. There is a need to convey some good communications channels to do away with breakdown of information or the lack of proper feedback mechanisms—which if not checked may erode the legitimacy of companies to operate.

### 7.2.2 Recommendations for Future Research

The study is limited to the context of Tanzania. Therefore, there are still different geographical contexts in which studies like this one can be conducted. Looking at the context of African developing countries, the same studies could for example be conducted in the East African region countries on a single or many countries basis to determine and understand whether or not the same results can be found. Likewise, other research design approaches and different data collection methods such as questionnaires or secondary data approaches could be used to study the same or other variables of CSR.

## References


Albareda, L., Lozano, J. M., Tencati, A., Midtun, A., & Perrini, F. (2008). The changing role of governments in corporate social responsibility: Drivers and responses. *Business Ethics: A European Review, 17*(4), 347–363.

Albareda, L., Lozano, J. M., & Ysa, T. (2007). Public policies on corporate social responsibility: The role of government in Europe. *Journal of Business Ethics, 74*(4), 391–407.

Amalados, M. X., & Manohar, H. L. (2013). Communicating corporate social responsibility—A case of CSR communication in emerging economies. *Corporate Social Responsibility and Environmental Management, 20*(2), 65–80.

Ararat, M. (2008). A development perspective for corporate social responsibility: Case of Turkey. *Corporate Governance, 8*(3), 271–285.

Arenas, D., Lozano, J. M., & Albareda, L. (2009). The role of NGOs in CSR: Mutual perceptions among stakeholders. *Journal of Business Ethics, 88*(1), 175–197.

Babiak, K., & Trendafilova, S. (2011). CSR and environmental responsibility: Motives and pressures to adopt green management practices. *Corporate Social Responsibility and Environmental Management, 18*(1), 11–24.

Bartle, I., & Vaas, P. (2007). Self-regulation within the regulatory state; towards a new regulatory paradigm? *Public Administration, 85*(4), 885–905.

Baughn, C. C., Bodie, N. L., & McIntosh, J. C. (2007). Corporate social and environmental responsibility in Asian countries and other geographical regions. *Corporate Social Responsibility and Environmental Management, 14*(4), 189–205.

Bebbington, A. (2004). NGOs and uneven development: Geographies of development intervention. *Progress in Human Geography, 28*(6), 725–745.

Bebbington, A., & Farrington, J. (1993). Governments, NGOs and agricultural development: Perspectives on changing inter-organizational relationships. *Journal of Development Studies, 29*(2), 199–219.

Beschorner, T., & Muller, M. (2007). Social standards: Towards an active ethical involvement of businesses in developing countries. *Journal of Business Ethics, 73*(1), 11–20.

Blowfield, M., & Murray, A. (2011). *Corporate social responsibility* (2nd ed.). Oxford University Press.

Brammer, S., Jackson, G., & Matten, D. (2012). Corporate social responsibility and institutional theory: New perspectives on private governance. *Socio-Economic Review, 10*(1), 3–28.

Bryman, A. (2016). *Social research methods* (5th ed.). Oxford University Press.

Campbell, J. (2006). Institutional analysis and the paradox of corporate social responsibility. *The American Behavioural Scientist, 49*(7), 925–938.

Carini, C., Rocca, L., Veneziani, M., & Teodori, C. (2018). Ex-ante impact assessment of sustainability information–the Directive 2014/95. *Sustainability, 10*(2), 560–584.

Christmann, P. (2004). Multinational companies and the natural environment: Determinants of global environmental policy standardization. *Academy of Management Journal, 47*(5), 747–760.

Collins, D. (2012). *Business ethics: How to design and manage ethical organizations*. John Wiley and Sons.

Demir, A., & Budur, T. (2019). Roles of leadership styles in corporate social responsibility to non-governmental organizations (NGOs). *International Journal of Social Sciences & Educational Studies, 5*(4), 174–183.

Dentchev, N. A., van Balen, M., & Haezendonck, E. (2015). On voluntarism and the role of governments in CSR: Towards a contingency approach. *Business Ethics: A European Review, 24*(4), 378–397.

Detomasi, D. A. (2008). The political roots of corporate social responsibility. *Journal of Business Ethics, 82*(4), 807–819.

Fawcett, B., & Pockett, R. (2015). *Turning ideas into research: Theory, design and practice*. Sage Publications.

Fiaschi, D., Giuliani, E., & Nieri, F. (2017). Overcoming the liability of origin by doing no-harm: Emerging country firms' social irresponsibility as they go global. *Journal of World Business, 52*(4), 546–563.

Flick, U. (2014). *An introduction to qualitative research* (5th ed.). Sage Publications.

Fox, T., Ward, H., & Howard, B. (2002). *Public sector roles in strengthening corporate social responsibility: a baseline study*. World Bank.

Frumkin, P., & Galaskiewicz, J. (2004). Institutional isomorphism and public sector organizations. *Journal of Public Administration Research and Theory, 14*(3), 283–307.

Galaskiewicz, J. (1997). An urban grants economy revisited: Corporate charitable contributions in the twin cities, 1979-81, 1987-89. *Administrative Science Quarterly, 42*(3), 445–471.

Gray, D. E. (2009). *Doing research in the real world* (2nd ed.). Sage Publications.

Greenwood, R., Suddaby, R., & Hinnings, C. R. (2002). Theorising change: The role of professional association in the transformation of institutionalised fields. *Academy of Management Journal, 45*(1), 58–80.

Griesse, M. A. (2007). The geographical, political, and economic context for corporate social responsibility in Brazil. *Journal of Business Ethics, 73*(1), 21–37.

Griseri, P., & Seppala, N. (2010). *Business ethics and corporate social responsibility* (1st ed.). South-Western Cengage Learning.

Guay, T., Doh, J. P., & Sinclair, G. (2004). Non-governmental organizations, shareholder activism, and social responsible investment: Ethical, strategic and governance implications. *Journal of Business Ethics, 52*(1), 125–139.

Guthrie, J., & Parker, L. D. (1989). Corporate social reporting: A rebuttal of legitimacy theory. *Accounting and Business Research, 19*(76), 343–352.

Hah, K., & Freeman, S. (2014). Multinational enterprise subsidiaries and their CSR: A conceptual framework of the management of CSR in smaller emerging economies. *Journal of Business Ethics, 122*(1), 125–136.

Høvring, C. M., Andersen, S. E., & Nielsen, A. E. (2018). Discursive tensions in CSR multistakeholder dialogue: A Foucauldian perspective. *Journal of Business Ethics, 152*(3), 627–645.

Idemudia, U., & Ite, U. E. (2006). Corporate-community relations in Nigeria's oil and industry: Challenges and imperatives. *Corporate Social Responsibility and Environmental Management, 13*(4), 194–206.

Khan, H. U. Z., Halabi, A. K., & Samy, M. (2009). Corporate Social Responsibility (CSR) reporting: A study of selected banking companies in Bangladesh. *Social Responsibility Journal, 5*(3), 344–357.

Khan, M., Lockhart, J., & Bathurst, R. (2021). The institutional analysis of CSR: Learnings from an emerging country. *Emerging Markets Review, 46*, 100752–100800. https://doi.org/10.1016/j.ememar.2020.100752

Kim, C., Kim, J., Marshall, R., & Afzali, H. (2018). Stakeholder influence, institutional duality, and CSR involvement of MNC subsidiaries. *Journal of Business Research, 91*, 40–47.

Koku, J. E., & Gustafsson, J. E. (2003). Local institutions and natural resources management in the South Tongu district of Ghana: A case study. *Sustainable Development, 11*(1), 17–35.

Kumi, E., Yeboah, T., & Kumi, Y. A. (2020). Private sector participation in advancing the sustainable development goals (SDGs) in Ghana: Experiences from the mining and telecommunications sectors. *The Extractive Industries and Society, 7*(1), 181–190.

Li, H., & Miao, Z. (2020). Corporate social responsibility and collaborative innovation: The role of government support. *Journal of Cleaner Production, 260*(12), 1028.

Liao, P. C., Xia, N. N., Wu, C. L., Zhang, X. L., & Yeh, J. L. (2017). Communicating the corporate social responsibility (CSR) of international contractors: Content analysis of CSR reporting. *Journal of Cleaner Production, 156*, 327–336.

Lin, X., McKenna, B., Ho, C. M., & Shen, G. Q. (2019). Stakeholders' influence strategies on social responsibility implementation in construction projects. *Journal of Cleaner Production, 235*, 348–358.

Long, W., Li, S., Wu, H., & Song, X. (2020). Corporate social responsibility and financial performance: The roles of government intervention and market competition. *Corporate Social Responsibility and Environmental Management, 27*(2), 525–541.

Magness, V. (2006). Strategic posture, financial performance and environmental disclosure: An empirical test of legitimacy theory. *Accounting, Auditing and Accountability Journal, 19*(4), 540–563.

Mahmood, M., & Humphrey, J. (2013). Stakeholders expectations of CSR practices: A study of local and multinational corporations in Kazakhstan. *Corporate Social Responsibility and Environmental Management, 20*(3), 168–181.

Maria, S. J. F. J., & Devuyst, E. (2011). CSR and development: A mining company of Africa. *Journal of Management Development, 30*(10), 955–967.

Matten, D., & Moon, J. (2008). Implicit and explicit CSR: A conceptual framework for a comparative understanding of corporate social responsibility. *The Academy of Management Review, 33*(2), 404–424.

Melubo, K., Lovelock, B., & Filep, S. (2019). Motivations and barriers for corporate social responsibility engagement: Evidence from the Tanzanian tourism industry. *Tourism and Hospitality Research, 19*(3), 284–295.

Merchant, K. A., & Van de Stede, W. A. (2006). Field-based research in accounting: Accomplishments and prospects. *Behavioural Research in Accounting, 18*(1), 117–134.

Michael, B. (2003). Corporate social responsibility in international development: An overview and critique. *Corporate Social Responsibility and Environmental Management, 10*(3), 115–128.

Moffat, K., Lacey, J., Zhang, A., & Leipold, S. (2016). The social licence to operate: A critical review. *Forestry: An International Journal of Forest Research, 89*(5), 477–488.

Mutti, D., Yakovleva, N., Vazquez-Brust, D., & Di Marco, M. H. (2012). Corporate social responsibility in the mining industry: Perspectives from stakeholder groups in Argentina. *Resources Policy, 37*(2), 212–222.

Nazir, O., & Islam, J. U. (2020). Influence of CSR-specific activities on work engagement and employees' innovative work behaviour: An empirical investigation. *Current Issues in Tourism, 23*(24), 3054–3072.

Okoye, A. (2012). Explaining the relationship between corporate social responsibility, law and development in an African context: Should government be responsible for ensuring corporate responsibility? *International Journal of Law and Management, 54*(5), 364–378.

Palazzo, G., & Scherer, A. G. (2006). Corporate legitimacy as deliberation: A communicative framework. *Journal of Business Ethics, 66*(1), 71–88.

Patriotta, G., Gond, J. P., & Schultz, F. (2011). Maintaining legitimacy: Controversies, orders of worth, and public justifications. *Journal of Management Studies, 48*(8), 1804–1836.

Paulus, T. M., Lester, J. N., & Dempster, P. G. (2014). *Digital tools for qualitative research*. Sage Publications.

Pfeffer, J., & Salancik, G. (1978). *The external control of organisations: A resource dependence perspective*. Harper and Row.

Punch, K. F. (2014). *Introduction to social research: Quantitative and qualitative approaches* (3rd ed.). Sage Publications.

Radaelli, C. M. (2000). Policy transfer in the European Union: Institutional isomorphism as a source of legitimacy. *Governance, an International Journal of Policy and Administration, 13*(1), 25–43.

Saleh, Y. A. B. (2019). Tax treatment methods of corporate monetary and in-kind donations and the extent of their compatibility with the content of the idea of corporate social responsibility: A justified view. *Administrative Management, 3*(1), 1034–1049.

Shauki, E. (2011). Perceptions of corporate social responsibility: A study in capturing public confidence. *Corporate Social Responsibility and Environmental Management, 18*(3), 200–208.

Simons, H. (2009). *Case study research in practice*. Sage Publications.

Sobhani, F. A., Amran, A., & Zainuddin, Y. (2009). Revisiting the practices of corporate social and environmental disclosure in Bangladesh. *Corporate Social Responsibility and Environmental Management, 16*(3), 167–183.

Steurer, R. (2010). The role of government in corporate social responsibility: Characterising public policies on CSR in Europe. Retrieved May 2, 2021, from http://ssrn.com/abstract=2342120.

Suchman, M. C. (1995). Managing legitimacy: Strategic and institutional approaches. *The Academy of Management Review, 20*(3), 571–610.

Tan, J. (2009). Institutional structures and firm social financial performance in transitional economies: Evidence of multinational corporations in China. *Journal of Business Ethics, 82*(2), 171–189.

Wu, J., & Davidson, D. K. (2011). The business-government-society relationship: A comparison between China and the US. *Journal of Management Development, 30*(1), 112–125.

Xu, S., & Liu, D. (2020). Political connections and corporate social responsibility: Political incentives in China. *Business Ethics: A European Review, 29*(4), 664–693.

Yang, K. (2010). *Making sense of statistical methods in social research*. Sage Publications.

Yang, X., & Rivers, C. (2009). Antecedents of CSR practices in MNC's subsidiaries: A stakeholder and institutional perspective. *Journal of Business Ethics, 86*(2), 155–169.

Yin, R. K. (2012). *Applications of case study research* (3rd ed.). Sage Publications.

**Ibrahim Ramadhani Kikwiye** is Senior Lecturer in Accounting and Finance at the Department of Accounting and Finance of the Institute of Finance Management (IFM) in Tanzania. Currently, he is located at Simiyu Campus. He holds a Ph.D. in Accounting and Finance from Anglia Ruskin University in the UK. He has a Master's in Business Administration majoring in Banking and Finance from Mzumbe University in Tanzania. Also, he has a Professional Accounting qualification of Certified Public Accountant (CPA) and an Advanced Diploma in Certified Accountancy from the then Institute of Development Management now Mzumbe University in Tanzania. His research interests include corporate social responsibility, sustainability reporting, corporate governance, finance, auditing, and public sector accounting.

# Implications of Fluidity in Compliance Regime for India's Compulsory CSR Law

Debanjali Saha, Rinkesh Dharod, and Raj Janagam

## 1 Introduction

> The joy of CSR is taken away under the law.—Mr. Noshir Dadrawala, CEO, Center for Advancement of Philanthropy (N Dadrawala, personal communication, 6th June, 2021)

India has a unique position in the CSR world being the third country to have a provision of "mandated CSR" under Section 135 of India's Companies Act 2013 (Chatterjee & Mitra, 2017, p. 1). Section 135 (1) of the Act illustrated the following:

> Every company having net worth of rupees five hundred crore or more, or turnover of rupees one thousand crore or more or a net profit of rupees five crore or more during any financial year shall . . . spends, in every financial year, at least two per cent of the average net profits of the company made during the three immediately preceding financial years, in pursuance of its Corporate Social Responsibility Policy—Ministry of Corporate Affairs (MCA), 2013, p. 87

At the 2016 UN Forum on Business and Human Rights, Dr. Bhaskar Chatterjee, Director General and CEO of Indian Institute of Corporate Affairs (IICA) and hailed as "'Father of CSR in India," had stated that India's private sector had a lot of vitality and innovation and the intent of the law was to "harness some of that through our CSR process and help it to complement and supplement what the Government is doing, then we can seriously upscale and update all our developmental efforts"

D. Saha (✉) · R. Janagam
Surge Impact Foundation, Hyderabad, India
e-mail: debanjali@surgeimpact.org; raj@surgeimpact.org

R. Dharod
AIC-IIITH Foundation, Hyderabad, India
e-mail: rinkesh.dharod@iiit.ac.in

 
I. G. Arraiano et al. (eds.), *Corporate Social Responsibility in a Dynamic Global Environment*, CSR, Sustainability, Ethics & Governance,
https://doi.org/10.1007/978-3-031-24647-0_7

(Chatterjee, 2018). This law was designed on "comply or explain" principle (Chatterjee, 2018). This was reflected in the law and rules drafted to implement the law (MCA, 2014a).

However, over time there has been a perceptible shift in this approach. The aim of this paper is to highlight that with time the regulatory regime has become very unstable and arbitrary, with constant changes which make identifying permitted CSR activities a challenge. This is further compounded by two parallel trends:

1. Changing nature and increasing quanta of compliance required under the law
2. Additional rules and red-tapeism imposed by corporations themselves in terms of what constitutes as CSR activity and the accompanying implementation process

This combination decelerates CSR activity in India making it expensive and restrictive and stifles the possibility of harnessing vitality and innovation that was at the heart of the law. Paradoxically, it facilitates use of CSR law to achieve short-term political aims and/or legalistic creativity for misapplication of CSR funds.

## 2 Methodology

This study is qualitative in nature where the primary sources of information are as follows:

a. Legal documents of Government of India
b. Expert opinion of Mr. Noshir Dadrawala, CEO, Center for Advancement of Philanthropy
c. Experiential narratives of T-Social Impact Group (T-SIG)[1]
d. Articles, videos, and blogs of CSR consultants and area experts

Qualitative study was chosen because it was felt that, only through a detailed analysis of the legal documents and available public documents and substantiating those with T-SIG's experiences, the fluidity of Indian CSR law, along with its varied contradictions, can be properly analyzed. However, none of the companies have been mentioned by name in the article due to confidentiality clauses with T-SIG.

[1] T-SIG, the official CSR cell of Telangana, has been initiated by the Department of Industries and IT, government of Telangana. Surge Impact Foundation, a Section 8 nonprofit, acts as the project management unit (PMU) to the department to manage this platform. The objective of the platform is to make CSR interventions across the state effective and sustainable by connecting all relevant stakeholders and creating synergies.

## 3 Limitations of the Study

Although private companies reveal data to MCA, which is showcased at the National CSR Portal, data for numerous companies is either nonexistent or contradictory. Moreover, CSR implementation agencies are often bound by nondisclosure agreements with companies. Furthermore, from T-SIG's experience, multinational private companies have shied away from nominating themselves for CSR Awards conducted by the MCA to avoid the basic question "What is the company's annual CSR budget?" Hence, the core limitation of this study is the lack of information regarding CSR activities of private companies.

Secondly, the second wave of COVID in India highly impacted the participation of civil society organizations in this study. Although the researchers had intended to approach CSR stakeholders like NGOs, companies, and retired government officials for semi-structured interviews of 30–45 min, most of these organizations and individuals were engaged in conducting on-ground emergency relief activities. This paper would have been surely been enriched with responses of such participants.

## 4 Evolution of CSR Regulatory Framework

As mentioned above, CSR law in India was brought in with a specific underlying philosophy in mind. While the CSR spending was legally mandated, the whole idea was based on the 'comply or explain' principle. This was referred to by Dr. Bhaskar Chatterjee as the 'Chatterjee Model' (Chatterjee & Mitra, 2017, p. 1) which is different from other CSR models because it is *focused on the social developmental agenda of the nation* (Chatterjee & Mitra, 2017, p. 4). This model is relevant because of Dr. Chatterjee's role in the formulation of the 'mandated CSR' regime in India whereby this model, launched in 2010 as CSR Guidelines for the Public Sector Enterprises of India, gives an insight into the original vision and ideas that underpin the CSR regime in India. Later Dr. Chatterjee was instrumental in ensuring that Section 135 became part of India's Companies Act 2013 and in formulation of the rules therein (Mitra, 2017).

The following are the elements of the models as laid down by Chatterjee and Mitra (2017, pp. 4–5):

1. A certain percentage of CSR spent to be done by large stable corporations.
2. Spending must be projectivized.
3. Spending must be beyond the core business areas of the corporation.
4. Areas of CSR activities must be in accordance with national priorities.
5. Activities must be outcome oriented and measurable with clearly budget allocated and money spent quantified.
6. Results must be measured, reported, and audited with respect to accounts and outcomes.

7. CSR activities must harness the strength of the corporation with aid of civil society.
8. CSR activities must be led by the board. Three board members including one independent board member are to authorize and monitor CSR activities.

Moreover, the compliance would be achieved not by penal provisions but through simple regulatory reporting by the stated principle of "comply or explain." This is explained as in case a company does not spend the mandated CSR amount in a financial year, it is "called upon to provide an 'explanation' on its website and also in the Annual Director's Report. The government does not stand in judgment on the quality or the validity of the explanation." Penal provisions would apply only if no explanation is provided (Chatterjee & Mitra, 2017, pp. 5–6).

There is an inherent contradiction in the design insofar that the law requires spending of a certain amount of money by companies on CSR, but enforcement was based on essential honor system of "comply or explain." As enumerated in this article, contradictions are part of the reason why the CSR ecosystem in India is suboptimal. Additionally, the involvement of board leadership in planning and implementation processes makes CSR very top-heavy.

# 5 Recent Changes in Framework

## *5.1 Basic Framework*

"Mandated CSR" in India is part of its Companies Act 2013 which governs the corporations in India. Section 135 of the law lays out the mandates that companies meeting certain criteria have to spend 2% of their income along with implementation modalities. Furthermore, areas in which CSR is carried out are laid out in Schedule VII of the Act.

India is a common law jurisdiction and relies heavily on the principles of delegated legislation. What this means is that the legislature generally lays down the basic framework of the law and executive frames the details of the "rules" on how the law is implemented. Many times, these rules not only provide the definition of terms but also can substantially change or modify the way laws are implemented.

In the case of CSR, the nature of implementation of the law is determined by the rules framed by the government for Section 135 and Schedule VII, and it is here that we see all the changes, and changing nature of the law and its contradictions and consequences play out.

Section 135 has itself been amended only three times, but rules governing it have been changed multiple times. Case with Schedule VII is even more gregarious. The entire schedule has been amended and even replaced multiple times by just changing the rules. To add to these, there have been multiple circulars and clarification issued by the MCA, which while not having statutory value are important factors to consider as discussed below.

## 5.2 *Story of Schedule VII*

As the authors have argued in an earlier paper, constant changes in rules related to CSR activities have made it difficult to decipher the meaning and spirit of the law. This has made genuine high-impact innovation in CSR difficult while making questionable use of CSR funds easier with legalistic creativity (Saha et al., 2020). Those findings have been further expanded here.

Schedule VII laid out the areas in which CSR activities are allowed. This section has been amended over ten times (on an average, one amendment per year) (MCA, 2014b). In addition, the government has introduced multiple clarifications and FAQs to clarify issues related to problematic drafting of Schedule VII. The most important is the general circular No. 21/2014 (June 18, 2014) which stated that "...entries in the said Schedule VII must be **interpreted liberally** so as to capture the essence of the subjects enumerated in the said Schedule" (emphasis original). This point was further emphasized by this curricular by including illustration like effective consumer grievance redressal mechanism under the category of "promoting education." By allowing for such a nonstandard and arbitrary interpretation, it opens the gates for similar such interpretation by others as seen below (MCA, 2014c).

Moreover, despite multiple changes, the categories of the schedule remain an eclectic mix of very broad categories like eradicating hunger, poverty, and malnutrition and very specific subjects like listing specific govt. institutions like IITs which can be funded for "conducting research in science, technology, engineering and medicine aimed at promoting Sustainable Development Goals (SDGs)" (MCA, 2014b) (see Annexure I). In fact, this line is the only place where global standards of SDGs are recognized. This mix has, in the authors' experience, created great confusion regarding the allowance of certain projects under the CSR law. Moreover, in many cases, it has seriously restricted the scope of activities and even invalidated some models, for example, in case of entrepreneurship/start-ups/social enterprises, funding is only allowed to incubators already funded by the government. This means nonprofit incubators unsupported by the government cannot receive CSR support.

Question arises then why such a situation exists. While there is little public documentation or explanation for the reason of such changes, one of the reasons for this instability can be routed back to the design of Schedule VII originating in the "Chatterjee model." The model clearly stipulates that CSR activities must be conducted in "areas as laid down in accordance with the national priorities" (Chatterjee & Mitra, 2017, p. 4). This means that area of the CSR activities is to be decided by the ruling government and not by corporations themselves. This also explains the need for the existence of Schedule VII instead of globally accepted standards like UN SDGs or allowing the corporations to define their own based on beneficiaries and/or outcomes. While referring to the last point in original 2013 version of Schedule VII which gave government power to add "such other matters as may be prescribed" (Ministry of Corporate Affairs, 2013, p. 87), Dr. Chatterjee states, "hence if the government in its enormous wisdom prescribes any other thing (as permitted CSR activities) then we guys can take, if they don't prescribe too bad" (NGOBOX, 2013).

Governments in democracy change. And priorities of the governments change. This means that CSR areas of activities are subject to changes which are "reactive, ad-hoc, arbitrary and do not consider any specific direction or underlying philosophies" (Saha et al., 2020). Apart from poor drafting, the primary causes of the changes are ever shifting and at times short-term political and policy priorities of the government of the day. Also, changes are a way to be in the good books of the government.

A good example of this is the category of entrepreneurship. Originally under the heading of "social business projects," it is now very restrictive and lines up with the government policy of Startup India via schemes like Atal Innovation Mission (AIM) (Department for Promotion of Industry and Internal Trade, n.d.). Under AIM, the government has funded 59 new incubators (NITI Aayog, n.d.). In the long term, this and other government funded incubators will also need CSR money. Thus, restricting CSR to only government funded incubators allows for directing CSR to meet government objectives.

More gregarious examples are of Statue of Unity and PM Cares Fund. State Unity, a political project of the government and the largest statue in the world, is the statue of Indian freedom fighter Sardar Vallabhbhai Patel. This was partly funded by the public sector units including the Oil and Natural Gas Corporation and Indian Oil Corporation with approximately US$ 19.84 million from their CSR funds under the category of "protection of national heritage" (Indian Audit and Accounts Department, 2018; Saha et al., 2020). The Comptroller and Auditor General of India (CAG) pointed out "[c]ontribution towards this project did not qualify as CSR activity as per schedule VII of the Companies Act 2013 as it was not a heritage asset" (Indian Audit and Accounts Department, 2018, p. 75). However, when asked by CAG for explanation, the companies mentioned that they relied on the "liberal interpretation" of the said provision (Indian Audit and Accounts Department, 2018, p. 76). This phrasing is derived from general circular No. 21/2014 (June 18, 2014) (Fig. 1).

The entire issue with the PM Cares Fund resonates with the arbitrary decision-making and changes in Schedule VII to meet short-term government objectives. With the start of the COVID pandemic, the central government created a PM relief fund for COVID relief activities and via a clarification dated March 28, 2020, declared contributions to it eligible for CSR (MCA, 2020a). At same time, via another General Circular No. 15/2020 dated April 10, 2020, the government declared that while PM Cares Fund is eligible for CSR, similar state government funds like Chief Minister's Relief Fund or State Relief Fund were ineligible for CSR donations (MCA, 2020b).[2] Realizing the circular was not adequate; the Schedule VII

[2]T-SIG, the official CSR cell of Telangana, has been initiated by Department of Industries and IT, government of Telangana. Surge Impact Foundation, a Section 8 nonprofit, acts as the project management unit (PMU) to the department to manage this platform. The objective of the platform is to make CSR interventions across the state effective and sustainable by connecting all relevant stakeholders and creating synergies.

**Fig. 1** PMO (2018)

itself was retrospectively amended to include PM Cares Fund as valid CSR funds on May 26, 2020 (MCA, 2020c). Previously, similar funds created for other central government projects like Swachh Bharat Kosh and Clean Ganga Fund have been added to Schedule VII through amendments (MCA, 2014b).

These examples show that aligning CSR activities with national priorities has allowed the ruling government to use CSR to fund its political priorities. Moreover, this arbitrary approach also violates the principle of the model that CSR spending should be in a projected mode (Rule 5.2, MCA, 2021 p. 15; MCA, 2014c).

Moreover, these constant changes have led to general confusion on interpretation of the law and rules leading to issuance of multiple clarifications, circulars, and FAQs since 2014. "Disaster Relief" was not even mentioned in the original schedule and was introduced via circular dated June 18, 2014, and consequent amendments of Schedule VII in May 2019 (MCA, 2014b, 2014c). Despite this, during COVID crisis, the MCA has had to issue multiple clarifications on utilization of CSR funds for COVID relief (at latest count 5), the last being on May 5, 2021.

Moreover, as stated in the earlier paper, "In a situation where laws and regulation are constantly changing; are simultaneously vague and restrictive; along with being driven by arbitrary political choices, is it possible to know what is the spirit of the

Circular made to the Contribution State Disaster Management Authority eligible, but in some states, these authorities did not even have a bank account at that time.

CSR law? Is it even possible to know where the letter of CSR law stands and what is the ordinary meaning of the text?" (Saha et al., 2020).

## 5.3 *Section 135: Rules, Intentions, and Confusion*

While Schedule VII (with all amendments and clarifications) lays out areas or subjects in which CSR activities can be undertaken by a company, Section 135 of the Companies Act 2013 lays down the mandate for compulsory CSR and various definitions and modalities for execution of the mandate (MCA, 2014d). Like Schedule VII, this section and its rules have regular amendments and clarifications issued by circular, though to a lesser degree. The Section itself has been amended three times since its enactment, and rules have been changed six times till date (MCA, 2014d, 2014e, 2014f). The primary focus of this paper is these amendments and most recent change to rules in January 2021 which significantly overhauls the rules governing CSR implementation.

The original rules as laid out in 2014 mirror the "Chatterjee model" very closely and are, in effect, translation of the model into a legal framework (MCA, 2014a). However, this translation gave rise to questions and confusion. A good example of this confusion is that based on rules in place at times in a Q and A session in 2014, Dr. Chatterjee said that activities related to road safety do not fall under preview of CSR activities (NGOBOX, 2014). However, in aforementioned Circular No. 21/2014 dated June 18, 2014, by MCA, road safety is allowed as CSR activities (MCA, 2014c).[3]

Moreover, the MCA has felt the need to issue detailed clarification repeatedly on various issues, clearly highlighting that confusion in the minds of the stakeholders [see, e.g., MCA 2014c, 2016, 2020b]. While some of the clarifications were needed as some companies were clearly seeing if they can use CSR funds for other legal mandate activities like expenditure under labor laws or land acquisitions, the core of the confusion was in way the act was designed.

This was compounded by the fact that multiple times some clarifications were contradictory. In the section on Apprentices Act, 1961, such issues are discussed in further details.

[3] One thing to note is that while references have been made to Dr. Bhaskar Chatterjee, the aim of the paper is not to be critical of him personally or his contribution. This is due to the nature of the evidence as he is in many cases only public available source of aim, intention, and understanding of the government when the law was formulated. He is of course not responsible for its subsequent implementation and evolution.

## 5.4 New Rules in Old Bottle

In January 2020, the rules were completely overhauled with more definitions laid down for clarity. New rules related to assets created out of CSR funds, use of CSR funds in multiyear projects, and allowing companies to "engage international organisations for designing, monitoring and evaluation of the CSR projects or programmes" are also laid out (MCA, 2021, pp. 14–17). Most significantly, for the first time, the CSR rules gave an exclusive definition on what will not be considered as CSR activities (MCA, 2021, pp. 13–14; N Dadrawala, personal communication, 6th June, 2021) which codified into rules some of the points of clarification issued earlier including FAQs on CSR, General Circular No. 01/2016 dated January 16, 2016 (MCA, 2016). Although there is confusion among the stakeholders regarding the new rules and their implications, the government is yet to come up with clarifications (N Dadrawala, personal communication, 6th June, 2021).

Some of the new rules which signify a massive shift from the underlying philosophy of the 2014 CSR rules are as follows:

1. **Nature of CSR:** Originally it was emphasized that the CSR money was not meant for government projects. This was reiterated by various actors including Dr. Chatterjee and the Union Minister of State for Finance and Corporate Affairs Anurag Thakur (NGOBOX, 2014; Press Trust of India, 2021). The aforementioned FAQ on CSR, General Circular No. 01/2016 dated January 16, 2016, stated, "CSR should not be interpreted as a source of financing the resource gaps in Government Schemes" (MCA, 2016). However, there was a slow shift away from this toward inclusion of government schemes and funds in Schedule VII. This has now been codified in form of rule 4 (1) (b) and (c) which allows CSR to be undertaken by "a company established under section 8 of the Act or a registered trust or a registered society, established by the Central Government or State Government" or "any entity established under an Act of Parliament or a State legislature" (MCA, 2021). Unlike NGOs, these entities are not required to show any track record of previous work. These rules are also seen as an attempt to prevent further controversies related to PM Cares Fund by allowing state governments to procure similar funds as an appeasement policy (N Dadrawala, personal communication, 6th June, 2021).
2. Even more importantly, the new amendments to the Companies Act in 2019 required that any funds not spent for 3 years be transferred "to a Fund specified in Schedule VII" (Ministry of Law and Justice, 2019, pp. 6). Clearly the government is looking at CSR funds as a source of financing to fill its own resource gap. As Mr. Dadrawala puts it, "PM Care and Swacch Bharat and all that are an attempt by. . . .the central government to kind of lay their hands on CSR funding" (N Dadrawala, personal communication, 6th June, 2021).
3. **Increase in compliance cost:** The new amendments and rules have drastically increased the rules of compliance and cost associated with it. Earlier the "main thrust and spirit of the law is not to monitor but to generate a conducive

environment. . . .Government has no role to play in monitoring the implementation of CSR by companies" (MCA, 2016).

However, currently, submission of annual action plans for CSR policies with details like manner of execution, modalities of utilization of funds, implementation schedules, monitoring and reporting mechanism for the projects or programs, and requirement of impact assessment, through an independent agency for large projects, is required under these new CSR rules. The detailed reporting has also increased with new reporting tables given in annexures of the rules (MCA, 2021, pp. 15–20).

Also, any amount not spent in a financial year has to be transferred to a separate bank account under "Unspent Corporate Social Responsibility Account" and reported accordingly (Ministry of Law and Justice, 2019, pp. 6). New rules also require certification by the Company Board and Chief Financial Officer or any senior executive person that funds were "utilised for the purposes and in the manner as approved" (MCA, 2021, p. 15). All these new rules are a clear shift away from the original design of comply or report with minimal government intervention which increases compliance costs. They also increase the risk of legal consequences to companies and their office bearers (N Dadrawala, personal communication, 6th June, 2021).

4. **From explain to punish:** As mentioned, both the Chatterjee model and old rules were designed on the principle of "comply or explain." In fact, under the old rules, penalty was not imposed in case of simply noncompliance but only when the noncompliance went unreported (MCA, 2014a, p. 8; read with Section 134 (3) (o) MCA, 2013, p. 87). However, the 2019 amendment drastically shifted away from these principles. It imposed not only a financial penalty on the companies for noncompliance but also penalty and jail terms for responsible officers of the company (MCA, 2019, pp. 6–7). After a massive uproar in corporate sectors, the jail term provisions were not implemented (Financial Express, 2019; Business Today, 2019). These were repealed in 2020 and replaced with modified provisions of financial penalty for the company and its officers (Ministry of Law and Justice, 2020, p. 6). In addition, as Mr. Dadrawala points out, new requirements of audits carry additional risk of harassment and loss of license for accountants auditing the activities (N Dadrawala, personal communication, 6th June, 2021).

Clearly these new rules represent a significant shift in terms of the nature of CSR and associated regulatory framework which will increase cost and shift focus to compliance. Many companies may find it more efficacious to simply donate the money to various approved government funds defeating the entire purpose of the law and converting it into a form of corporate tax. To quote Mr. Noshir Dadrawala, "As companies proudly say, it is an additional charity tax that the government has imposed upon us" (N Dadrawala, personal communication, 6th June, 2021). Mr. Dadrawala also mentioned that there is a feeling among some corporates that paying extra tax would be better than mandating CSR with its additional compliance requirements (N Dadrawala, personal communication, 6th June, 2021).

One of the reasons for this is that the original CSR law was designed in terms of money spent on activities approved by the government and not in terms of impact or beneficiaries. Dr. Chatterjee defined CSR with respect to hard cash spent by the company whereupon "if it is not rupee measurable it is not CSR" (NGOBOX, 2013). However, as Mr. Dadrawala pointed out, currently "CSR has been reduced to arithmetic" (N Dadrawala, personal communication, 6th June, 2021). This not only sidelines other elements of CSR like employee engagement and social and environmental responsibility but also drives the regulatory regime in a penal direction. If the mandate is to spend money, then the unspent funds are seen as financial impropriety and corruption. Requirements for penal provision follow as simple derivation.

## 5.5 *Entanglement of CSR with Other Laws*

Increasing regulatory requirements and ad hoc changes in rules and contradictory clarifications have meant that interaction of CSR with other laws lacks clarity. A few of these issues are exemplified below:

1. **Story of Apprentices Act:** Question on whether CSR funds can be used for requirement under Apprenticeship Act bounced repeatedly in government circles including a referral to Prime Minister's Office leading to multiple confusing and contradictory clarifications (Saha et al., 2020). To summarize, multiple curricular clearly stated that "Expenses incurred by companies for the fulfilment of any Act/ Statute of regulations (such as Labour Laws, Land Acquisition Act etc.) would not count as CSR expenditure under the Companies Act," directly referring to Apprenticeship Act (MCA, 2014c, 2016). Yet since skilling and promotion of apprenticeship was a policy objective of the government, the new circular issued by the Union Ministry of Skill Development on March 11, 2020, allowed use of CSR funds for apprenticeship under certain conditions (Ministry of Skill Development and Entrepreneurship: Government of India, 2016; Ministry of Skill Development and Entrepreneurship, 2020). However, it needs to be noted that what was earlier a clarification is now rules. Rule 2(d) (vi) clearly states that the "activities carried out for fulfilment of any other statutory obligations under any law in force in India" (MCA, 2021, p. 14) cannot be considered as CSR activities. It is not clear if the clarification related to Apprentices Act or any other clarifications issued before January 2021 are still applicable.
2. **CSR-1 Form and IT Act:** An interesting change in the new rules is the introduction of CSR-1 Form which is mandatory to be filled by any organization wishing to access CSR funds. This form is technically for registration with MCA, thereby increasing the credibility of the organization (MCA, 2021, pp. 15, 20). Hence, even though it increases compliance, the simple form brought clarity to corporates and other stakeholders. However, the form was not mapped with IT laws. The form requires that any independent nonprofit organizations have 12A

and 80G certificates to be able to fill the form (MCA, 2021, pp. 15, 20). 80G was not required earlier to access CSR funds. Moreover, as Mr. Dadrawala pointed out, many organizations only possess 12A and not 80G, if they were not donor dependent. Also, many charitable schools and hospitals do not fall under 80G but under Section 10(23) of the IT Act. Similarly, research organizations have better deduction under Section 35 of the IT Act. The compulsion of possessing an 80G certification disables all these organizations from filling the form and, consequently, receiving CSR funds from financial year 2021–2022 (N Dadrawala, personal communication, 6th June, 2021). Clearly, more care must be taken in drafting these compliance rules.

3. **Foreign Contributions Regulation Act (FCRA):** The FCRA 2010 regulates the acceptance and utilization of foreign contributions by individuals, organizations, and companies to Indian NGOs. The amendments introduced to this Act in September 2020 disallowed multiple small and medium grassroots organizations to accept funds from large FCRA mandated organizations which, in turn, have CSR donors. Thus, collaborative social development work with smaller organizations has come to a screeching halt.

   In effect, arbitrary reduction in the utilization of foreign funds under administrative heads from 50% to 20% also deters capacity building of organizations. This also means stifled research collaboration with international or national partners for the larger NGOs. Furthermore, all FCRA NGOs across India were mandated to establish FCRA bank accounts with the largest nationalized bank (State Bank of India) in New Delhi wherein they will receive foreign funds to ensure better transparency. Initially restricted till March 2021, these NGOs were provided extended time (till September 2021) to open their accounts in light of the COVID second wave. Although the ramifications of the FCRA Amendment 2020 on the CSR law are comparatively less, these amendments again increase compliance costs for all stakeholders.

## 5.6 Corporate Bureaucracy and Deceleration in Decision-Making Process

In light of strict compliance associated with the CSR laws in India, companies have also initiated internal processes to ascertain proper utilization of the CSR funds. However, such processes, often viewed as arbitrary and ad hoc, impose further challenges to NGOs through long drawn out and risk-averse selection processes of CSR projects. Two of the authors, as part of the official CSR cell of Telangana, T-Social Impact Group, which acts as an intermediary between corporates, government bodies, and civil society organizations to efficient CSR collaborations in Telangana, have had experiences which attest to the bureaucratic processes of corporates in addition to the ever-changing laws of the government. Multiple companies have constructed internal criteria, in addition to the CSR laws, to select an NGO partner and its project. These eligibility criteria often result in the companies being left with a highly reduced pool of NGOs and, consequently, mostly traditional CSR projects to select from. Corporate bureaucracy as experienced by the T-SIG platform can be primarily divided into parts – (a) internal policies and (b) decision-making processes. Some of these issues have been exemplified below:

### 5.6.1 Internal Policies

Company A has two different business arms; say Company A1 and Company A2. As part of their internal policies, the Executive Committee of the company refrains the two arms from supporting the same NGO (even if the projects supported are different). Thus, if a credible and impactful NGO is already supported by Company A1, it cannot be supported by Company A2. Moreover, once the CSR project of an NGO is selected by Company A, internal policy only allows for providing 15% of the NGO's income for the previous financial year as CSR funds. Hence, the provision of CSR funds, in this case, is not related to the relevance, impact, or costs associated with the implementation of the CSR project but to the size of the NGO. Such policies, added to the legal requirements mandated by the government, reduce the pool of NGOs from whom companies can select projects.

In fact, in a quest of working with multiple partners, some companies do not wish to work with the same NGO, post completion of a 3-year CSR project. India's CSR law has never mandated till date that a company must work with new NGOs every 3 years, but this has become a norm among some multinational companies, as experienced by T-SIG. Hence, evidently, not only does this approach hamper the sustainability of ongoing projects and increase administrative costs of all stakeholders but also shorten the list of eligible NGOs for companies.

### 5.6.2 Decision-Making Processes

Slow decision-making process with multiple levels of approval-seeking coupled with innumerable rounds of conducting due diligence is another form of bureaucracy to which NGOs are subjected to. In fact, attempts of companies to form CSR committees and departments with personnel from social sector background have largely remained ineffective in understanding the workings of the social sector. Parekh (2021a) mentions, "(w)hen it comes to companies with a large manufacturing footprint, a further layer of decision making is involved since most factories or plants also have a local CSR structure, made up of plant leaders and support services at the plant level." Parekh elaborates how in conglomerates, defending the choice of a particular CSR project is similar to defending a doctoral thesis by putting forth a clear picture (Fig. 2) of the multiple approvals required to finalize a CSR project. She delves deeper into how within each company, different verticals or departments such as procurement team, legal team, company secretariat, and finance team also partake in finalizing CSR projects, where each team prioritizes the well-being and compliance of the company more than deliberating on the social impact of the CSR project.

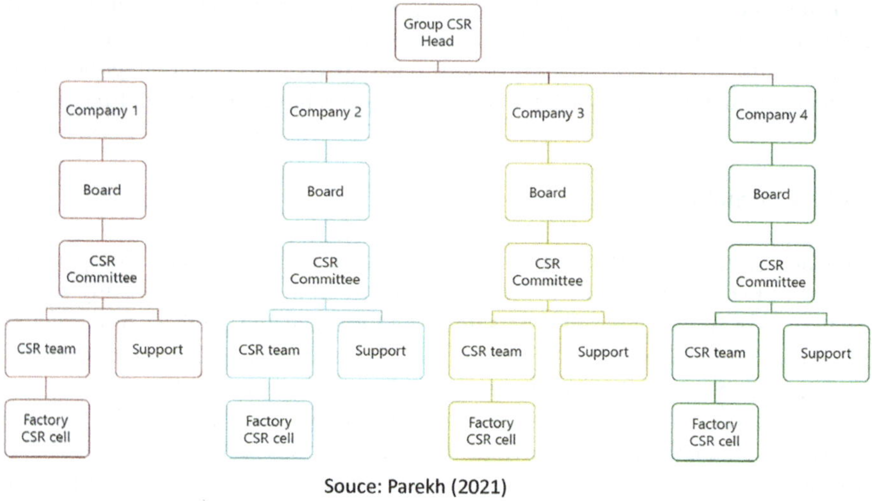


**Fig. 2** Structure of CSR decision-making in companies in India. Source: Parekh (2021a)

> They (corporate) therefore prefer to opt for the letter, rather than the spirit of the law, focusing on compliance and risk avoidance and trying to transpose corporate processes, templates, and contracts to social programmes. (Parekh, 2021b)

Even during emergency crisis periods such as COVID where ground realities are dynamic and beneficiary needs are constantly evolving, companies have remained true to their bureaucratic nature. Rather than expediting disbursement of CSR funds to be swiftly utilized by grassroots organizations for immediate requirements like food, ration, and protective gear, T-SIG has encountered companies who have taken more than a month to disburse funds.

But as Mr. Noshir Dadrawala rightly mentions, the corporate bureaucracy is a result of the "top-down approach of the CSR law" (personal communication, 6th June, 2021) and cannot be seen in isolation. With increasing compliance on part of the companies, it is understandable that the companies wish to be careful about their CSR expenditure while reporting to the MCA. This leads companies to take extra caution, and inadvertently, the NGOs are at the receiving end of this governmental pressure.

On the other hand, for companies which are public sector units (PSUs), corporate bureaucracy is evident through different methods. Through T-SIG's experience, it has been perceived that these PSUs primarily prefer to donate their CSR funds to a trust or society registered under a certain government department, for example, municipal corporation and healthcare or education departments. More often than not, such CSR projects are halted midway or delayed due to multiple reasons such as government bandwidth, change in government focus areas, or even change in government officials. This, in turn, leads to issues related to project implementation, reporting, and monitoring and even impact assessment which could be easily negated if more PSUs worked with credible NGOs. In fact, the CSR expenditure by PSUs on the Statue of Unity, as illustrated above, is the perfect example of how PSUs fund government schemes and initiative through their CSR.

## 5.7 *Impact of Indian CSR on COVID Emergency*

The structural issues in the CSR law and its reflection in the operation of the law became abundantly clear during the first and the second waves of the coronavirus pandemic in India. In the first wave, the obscurity of what constitutes as CSR activity was primarily refrained to the debate of PM Cares Fund and CM Relief Fund. In spite of all the upheaval created with the inception of PM Cares Fund and the repeated right to information (RTIs) for transparency regarding the expenditure of the funds, little action has been initiated by the central government (Jebaraj, 2020). In fact, the Ministry of Home Affairs has exempted PM Cares Fund from FCRA, while small NGOs have to follow all the rules even during COVID pandemic.

> In the last few days of March, the Fund collected Rs 39.68 lakhs [US$ 53,927.92] in foreign currency donations. There is no clarity on how much of this is from foreign donors and how much from domestic donors. (The Wire, 2020)

However, in the second wave, with the new CSR amendments in January 2021, the confusion was also expanded to which NGOs could receive CSR funds. The mandatory CSR Form 1 removed any NGO who had not registered with the MCA, outside the ambit of receiving any CSR funds, thereby removing a huge quantum of NGOs from consideration. Even in the case of receiving foreign funds, many NGOs who were in the process of opening a bank account in SBI, New Delhi, were left in a lurch. With the pandemic showcasing its ugliest form in the country's capital and the consequent lockdown of Delhi, many NGOs were in a limbo. Although these NGOs had a preexisting network of companies with a corpus of foreign funds, the technicality of not having a bank account in the government specified bank deterred these companies from donating foreign funds to the NGOs. Most of these NGOs had to resort to platforms like Give India for foreign fundraising and pay an additional 5% administrative charges to receive funds even though they possessed FCRA certification.

Another issue highlighted by Mr. Dadrawala (personal communication, 6th June, 2021) was not considering donation of products by manufacturing companies as part of the CSR laws. The CSR law in India focuses strictly on utilization of CSR funds to be considered as CSR activity.

It does not consider cases where manufacturing companies can donate their products as part of relief efforts as legitimate CSR activities. This creates a roundabout loop where companies disburse funds to NGOs to procure their own products and then distribute them. In any crisis where time is of value, this process can be reduced if companies' product donation to NGOs was considered as CSR activity.

Through Surge Impact Foundation's experience of working on-ground during the second wave of COVID, similar experiences of delay and slow decision-making processes were faced while fundraising for emergency CSR projects. Company B had agreed to finance a COVID CSR project to provide protective gear to the youth of a slum area in Hyderabad, Telangana, who are acting as frontline workers on April 26, 2021. The project was to be implemented by Surge Impact Foundation and costs US$ 3,721.77. Consequently, the due diligence process was initiated whereby the NGO submitted all relevant documents and declarations. However, till June 21 (57 days), the CSR funds were not disbursed even after multiple follow-ups. One of the main reasons for this was the normal process of finalizing a CSR project being followed during emergency crisis situations. Instead of understanding the immediate on-ground requirements, Company B was primarily involved in ensuring that all compliance rules and regulations were followed. By June 18, 2021 (the funds were yet not disbursed), the requirement of such protective gear for the youth was not valid with the number of COVID cases reducing in the city. Upon inquiry regarding the utilization of the agreed upon CSR funds for another pressing issue (like the payment of hospital bills for COVID patients from below poverty line), Company B mentioned that as per the company's internal policies, the funds cannot be directed for any other cause and only the location of project implementation could be revised. As a result, instead of directing the funds to be utilized in a more efficient manner, $ 3,721.77 could strictly be utilized for the procurement of protective gear, even if its requirement had become redundant.

In such emergency situations, the duty of the government is to make rules and regulations more lenient to ensure that civil society organizations can work in tandem with the government bodies to provide relief. Even Mr. Noshir Dadrawala mentioned, "Financial Years 2020–2021 and 2021–2022 were not years for implementation [of new rules of CSR laws and FCRA] but years to be benevolent" (N Dadrawala, personal communication, 6th June, 2021).

## 6 Conclusion

Thus, one can conclude that CSR in India suffers from double bureaucracy which seriously hobbles its impact. At one end, an unstable regulatory regime and increasing government compliance have made CSR expensive and burdensome for the stakeholders. The regulatory framework suffers from arbitrary choice made with short-term objectives which make innovation difficult but do not curb questionable use of CSR funds. As argued in the paper, this partly arises from the inherent nature of the Chatterjee model which defines CSR in terms CSR funds spent on government priorities. This enables arbitrary behavior and red-tapeism on part of the government. For a country that is looking to showcase ease of doing business, mandated CSR with heavy and penal regulatory framework is discouraging. However, it must be reiterated that while all models have limitation, the core issues in India's CSR regime arises from political and policy decisions made by the power that be to not only exploit the limitation of the model but also consciously or unconsciously move away from its underlining philosophy.

On the other hand, private corporates have also created their own internal policies and eligibility criteria for the selection of NGOs and CSR projects. Although these measures can be traced back to compliance issues and reporting of companies to MCA, the ultimate burden falls on the NGOs who have to justify their organizational structure and administrative overheads along with the relevance of their project and its intended outcomes. Already under the strict surveillance of the government, the additional legal documents sought by companies restrict numerous small and medium-sized credible NGOs from entering the CSR arena.

One solution would be to go back to the original principle of "comply and explain" while redefining CSR in terms of impact using international standards like UN SDGs which will reduce scope of arbitrary changes with short-term goals. Furthermore, compliance should be kept at minimum. The government should see CSR as a force multiplier for social development and not as a source of financing where its sole role is to provide a clear and stale regulatory framework. Companies and civil society organizations must be provided with necessary freedom to innovate and harness the strength for social impact. This will also reduce the cautionary approach undertaken by companies to finalize their CSR projects, thereby reducing the second level of bureaucracy faced by NGOs. A more sensitive, collaborative, and responsive approach to rule-making would be a good place to start.

Lastly, there is need to explore whether the compulsory CSR is even the most efficient way to achieve the underlining aims, given the tendency of the government to treat CSR as a revenue source.

## Annex I

Timeline of changes in Schedule VII related to healthcare and entrepreneurship (Saha et al., 2020)

| | Healthcare | Entrepreneurship |
|---|---|---|
| As in act passed by parliament | (v) Combating human immunodeficiency virus, acquired immune deficiency syndrome, malaria, and other diseases | (viii) Social business projects |
| February 27, 2014 | (i) Eradicating hunger, poverty, and malnutrition, promoting preventive healthcare and sanitation, and making available safe drinking water | (ix) Contributions or funds provided to technology incubators located within academic institutions which are approved by the central government |
| March 31, 2014 | For "promoting preventive health care" read "promoting health care including preventive health care". | |
| October 11, 2019 | | (ix) Contribution to incubators funded by central government or state government or any agency or public sector undertaking of central government or state government and contributions to public funded universities, the Indian |

(continued)

| | Healthcare | Entrepreneurship |
|---|---|---|
| | | Institute of Technology (IITs), national laboratories, and autonomous bodies (established under the auspices of the Indian Council of Agricultural Research (ICAR), Indian Council of Medical Research (ICMR), Council of Scientific and Industrial Research (CSIR), Department of Atomic Energy (DAE), Defence Research and Development Organisation (DRDO), 7 [Department of Biotechnology (DBT)], Department of Science and Technology (DST), Ministry of Electronics and Information Technology) engaged in conducting research in science, technology, engineering, and medicine aimed at promoting Sustainable Development Goals (SDGs) |
| August 24, 2020 | | (ix) (a) Contribution to incubators or research and development projects in the field of science, technology, engineering, and medicine, funded by the central government or state government or public sector undertaking or any agency of the central government or state government<br>(b) Contributions to public funded universities; the Indian Institute of Technology (IIT); national laboratories and autonomous bodies established under the Department of Atomic Energy (DAE); Department of Biotechnology (DBT); Department of Science and Technology (DST); Department of Pharmaceuticals; Ministry of Ayurveda, Yoga and Naturopathy, Unani, Siddha and Homoeopathy (AYUSH); Ministry of Electronics and Information Technology and other bodies, namely, Defence Research and Development Organisation (DRDO); Indian Council of Agricultural Research (ICAR); and Indian Council of Medical Research (ICMR) and Council of Scientific and Industrial Research (CSIR), engaged in conducting research in science, technology, engineering, and medicine aimed at promoting Sustainable Development Goals (SDGs) |

## References

Business Today. (2019, August 14). *Govt likely to do away with plan to jail officials under new CSR norms.* Retrieved June 21, 2021, from https://www.businesstoday.in/current/economy-politics/govt-likely-to-do-away-with-plan-to-jail-officials-under-new-csr-norms/story/372507.html

Chatterjee, B. (2018, June 29). *'Father of CSR' in India Addresses U.N. Forum on Business and Human Rights.* Retrieved June 20, 2021, from https://www.prnewswire.com/news-releases/father-of-csr-in-india-addresses-un-forum-on-business-and-human-rights-300365087.html#continue-jump

Chatterjee, B., & Mitra, N. (2017). CSR should contribute to the national agenda in emerging economies - The 'Chatterjee Model'. *International Journal of Corporate Social Responsibility, 2*(1). https://doi.org/10.1186/s40991-017-0012-1

Department for Promotion of Industry and Internal Trade. (n.d.). *Start-up India.* Retrieved from https://www.startupindia.gov.in/content/sih/en/home-page.html

Financial Express. (2019, August 1). *CSR Violations: India Inc feels jail term too harsh, mandatory spend akin to impost.* Retrieved June 21, 2021, from https://www.financialexpress.com/industry/csr-violations-india-inc-feels-jail-term-too-harsh-mandatory-spend-akin-to-impost/1662535/

Indian Audit and Accounts Department. (2018). *Report of the Comptroller and Auditor General of India for the year ended 31 March 2017. Report No. 8 of 2018.* Union Government (Commercial)

Jebaraj, P. (2020, August 17). *PMO denies RTI plea seeking info on PM-CARES.* Retrieved from https://www.thehindu.com/news/national/pmo-denies-rti-plea-seeking-info-on-pm-cares/article32369180.ece on 16th June, 2021.

Ministry of Corporate Affairs. (2013). *The Companies Act, 2013.* Government of India. Retrieved from http://ebook.mca.gov.in/notificationdetail.aspx?acturl=6CoJDC4uKVUR7C9Fl4rZdatyDbeJTqg3F94LXwPqb1E+n2xkV265F/0eSYr1zaxEYLeJ4JSYA+w/NG9H7RPU3w==

Ministry of Corporate Affairs. (2014a, February 27). *Ministry of Corporate Affairs Notification New Delhi, the 27th February' 2014.* Retrieved from https://www.mca.gov.in/Ministry/pdf/CompaniesActNotification2_2014.pdf

Ministry of Corporate Affairs. (2014b, February 27). *SCHEDULE VII.* Retrieved from http://ebook.mca.gov.in/Actpagedisplay.aspx?PAGENAME=17923

Ministry of Corporate Affairs. (2014c, June 18). *Clarifications with regard to provisions of Corporate Social Responsibility under section 135 of the Companies Act, 2013. General Circular No. 21/2014. Government of India.* Retrieved from https://www.mca.gov.in/Ministry/pdf/General_Circular_21_2014.pdf

Ministry of Corporate Affairs. (2014d, April 1). *Corporate Social Responsibility.* Retrieved June 21, 2021, from http://ebook.mca.gov.in/Actpagedisplay.aspx?PAGENAME=17518

Ministry of Corporate Affairs. (2014e, April 1). *Definitions.* Retrieved June 21, 2021, from http://ebook.mca.gov.in/Actpagedisplay.aspx?PAGENAME=29016

Ministry of Corporate Affairs. (2014f, April 1). *CSR Activities.* Retrieved June 21, 2021, from http://ebook.mca.gov.in/Actpagedisplay.aspx?PAGENAME=29017

Ministry of Corporate Affairs. (2016, January 12). *FAQ CSR, General Circular 01/2016.* Retrieved from https://www.mca.gov.in/Ministry/pdf/FAQ_CSR.pdf

Ministry of Corporate Affairs. (2019, July 31). *The Companies (Amendment) Act, 29 No. 22 Of 2019.* Retrieved from https://www.mca.gov.in/Ministry/pdf/AMENDMENTACT_01082019.pdf

Ministry of Corporate Affairs. (2020a, March 28). *Clarification on contribution to PM CARES Fund as eligible CSR activity under item no. (viii) of the Schedule VII of Companies Act, 2013. eF. No. CSR-05/1/2020-CSR-MCA.* Government of India. https://www.mca.gov.in/Ministry/pdf/Circular25_20062020.pdf

Ministry of Corporate Affairs. (2020b, April 10). *COVID-19 related Frequently Asked Questions (FAQs) on Corporate Social Responsibility (CSR)*. General Circular No. 15 /2020. Government of India. https://www.webtel.in/Image/General%20Circular%2014april%202020.pdf

Ministry of Corporate Affairs. (2020c, May 26) Notification, http://ebook.mca.gov.in/notificationdetail.aspx?acturl=6CoJDC4uKVUR7C9Fl4rZdatyDbeJTqg3HE7Xx+PAr2tIpnrftKCE2ZF9raArVVO+

Ministry of Corporate Affairs. (2021, January 22) *Notification CSR Amendment Rules 2021*. https://www.mca.gov.in/Ministry/pdf/CSRAmendmentRules_23012021.pdf

Ministry of Law and Justice. (2019, July 31). The Companies (Amendment) Act, 2019. Retrieved from http://ebook.mca.gov.in/notificationdetail.aspx?acturl=6CoJDC4uKVUR7C9Fl4rZdatyDbeJTqg3c+9Unnk2xevZ+9h/lvkVY3TGLaBTBGrakRRKnu3V8Do=

Ministry of Law and Justice. (2020, September 28). *The Foreign Contribution (Regulation) Amendment Act, 2020*. Retrieved from https://fcraonline.nic.in/home/PDF_Doc/fc_amend_07102020_1.pdf on 12th June, 2021

Ministry of Law and Justice. (2020, September 28). *The Companies (Amendment) Act, 2020*. Retrieved June 21, 2021, from http://ebook.mca.gov.in/notificationdetail.aspx?acturl=6CoJDC4uKVUR7C9Fl4rZdatyDbeJTqg30ApO1EAOX4LB6NIr39q6zNZ3uzmI1wj8sclx2dyU4aU=

Ministry of Skill Development and Entrepreneurship. (2016). N*ational Apprenticeship Promotion Scheme (NAPS): Ministry of Skill Development and Entrepreneurship: Government of India*. Retrieved June 21, 2021, from https://msde.gov.in/en/schemes-initiatives/apprenticeship-training/naps

Ministry of Skill Development and Entrepreneurship. (2020, March 11). *Office Memorandum, 11 March 2020*. Retrieved June 21, 2021, from https://naps-cdn.s3.ap-south-1.amazonaws.com/CSR_funds_utilization_under_Apprenticeship_Act.pdf

Mitra, N. (2017, January 01). *Believe that CSR can change your nation, Says Dr. Bhaskar Chatterjee*. Retrieved from https://indiacsr.in/believe-that-csr-can-change-your-nation-says-dr-bhaskar-chatterjee/

NGOBOX. (2013, October 05). *Dr Bhaskar Chatterjee on CSR Rules in India Part I*. YouTube. Retrieved June 20, 2021, from https://www.youtube.com/watch?v=PG4AQ2NB-kM

NGOBOX. (2014, September 25). *Dr. Bhaskar Chatterjee, DG & CEO of IICA in India CSR Summit 2014 on CSR Rules*. YouTube. Retrieved June 21, 2021, from https://www.youtube.com/watch?v=t4S486lNJlA

NITI Aayog. (n.d.). *Atal Innovation Mission*. Retrieved from https://aim.gov.in/atal-incubation-centres.php

Parekh A. (2021a, June 17). *How CSR Decisions Get Taken in India*. Retrieved from https://idronline.org/article/philanthropy-csr/how-csr-decisions-get-taken-in-india/ on 17th June, 2021

Parekh, A. (2021b, June 17). *Untitled [Graph]. How CSR Decisions Get Taken in India*. Retrieved from https://idronline.org/article/philanthropy-csr/how-csr-decisions-get-taken-in-india/ on 17th June, 2021

PMO India. (2018, October 31). *Statue of Unity [Digital image]*. Retrieved June 21, 2021, from https://bsmedia.business-standard.com/_media/bs/img/article/2018-10/31/full/1540957771-728.jpg

Press Trust of India. (2021, March 22). *CSR funds not used to run government schemes, says Anurag Thakur*. Retrieved from https://www.business-standard.com/article/economy-policy/csr-funds-not-used-to-run-government-schemes-says-anurag-thakur-121032200630_1.html

Saha, D., et al. (2020, December 10). *Missing Spirit of CSR: Structural Causes of the Challenge*. Down to Earth. Retrieved May 31, 2021, from https://www.downtoearth.org.in/blog/governance/missing-spirit-of-csr-structural-causes-of-thechallenge-74590

The Wire Staff. (2020, September 12). *PM-CARES Fund Exempted From FCRA, MHA Reveals Under RTI*. Retrieved from https://thewire.in/government/pm-cares-fcra-exemption on 16th June, 2021

**Debanjali Saha**, a postgraduate in Women's Studies from Tata Institute of Social Sciences, Hyderabad, is the Head of Business Development for Surge Impact Foundation, a Section 8 nonprofit organization based out of Hyderabad, India. Her experience includes conducting field research in areas of corporate social responsibility, refugees, girl child education, and women's rights.

Moreover, Debanjali, in her previous capacity, has been deeply involved in CSR as the Program Manager for the Telangana state-initiated and funded official CSR platform, "T-Social Impact Group" (T-SIG) and in organizing focus group discussions between corporates and government officials, content creation, and research work for the T-SIG platform. She is also the lead author for Surge Impact Foundation's research paper "Missing Spirit of CSR: Structural Causes of the Challenge" published by *Down to Earth* journal. Having previously worked with organizations such as IT for Change and Save the Children, Debanjali has an avid interest in conducting action research on contemporary socioeconomic issues. Her research skills have also been honed as a result of working with the University of Berkeley, California, and Jadavpur University.

**Rinkesh Dharod** has done his MA in History and LLB from Mumbai University and has interest in history and politics of modern and contemporary India. His past work is primarily in area of philanthropy and CSR. He has worked as a Research Assistant at K R Cama Oriental Institute on their project "Genealogy of the Seth and Sethna Family: 1930–2010" and coauthored two volumes on the same.

Rinkesh has worked as a History and Theory of Knowledge Teacher for IBDP and IGSCE syllabus for 5 years. At present, he is working in field of start-up incubation, CSR, and social enterprise and is a Knowledge Manager, AIC-IIITH Foundation, which is an incubator for social enterprises.

Rinkesh has coauthored two volumes: *The Sett and Sethna Family of Rustom Maneck (17th to 1st century) Vol I: Genealogy and History* and *The Sett and Sethna Family of Rustom Maneck (18th to 21st century) Vol II: Institutional Philanthropy*. He has also coauthored a paper on CSR in *Down to Earth* magazine titled, "Missing Spirit of CSR: Structural Causes of the Challenge." He has also presented paper on "Viewing Design: Changing Perceptions of Khadi" at "Towards Global Design Histories: Postcolonial Perspectives" – the Design History Society Annual Conference 2013.

**Raj Janagam** works on high-impact businesses in the capacity as a business advisor, coach, and consultant. He has coached over 85 start-ups since 2013 and is the CEO of the Atal Incubation Centre-IIIT Hyderabad Foundation to support early-stage social enterprises leveraging technology to achieve SDGs. In his earlier avatar, he cofounded Surge Impact Foundation and founded UnLtd Hyderabad. He advises the Telangana Social Impact Group which is the CSR cell of the government of Telangana. At Global Shapers Community, he was a part of the Advisory Council on Governance and Accountability for hubs across the world and is also an alumnus of the Hyderabad Hub which he led as a curator in 2018–2019. Raj is a partner at Camp CeLiberate – an art company promoting modern arts and artists from India.

Raj brings in the experience of working across sectors of the government, healthcare, sports, celebrity philanthropy, research and advocacy, politics, skilling, modern arts, nonprofit, and social enterprise start-ups.

He is a philosophy graduate and an MBA in social entrepreneurship. From being an all-round best student across universities, Raj has won much national and international recognition and coverage for his work and ideas. Raj is a public speaker with many keynotes like TedX and Josh Talk events including sharing about SDG progress in Telangana at some key events like Davos 2019 joined by political and business leaders from across the world.

# Part II
# CSR Experiences in Different Industries

# Promoting Decent Work in Global Production: Lessons Learned from the Jordanian Garment Industry

**Britta Holzberg**

## 1 Introduction

In 2019, the International Labour Organization (ILO) found that despite CSR improvements, the current rate of progress is insufficient to achieve "decent work for all" (SDG 8) by 2030 (ILO, 2019). In particular, the global garment production network has been highlighted as a prime case of corporate social irresponsibility (Luque & Herrero-García, 2019). Many garment workers encounter excessive and forced overtime, pay below living wages, health and safety concerns, repression of collective bargaining rights, discrimination, and sexual harassment. One strategy to support the enhancement of CSR efforts is based on the analysis of cases in which significant decent work progress has been made. Case studies foster understanding of how complex factors interact in a real-life environment (Yin, 2009). In this paper, the case of the garment export industry in Jordan is examined, where significant decent work improvements have been realized, although the end goal has not yet been achieved (Better Work Jordan, 2019b; Business and Human Rights Resource Center, 2018; Kolben, 2019).

The current paper builds on previous research regarding Jordan. Early articles highlighted the precarious working conditions in the garment export industry and linked them to the trade regime (Kolben, 2013), the dominance of migrant labor (Francoise De Bel-Air & Dergarabedian, 2006), and the questionable economic viability of the industry (Domat et al., 2012). Jordan had no natural competitive advantage in garment production, little experience in garment exports, and few

B. Holzberg (✉)
Institute of Development Research and Development Policy (IEE), Ruhr University Bochum, Bochum, Germany

International Institute of Social Studies (ISS), Erasmus University Rotterdam, The Hague, The Netherlands

 
I. G. Arraiano et al. (eds.), *Corporate Social Responsibility in a Dynamic Global Environment*, CSR, Sustainability, Ethics & Governance,
https://doi.org/10.1007/978-3-031-24647-0_8

qualified garment workers when the country entered a preferential trade agreement with the United States. Foreign companies subsequently set up factories in Jordan to benefit from the agreement, but they remained disconnected from the local economy and employed mostly vulnerable migrant labor. This situation nourished the ground for labor abuse (Azmeh, 2014; Kolben, 2013). More recently, studies have addressed efforts for improving working conditions and identified success factors. Kolben (2015) discusses the interaction of private and public labor regulation to enforce labor standard compliance in Jordan. Aissi et al. (2018) find that the heightened monitoring and intergovernmental dialogue driven by the Jordan-United States free trade agreement benefitted labor rights. Kolben (2019) traces how Better Work Jordan (BWJ), a joint program of the ILO and the International Finance Corporation, supports the development of the garment industry and decent work. Robertson (2019) examines one BWJ project in particular and links the BWJ transparency portal to improved labor standard compliance.

Although the early observations of unacceptable working conditions in Jordan are not unusual for global garment production (Anner, 2019), the identified success factors are not necessarily universally applicable. Free trade agreements were not found to benefit labor rights in South Korea (van Roozendaal, 2017), and Bair (2017) finds the effectiveness of the Better Work (BW) program to be contingent on the local context. Such findings underline the need to account for contextual specifics when discussing enablers and inhibitors of decent work progress. The current scientific debate features an increasing consensus that no single factor alone can explain decent work outcomes. The interaction of different global, local, public, and private factors shapes working conditions in global production (Lund-Thomsen et al., 2016; Niforou, 2015). Therefore, current research on individual drivers of decent work needs to be accompanied by case studies that illustrate the interplay of drivers and highlight contingencies. Holzberg (2019) theoretically systematized factors that interact and influence labor-related CSR decisions in supplier firms. She argues that the interaction of global competitive forces (e.g., price pressure, lead time pressure) and global public and private regulatory efforts (e.g., labor standards, audit regimes), combined with local institutional factors (e.g., legal institutions, worker institutions), sociocultural factors (e.g., work ethics, gender norms), and industry factors (e.g., labor markets, production sophistication), must be considered when trying to understand decent work outcomes. Global and local factors "crossverge" at the point of the supplier and through their interaction shape decent work realities on the factory floor (Holzberg, 2019). As the framework mirrors the complexity of the real-life environment of production networks, it serves as the theoretical starting point for the design of this case study.

The case study analyzes the interaction of global, local, public, and private factors that shape working conditions in the Jordanian garment export industry to answer two research questions:

(1) Which factors and factor interactions enabled decent work progress in the garment export industry in Jordan between 2008 and 2018?
(2) Which factors and factor interactions inhibit further decent work progress in the garment export industry in Jordan in 2018?

The background of the Jordanian case and the methodological research design are outlined in the next section of the paper. Case study findings are presented and discussed subsequently. The concluding section sums up the study's main research contributions and their implications for the Jordanian and the global garment production networks.

## 2 Jordanian Garment Export Industry: Background and Decent Work Developments

By producing 25% of the country's total exports and employing 68,300 workers, the garment export industry is central to the Jordanian trade and labor market (Better Work Jordan, 2019a, data: 2018). Approximately 95 factories operate in Jordan, producing mainly for the United States (Better Work Jordan, 2019a, data: 2018). A distinguishing feature of the industry is its strong international linkages. The garment export industry developed after a preferential trade agreement with the United States came into effect in 1997. Asian manufacturers in particular invested in setting up garment factories in Jordan (Azmeh, 2014). Currently, almost all factories are foreign-owned and 75% of the workforce, along with most owners and managers, are migrants, especially from Bangladesh, India, Sri Lanka, and Nepal (Better Work Jordan, 2019a, data: 2018). Jordanian labor law requires a 25% share of Jordanian workers, which is supported by the creation of satellite factories in rural areas.

Poor working conditions in the industry first drew attention in 2006 owing to a report by the National Labor Committee (2006) (NLC). The report included testimony by migrant workers about working hours exceeding 20 hours per shift on 6–7 days per week, beatings and torture by supervisors, sexual abuse, and wages being withheld or severely underpaid. The accounts portrayed inhumane living conditions, with insufficient living space, food, and water. The information sparked an international outcry. Buyers retreated from Jordan, and several of the factories mentioned in the NLC report closed (Kolben, 2013). Jordan was placed on the US Department of Labor's (USDOL) forced labor list in 2009, and the US government started providing funding and technical expertise to aid labor standard improvement. Agreement was reached to institutionalize the BW program in Jordan. The program aims to promote labor standards and productivity in the global garment industry.

In 2019, international observers reported significant improvements and comparably favorable working conditions (Aissi et al., 2018; Kolben, 2019; Robertson, 2019). Jordan was removed from the USDOL's forced labor list in 2016, and a 2019 report issued in the name of the Dutch Ministry of Foreign Affairs noted "the excellent compliance with environmental and social standards" as a core benefit of the industry (Enclude BV & CMC, 2019, p. 4). In particular, at 220 JOD (approximately 310 USD in 2018), the monthly minimum wage was more than three times higher than in Bangladesh and could cover living expenses in most rural areas of Jordan. Workers worked on average 53.5 hours per week, and excessive hours, forced labor, physical violence by supervisors, and unauthorized wage deductions were found to be exceptions rather than the norm (Better Work Jordan, 2019a, data: 2018). Nevertheless, decent work for all has not yet been achieved (Business and Human Rights Resource Center, 2018). Especially at subcontractors, working conditions remain substandard, for example, with working hours of up to 97 hours per week (Better Work Jordan, 2019a, data: 2018). Furthermore, Jordan has a single trade union structure and therewith remains noncompliant with the ILO's Freedom of Association convention. In addition, BWJ criticizes occupational safety and health breaches, especially in dormitories, and acknowledges shortcomings in tracking and preventing incidents of rape and sexual harassment.

## 3 Methodology of Evaluation and Data

The decent work developments in the Jordanian garment export industry were analyzed with an explanatory case study design. The analysis aimed to explain which developments enabled decent work improvements, which factors hindered further progress, and how interactions between different drivers shaped the current decent work situation.

Evidence was collected from multiple sources (see Table 1 for an overview and Annex 1, 2, and 3 for further details): Between April and May 2018, 7 industry and labor experts and 13 factory owners-managers were interviewed. A maximum variation logic guided the sampling of respondents as common themes that are discovered across a diverse range of respondents are likely to reflect core and shared

**Table 1** Overview of collected data

| Supplier firm interviews | Factory owners-managers | | Expert interviews | Experts | |
|---|---|---|---|---|---|
| | *Jordanian* | *Non-Jordanian* | | *Jordanian* | *Non-Jordanian* |
| *Factory size* | | | *Area of expertise* | | |
| 0–499 | 4 | 3 | Industry expert | 3 | |
| 500–1499 | 2 | 1 | Labor expert | 2 | 2 |
| 1500–5500 | 2 | 1 | | | |
| **Worker group interview** | **Workers** | | **Document analysis** | | |
| | *Jordanian* | *Non-Jordanian* | | *Documents* | |
| *Gender* | | | *Type* | | |
| Female | 1 | 11 | Legal information | >10 | |
| Male | | 8 | Report/study | >40 | |
| | | | News article | >60 | |

experiences of the overall research case (Patton, 2002). Furthermore, 20 workers participated in group interviews, and the author observed three industry and workers' events. During the interviews, respondents reported on current working conditions and reflected on changes over time, reasons for improvements, and remaining obstacles. Interviews lasted 60–90 min, and most were audio-recorded and transcribed. If respondents did not consent to recording, the interviews were documented in detail. The interviews and observations were primarily conducted in English or German. Selected interviews and observations were aided by translators (Arabic, Bengali, Hindi, Nepali, and Sinhala). All interviews were labeled with a code that specified the type of interview (Firm, Expert, or Worker). The 18th interview in Jordan, which was the 12th firm interview, was labeled J18F12. Complementing the field research, over 60 international and local publications were analyzed to obtain insights on the case of Jordan between 2006 and 2018.

The data were analyzed with the help of Yin's (2009) explanation building technique. Holzberg's (2019) crossvergence framework provided the initial theoretical propositions. It proposes that global competitive pressures (price, lead time, flexibility, quality pressure), global public labor market regulation (labor standards set and promoted by international institutions), global private labor market regulation (labor standards set and promoted by buyers and private multi-stakeholder initiatives), local institutions (labor laws, business, labor and civil society institutions), local socioculture (local values, norms, and decent work beliefs), and the local industry environment (technological, business, and labor characteristics of local industrial clusters) shape work outcomes in global garment production. The propositions were revised until a coherent explanation for the developments in Jordan was found. Interviews and documents were coded following the steps of a qualitative content analysis as outlined by Schreier (2014). The final category system is presented in Annex 4, and Annex 5 shows an excerpt from the coding frame. The interviews with owners-managers served to identify which factors influenced the design of working conditions and (de-)motivated decent work improvements. Insights from experts helped to elucidate the factors that were addressed and to evaluate the extent to which the interview insights were common for the industry. The documents primarily served to trace developments over time, and along with the worker interviews, they were used to triangulate the data following Yin's (2009) suggestion to use multiple sources of evidence to ensure validity. Using the iterative process of the explanation building technique was intended to ensure the internal validity of the case study. The analysis was done twice and documented in MAXQDA to foster reliability.

## 4 Case Study Results

The analysis confirmed that an interplay of multiple factors enabled decent work progress in Jordan. As proposed by previous authors (Holzberg, 2019; Lund-Thomsen et al., 2016; Soundararajan & Brown, 2016), global and local industry factors interacted at the factory level and needed to be considered simultaneously to understand decent work developments. Thereby, categorization according to level was insufficient for structuring the multiple factors detected. As suggested by the explanation building technique, different theoretical logics were considered before eventually finding that an actor-oriented subcategorization suited the data. Enabling factors on each level could be attributed to the influence of business, labor, or public stakeholders, whereby labor-oriented societal actors such as NGOs or local clans were included among labor stakeholders. Figure 1 depicts the different influences that shaped decent work realities in Jordan. Theoretically, it mirrors Gereffi and Lee's (2016) proposal that private (business), social (labor), and public governance on global and local levels affect social upgrading in global supply chains and extends it by adding the factory level. Thereby, in the case of Jordan, stakeholders on each level not only governed decent work efforts but actively supported their implementation to various degrees. Moreover, the analysis revealed that the coordinated interaction of different actors was crucial for decent work progress. The work of BWJ played a particularly important role in this regard and was placed in its own category as a cross-level influence in the analysis. The following sections outline the enablers of decent work progress in Jordan by level and highlight their interactions and the role of BWJ in facilitating them. The section on inhibitors identifies remaining issues in Jordan in 2018 that contributed to inhibiting or even deteriorating progress.

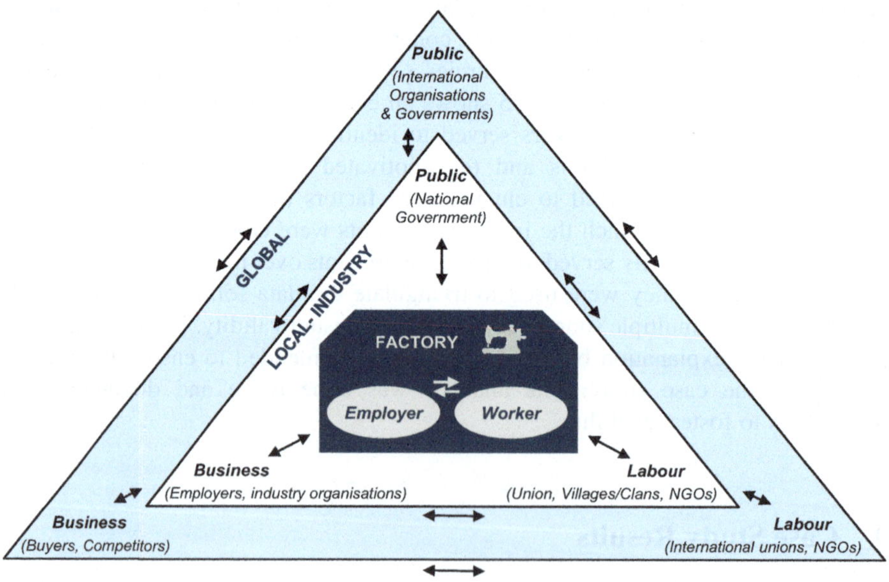


**Fig. 1** Map of stakeholder influence on decent work in the garment industry in Jordan

## 4.1 *Enablers of Decent Work Progress*

### 4.1.1 Global-Level Developments

On a global level, public, business, and labor stakeholders have put pressure on the Jordanian garment export industry to improve working conditions.

With regard to public governance and support, North American governments in particular, and since 2016 also the EU, had an interest in the industry. US governmental actors used trade agreements to demand decent work and put Jordan on the USDOL's forced labor list from 2009 to 2016. The EU-Jordan agreement financially motivated suppliers to employ Syrian refugees [BWJ-report-2016], and the industry received technical and financial support to fulfill these demands. US governmental actors have repeatedly acted as mediators in local decent work debates, and the United States, Canada, and the EU have substantially helped finance BWJ to monitor and support decent work progress in Jordan.

In terms of business governance and support, international buyers put pressure on suppliers by specifying codes of conduct or by requiring international certificates that (partially) go beyond the Jordanian law and ILO core labor standards. The characterization of buyer-supplier relationships thereby differed between respondents. While some characterized their relationship as a "partnership" [J18F12], others highlighted discrepancies between buyers' codes of conducts and the high pressure on prices, as will be further discussed in the inhibitors section. All respondents agreed that the work with BWJ substantially helped to improve understanding between buyers and suppliers. "Better Work teach us 90% from what [buyer] auditors need" [J08F05].

Finally, with regard to labor governance and support, the international union Industri-All supported the Jordanian garment union, for instance, by providing trainings to union representatives. Furthermore, international NGOs continued to monitor the industry. The Institute for Global Labour and Human Rights, for instance, has regularly published reports on the situation in Jordan [IGLHR-2011a, IGLHR-2012, IGLHR-2013].

### 4.1.2 Local Industry-Level Developments

On the local industry level, Jordan experienced substantial changes driven by public, business, and labor stakeholders. Thereby, stakeholder collaboration was key to promoting decent work.

Several decent work improvements were linked to changes on the public level, especially legislative changes. The minimum wage was frequently raised for Jordanians (see Fig. 2). Furthermore, progress toward the support of migrant labor was made in 2010 when the workers received the legal right to join trade unions. Even though compliance with labor laws was still mainly assessed by BWJ in 2018, resources and capabilities of the local labor inspectorate were incrementally

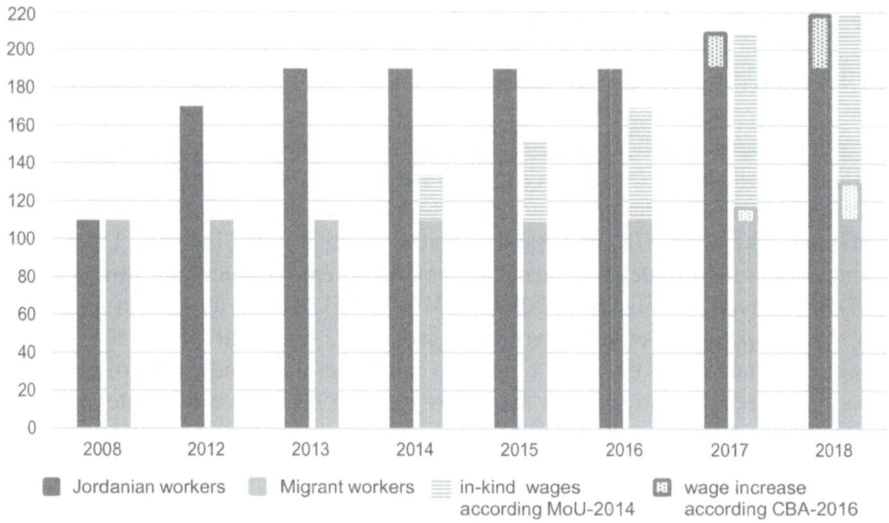


**Fig. 2** Legal minimum wages in the garment export industry in Jordan [in JOD]

developed to eventually take over factory assessments. Generally, a pattern was discerned indicating that governmental efforts oftentimes (initially) focused on Jordanian workers but subsequently also benefitted migrant workers as the garment export industry is mandated to not discriminate between nationalities, with the mandate being strictly controlled by BWJ and other stakeholders.

For business and labor organizations, documents and interviews showed a marked increase in decent work sensitivity and capabilities. All interviewees from these organizations underlined the necessity to focus on decent work for the survival of the industry. Employer representatives placed this need in the context of the need to remain competitive, whereby the focus on decent work was frequently considered as the strategic edge of Jordan. While the treatment of migrant labor in particular was a black box for union representatives in the past, interviews and observations showed that investments in infrastructure and capabilities enabled the union to represent both local and migrant members in 2018.

Along with decent work sensitivity and capabilities, collaboration between employers and employees improved. The introduction of a collective bargaining process, which resulted in the first collective bargaining agreement (CBA) in 2013, was highlighted by industry and labor experts as a milestone. The process was intensely supported by BWJ and was described as functioning by experts in 2018: "We work every two years, we have a new collective bargaining agreement, we sit, we bargain, we communicate. The dialogue becomes very intense at one point or another, but we always reached the conclusion that both parties can agree upon. That is reasonable, acceptable, and workable" [J15E05]. Among others, CBAs reduced wage discrimination (see Fig. 2), set organizational safety and health standards for dormitories, and unified the contracts for migrant workers.

### 4.1.3 Factory-Level Developments

Interests and power structures on the factory level changed substantially between 2008 and 2018 and were essential for understanding decent work progress in Jordan. Employers' and workers' decent work resources developed, and dialog between these two groups was established in many factories.

Many interviewed owners-managers and all experts described compliance with labor standards as an economically beneficial necessity. Based on this awareness, decent work capabilities were continuously developed. Direct exporters especially possessed the necessary structures, policies, and management skills to promote decent work. As most direct exporters operating in Jordan belong to large Asian textile and garment firms, they additionally had the financial resources to invest in decent work. However, not all interviewed owners-managers possessed awareness, capabilities, and resources. Limitations, especially concerning subcontractors, are elaborated in the inhibitors section.

> J18F12: [decent] work is growing the industry [. . .]. We are focusing on much better issues today because the bar is rising in the industry because of the willingness, of the owners also, management also. That's the whole equation.

Workers initially started with a very weak bargaining position in Jordan. Being young, foreign, poor, and poorly educated, they were subject to the (good)will of their employers in 2008. These characteristics still hold true for migrant labor in 2018 (see Sect. 4.2.4), but their support structures and decent work capabilities have improved. The union supported workers and trained them in their rights and responsibilities and their negotiation skills. BWJ and other stakeholders gave workers a voice in their assessments:

> J19F13: During this audit, they will make interview with the worker. [. . .] Then you will [get a] feeling if they are happy or not. If they are not happy, [they will ask] why and what is the issue [. . .]. Better Work, they will do this, and Ministry of Labor, they will do this, plus the buyers, before issue of any order. He will be continual visit, every time visit.

Furthermore, the value of migrant labor for employers improved as the law currently requires employers to hire 25% Jordanian workers, who were said to be less productive and reliable. The integration of Jordanian workers additionally contributed to enhancing working conditions as their bargaining position is relatively strong. (Skilled) Jordanian labor was reported to be difficult to recruit, and particular care was reported to be taken to retain women, who commonly do not work in rural Jordan: "[. . .] this industry becoming not anymore attractive to the worker, especially the local workers. It's not easy to keep the workers in the company" [J09F06]. Individual workers also benefited from social and family support. Respondents reported cases in which all workers from a community left the factory if one of them felt mistreated.

When asked about dialog, respondents highlighted the Performance Improvement Consultative Committees (PICCs) in which representatives of all groups of workers interact and consult with their employers. Factories that had been working with BWJ since the beginning especially praised the benefits of having PICCs. Supervisor and management trainings initiated by BWJ furthermore focused on improving day-to-day communication on the factory floor. J15E05 emphasized that for decent work progress "it's always important to listen to them [workers], hear their stories and see what they need, what is missing."

### 4.1.4 Interactions

Developments on factory, local industry, and global levels are not independent of each other. Interview responses and documents did not indicate single factors that enabled change but rather described an interwoven process driven by various stakeholders on different levels. The following example illustrates this interaction.

The Jordanian government increased the minimum wage for local labor in 2012 but failed to protect migrant labor (see Fig. 2). Local stakeholders, in this case the local union supported by BWJ, admonished this shortcoming. They highlighted that different wages for local and migrant labor constituted an act of discrimination, which is prohibited by international labor standards and most buyers' codes of conducts. Local stakeholders struggled to solve this dilemma. In a mediation led by the United States, the Jordanian government and local business and labor organizations finally agreed that a sectoral agreement should remedy the breach. This agreement started a collective bargaining process on the local level that was again supported by global stakeholders, specifically the ILO in cooperation with BWJ. In the end, employer and employee representatives agreed that in-kind benefits provided to migrants could offset the difference between the in-cash minimum wages of local and migrant labor [CBA-2013]. Overtime pay of migrant workers, however, was still calculated based on their in-cash wages, a significant shortcoming in an industry environment where overtime is indispensable to make a living. A memorandum of understanding (MoU) remedied this shortcoming in 2014 [MoU-2014]. The MoU obligated suppliers to calculate an increasing amount of in-kind benefits into their overtime pay. By August 2017, overtime pay had to be calculated on the same basis for Jordanian and migrant workers. Despite these provisions, equal and minimum wages were not yet provided to all workers in 2017. BWJ detected inaccurate calculations of overtime wages in 27% of analyzed cases [BWJ-Report-2017]. Only after BWJ trained factory managers and the global and local union trained worker representatives on the correct calculation of overtime pay was this percentage reduced to 15% in 2018 [BWJ-Report-2018]. Hence, the contribution and interaction of multiple actors on global, local industry, and factory levels was necessary to improve wages for all workers.

### 4.1.5 Better Work Jordan

As already indicated, BWJ played a crucial role in promoting developments on global, local industry, and factory levels, particularly by coordinating interactions between the different stakeholder groups. Interview respondents highlighted that BWJ consulted, monitored, and facilitated across stakeholders.

BWJ worked with all industry stakeholders and provided advisory services, trainings, secondments, information resources, and/or project support. Services were portrayed as highly successful by respondents. Almost all changes on global, local industry, and factory levels were made with the support of BWJ.

BWJ factory assessments were especially characterized as instrumental for decent work progress. "Better Work is watching you. That is good, Better Work, it is good that we have them. [...] One needs someone, an external buddy, who says, this is right, this is wrong" [J02E01]. Suppliers and experts agreed that BWJ assessments were increasingly accepted by buyers as substitutes for their own audits, and thorough advisory services through the program enabled suppliers to develop compliance with global and local labor regulations. Assessments additionally helped BWJ to understand which underlying factors caused decent work deficits in Jordan. As described above, the detection of multiple cases of wage discrimination in factory assessments in 2012 helped to initiate the collective bargaining process to harmonize wages.

In addition to the advisory and assessment services, BWJ's facilitatory role was emphasized in interviews and documents. Different information portals, reports, and publications made the status quo of decent work in Jordanian factories transparent to buyers and the public. Building on increased transparency, BWJ stimulated dialog between stakeholders. On the factory level, owners-managers underlined that BWJ greatly helped employer-worker dialog by promoting PICCs and worker-management committees. "[PICCs existed] since long time with the labour union. They've got more organized with BWJ, when BWJ started. BWJ started in 2009. That's when it got official, more documented" [J07F04]. On the industry level, BWJ supported the negotiation of CBAs. Local stakeholders are connected in BWJ's Project Advisory Committee, and global buyers are connected with local stakeholders during the annual buyer forums. Furthermore, respondents highlighted that BWJ promotes understanding and solves problems across stakeholders in each training and advisory meeting.

As the BW program exists in many countries but is not equally successful in all (van Roozendaal, 2017), the analysis also examined which factors enabled BWJ to successfully serve as an advisor, assessor, and facilitator in Jordan. The obligation to participate and program acceptance across stakeholder groups were thereby shown to be of core importance. In 2010, the Jordan government mandated that suppliers exporting to the United States and Israel and their subcontractors had to participate in the program. Most BW programs in other countries are voluntary (Bair, 2017). Furthermore, documents and reports showed that global buyers, governments, and NGOs put pressure on local stakeholders to work with BWJ. Consequently, BWJ covered almost all exporting garment suppliers in Jordan in 2018. In addition, BWJ managed to win acceptance and respect across stakeholder groups. Interviewees repeatedly highlighted that BWJ balanced the interests of all stakeholders.

J15E05: When Better Work Jordan came to Jordan [. . .] we were worried. [. . .] Actually, it came up to be a very healthy relationship. They taught factories and they taught the sector how to improve working conditions, to have decent jobs, decent work, good working environment. They pinpoint on the relations, where we can improve, and we can work together to accomplish better results at the end.

## *4.2 Inhibitors of Decent Work Progress*

Despite progress, the garment export industry had not achieved decent work for all in 2018. Thus, the analysis also sought to identify which issues inhibited or even deteriorated progress.

### 4.2.1 Competitiveness of the Jordanian Export Industry

The quest to improve working conditions was frequently set against the need to remain competitive. Even though several interviewees highlighted that decent work helps suppliers improve competitiveness, this criterion coexisted with buyers' price, lead time, and quality demands. Decent work improvements affecting running costs (e.g., increasing wages or reducing working hours) were frequently described as a threat to competitiveness. When asked about the possibility of negotiating with buyers, one respondent explained: "The American clients, no. The Americans, they are coming, 'I have this jacket with this cost, take it or leave it' [. . .]" [J09F06].

Subcontractors especially criticized the immense price competition. The interviewed subcontractors were medium-sized firms with limited financial resources. They stressed their lack of bargaining power vis-à-vis direct exporters and expressed they could only be profitable with excessive overtime and a focus on migrant labor. Interviewed subcontractors had not been trained in decent work, and their management systems were much less developed than those of direct exporters. They had only recently started working with BWJ because they were legally obligated to and/or they considered decent work progress as a prerequisite to economically upgrade. At the time of the analysis, it remained unclear whether they would manage to economically and socially upgrade.

The competitive situation of most direct exporters looked more favorable. Their highly sophisticated production system and qualified (migrant) workers ensured outstanding productivity, and as direct exporters, they earned higher income from sales than subcontractors. Nevertheless, they also underlined that the future of the export industry in Jordan was questionable, and some showed reluctance to make substantial further investments. "If I cannot project my cost for the next three to five years, if I cannot project the laws that I'm working under for the next three to five years, it's a very sceptical concept" [J07F04]. The lack of a promising perspective for the garment export industry in Jordan was linked by owners-managers and experts to (1) the minimum wage, which was more than three times higher in Jordan than in Bangladesh; (2) the dependence on the United States, where the political climate in 2018 questioned the benefits of free trade; and (3) the repeated ad hoc changes in local labor regulations (minimum wages, tax exemptions, labor policies concerning local labor). As one way out of this dilemma, respondents contemplated winning European buyers. At the time of the interviews, the need to hire 15% of refugee labor to qualify for the EU-Jordan agreement and the lack of connections to European buyers prevented most firms from starting trade with Europe.

### 4.2.2 Institutional Environment in Jordan

When asked about what factory owners-managers wished to change to further develop socially and economically, a common theme involved government regulations and government cooperation with the export industry. This argument again referred to the link between decent work and competitiveness. Respondents outlined examples of intense bureaucracy and frequent changes in regulatory demands which impaired competitiveness. "They [Ministry of Labor] are supporting and they are making problems for us. Yearly they are changing the law. Yearly they are changing, so it's not workable actually" [J08F05]. Migrant managers especially felt that foreign investors were not welcome in Jordan.

Experts remarked that frequent changes in government officials impaired dialog and that they hoped for more support of the export industry from government side.

> J07F04: "this is a sector that recruits 70,000 people. There's big potential for the sector to grow. [...] Doubling the 70,000 to become 140,000 is a great opportunity in front of us. [...] We just need to see the future. [...] If the government would believe in the sector and would take the sector to the next level [...]".

Even though different steps toward more stakeholder dialog were taken, documents and interview responses showed that there was no joint perspective for the garment industry in 2018. While the mainly foreign-owned garment export industry was rooted in the Cut-Make-Trim business and relied heavily on migrant labor, the government was interested in moving the industry into higher value-added segments. It furthermore continued to enact regulations reducing the share of migrant labor to counter rising unemployment rates in Jordan. Notably, the government strategy was considered to be an inhibitor in the mainly foreign-owned export industry. For the local industry, it was viewed as enabling social and economic upgrading.

### 4.2.3 Limits of Social Auditing

Factories in Jordan are subject to the Jordanian labor law, ILO core labor standards, and buyers' codes of conduct. The factories are inspected by BWJ, local labor inspectors, and international auditors, whereby BWJ succeeded in substantially aligning these assessments. While most respondents considered BWJ assessments as just and helpful, respondents perceived local inspections to be a burden. Buyer inspections were considered acceptable at the time of the interviews, whereby improvements in buyer inspections were frequently attributed to BWJ. The responses showed that in 2018 a functioning inspection system depended on BWJ.

However, BWJ inspections were not perfect as BWJ interviews and reports showed. First, BWJ uses local legislation and ILO core labor standards for reference in their assessments and is thereby limited. Regarding Freedom of Association, the Jordanian law only allowed for one trade union in the industry, and factories therefore could not comply with this standard. Second, local and BWJ assessments held scope for deceit. Some workers reported cases in which they were asked to work overtime without it being recorded. Third, BWJ repeatedly noted difficulty in assessing some standards, particularly the prevention of sexual harassment which is a highly sensitive topic in Jordan. Fourth, BWJ assesses decent work issues which local stakeholders have limited power to change. The support of migrant-sending countries and foreign recruitment agencies is, for instance, needed to permanently eradicate child labor, forced labor, and the payment of recruitment fees.

### 4.2.4 Worker Vulnerability

Workers often belonged to vulnerable groups (young, poor, mainly female, poorly educated), and migrant workers had an especially weak bargaining position. They were isolated in industrial zones and commonly only stayed in Jordan for 2 years. They depended on local and international organizations to fight for their rights, and it is especially notable that migrant workers were not legally eligible to stand for union office in 2016 and their right to vote in union elections was not enforced in 2015 [BWJ-report-2015]. The single union structure prevented them from selecting another union. Although migrant workers were much better protected in 2018 than in 2008, they remained dependent on global actors for protection.

Additionally, many direct exporters struggled with the management of Jordanian labor. Interviews and documents revealed a sociocultural disconnect, especially between foreign managers and local workers. Interviews with Jordanian managers who successfully run satellite or locally owned firms showed that the needs of local and migrant workers differed. Improving formal labor standards had not necessarily made a workplace attractive for Jordanians. And even though Jordanians had more bargaining power than migrants vis-à-vis their employers, they still belonged to a vulnerable labor group and were sometimes induced to act against their will:

> J18F12: "Jordanians like to work only eight hours, predominantly. But [...] they want to make extra money by doing overtime, because the cost of living is going up. They might not like it but now the change is coming."

## 5 Conclusions and Perspectives

The garment export industry in Jordan went from being a prime case of social irresponsibility to a good practice case of social compliance in the garment supply chain. The analysis revealed that multiple actors on global, local industry, and factory levels contributed to this success. The identification of most of the decent work enablers and inhibitors confirmed previous research insights: Complementary private, social, and public governance (Gereffi & Lee, 2016), strengthened labour agency (Siegmann et al., 2016), more developed supplier resources (Soundararajan & Brown, 2016), and multi-stakeholder initiatives (MacDonald et al., 2019) can help decent work progress. Competitive pressures, worker vulnerability, and limits of social auditing often inhibit it in the garment supply chain (Anner, 2019). The bird's-eye perspective of the case study prevents an in-depth discussion of each identified enabler or inhibitor. This conclusion therefore focuses on two particular observations and their implications that were especially relevant for Jordan, plus the overall implications of the detected multiple-level, multiple-stakeholder interactions.

The case showed that global pressure and local institutional changes were a necessary but not yet sufficient condition for decent work progress. Capacity development, CSR sensitivity, and financial viability at the factory level were needed to support suppliers in their efforts to incrementally enhance labor standards. Factories with sophisticated management systems and high productivity rates managed to run profitable and responsible businesses. At the time of the analysis, subcontractors especially showed a substantial need for further development. Previous research has highlighted the need for technical capabilities and financial resources to promote decent work across garment production contexts (Nadvi & Raj-Reichert, 2015). This case study builds on this insight and extends it by showing how technical resources that enable decent work progress can be successfully developed: BWJ started work with suppliers in Jordan over 10 years ago, incrementally included a growing number of factories into their services, and adapted their services to the needs of the industry. This long-term investment tailored to the specific needs of local factories made the difference. However, such investments are difficult to finance. Many garment producers, especially medium-sized ones in the Global South, are not able to pay for such services. A model that obligates buyers, along with suppliers or international organizations, to pay for advisory services is therefore recommended.

While substantial decent work progress had been realized in Jordan, future progress remains uncertain. Several respondents were unsure how long it will remain financially viable to produce in Jordan. If the export industry moves elsewhere, decent work progress has been in vain, a fear shared with many other garment production contexts (Khan et al., 2019; Ruwanpura & Wrigley, 2011). Some experts contemplated strategically focusing on CSR-sensitive buyers and high-quality products to ensure competitiveness and to aid efforts to win European buyers. The current trend in Europe toward nearshoring and sustainability-sensitive consumerism (McKinsey & Company, 2018) could help suppliers in these efforts. To win the support of the Jordanian government, suppliers need to further invest in satellite units and local labor. By providing large-scale work opportunities to local and refugee workers, the garment export industry incentivizes local public actors to invest in its development. As this envisaged market is a niche market (McKinsey & Company, 2018), however, this strategy is not universally applicable.

The case study especially illustrated that no single factor alone can explain decent work progress. Multiple stakeholders interacted and jointly shaped decent work realities. While the complexity of global production networks and the multiple-stakeholder influence on decent work realities has been previously pointed out in literature (Lund-Thomsen et al., 2016; Niforou, 2015), this case study contributes to existing research by practically showing how this interaction plays out in practice. A crossvergence perspective (Holzberg, 2019), which assumes an interplay of global and local factors at the factory level, thereby proved useful to grasp decent work progress. It breaks with the common research practice to primarily focus on the influence of buyers on suppliers and shows global and local demands to equally shape decent work outcomes in supplier firms. Although case studies can illustrate the web of influences on suppliers, they cannot discuss individual enablers or inhibitors in depth. Therefore, future research is recommended to continue studying individual influences on suppliers while placing research findings more strongly in their external context. For this purpose, this study provides a map (see Fig. 1) systematizing decent work enablers and inhibitors by level (global, local industry, factory) and stakeholder group (public, business, labor) to the debate.

In practice, the interplay of public, business, and labor stakeholders on global, local industry, and factory levels emphasizes the need for cross-level stakeholder dialog and aligned actions to drive sustainable change. However, joint efforts are often compromised because stakeholder interests differ or because stakeholders lack a unified understanding as they oftentimes exist in very different economic, social, and cultural realities. This finding is in line with previous research highlighting the need for multi-stakeholder interaction and engagement to foster decent work progress (Baumann-Pauly et al., 2017; MacDonald et al., 2019). Contributing to this stream of literature, the case shows how multi-stakeholder engagement can be successfully coordinated. In Jordan, BWJ contributed substantially to facilitating dialog and aligned actions across stakeholders. Thereby, the case showed that the decent work journey is not necessarily a linear process. Changes in one node of the network (e.g., if interests or personnel of stakeholders' change) and changes in network linkages (e.g., through power shifts in global competition) can reverse progress and require renewed efforts to facilitate dialog and to support a joint agenda to promote decent work. Furthermore, limits of BWJ's influence were revealed; for example, BWJ has limited means to ease global competition or to relieve worker vulnerability. In conclusion, the work of BWJ should not be a one-time commitment; it needs to be institutionalized as a permanent institution in Jordan and further strengthen cooperation with actors that have access to BWJ's "blind spots." While Jordan's journey toward decent work cannot be copied point for point to other contexts, the idea of a multiple-level, multiple-stakeholder mediator that facilitates dialog and helps to align actions across public, business, and labor actors can.

> J18F12: Everybody is partnering with BWJ, Ministry of Labor, factories, owners, the buyers. It's basically a teamwork. Everybody is contributing and everybody is benefiting at the end of the day.

# Appendices

**Annex 1** Profiles of experts interviewed

| | Code | Type of expert | Type of organization | Local/international |
|---|---|---|---|---|
| 1 | J02E01 | Industry expert | Employer organization | Local |
| 2 | J03E02 | Industry expert | Employer organization | Local |
| 3 | J04E03 | Labor expert | Development and training organization | International |
| 4 | J11E04 | Labor expert | Development and training organization | Local |
| 5 | J15E05 | Industry expert | Employer organization | Local |
| 6 | J17E06 | Labor expert | Union | Local |
| 7 | J21E07 | Labor expert | Development and training organization | International |

**Annex 2** Profiles of factory owners-managers interviewed

| | Code | Role | Status | Firm type | Factory location | Approx. no. of workers |
|---|---|---|---|---|---|---|
| 1 | J05F02 | HR & Compliance | Migrant | Direct exporter | Industrial | 600 |
| 2 | J07F04 | General Manager | Jordanian | Direct exporter | Industrial | 2000 |
| 3 | J08F05 | HR Manager | Jordanian | Direct exporter | Industrial | 3000 |
| 4 | J09F06 | COO | Jordanian | Direct exporter | Residential | 800 |
| 5 | J16F11 | Factory Manager | Jordanian | Direct exporter | Residential | 450 |
| 6 | J18F12 | General Manager | Migrant | Direct exporter | Industrial | 1800 |
| 7 | J19F13 | HR & Compliance | Jordanian | Direct exporter | Industrial | 950 |
| 8 | J06F03 | Owner | Jordanian | Ex subcontractor | Industrial | 150 |
| 9 | J12F08 | Owner | Migrant | Subcontractor | Industrial | 200 |
| 10 | J13F09 | Owner | Migrant | Subcontractor | Industrial | 150 |
| 11 | J01F01 | Owner | Jordanian | Local market | Residential | 50 |
| 12 | J10F07 | Owner | Jordanian | Local market | Residential | 150 |
| 13 | J14F10 | Owner | Refugee | Local market | Residential | 40 |

**Annex 3** List of analyzed documents (selection of most relevant documents)

| | Code | Document |
|---|---|---|
| 1 | Law-2019 | BWJ (2019), Guide to Jordanian Labour Law for the Garment Industry, 3rd ed. |
| 2 | Law-2018 | BWJ (2018), Guide to Jordanian Labour Law for the Garment Industry, 2nd ed. |
| 3 | Law-2013 | BWJ (2013), Guide to Jordanian Labour Law for the Garment Industry, 1st ed. |
| 4 | Law-1996 | Jordan Labour Code, Law No. 8 of 1996. Dated 2nd March 1996 |
| 5 | CBA-2017 | Collective Labour Contract (2017) between the (1) Jordan Garments, Accessories & Textiles Exporters Association (JGATE); Association of Owners of Factories, Workshops and Garments (AOFWG); and (2) General Trade Union of Workers in Textile, Garment and Clothing Industries |
| 6 | CBA-2015 | Collective Agreement (2015) between the (1) Jordan Garments, Accessories & Textiles Exporters Association (JGATE); Association of Owners of Factories, Workshops and Garments (AOFWG); and (2) General Trade Union of Workers in Textile, Garment and Clothing Industries |
| 7 | Contract-2015 | A standard contract for non-Jordanian workers in Textile, Garment and Clothing Industry (2015), developed by the (1) Jordan Garments, Accessories & Textiles Exporters Association (JGATE); Association of Owners of Factories, Workshops and Garments (AOFWG); and (2) General Trade Union of Workers in Textile, Garment and Clothing Industries |
| 8 | MoU-2014 | Memorandum of Understanding on Equalizing Payment of Wages and Benefits for all Workers (2014) between the (1) Jordan Garments, Accessories & Textiles Exporters Association (JGATE); General Union of Garment Factories; and (2) General Trade Union of Workers in Textile, Garment and Clothing Industries |
| 9 | CBA-2013 | Collective Agreement (2013) between the (1) Jordan Garments, Accessories & Textiles Exporters Association (JGATE); Association of Owners of Factories, Workshops and Garments (AOFWG); and (2) General Trade Union of Workers in Textile, Garment and Clothing Industries |
| 10 | MoL-2007 | Ministry of Labor Jordan (2007), Labor Compliance in Jordan's Apparel Sector: Actions to date and next steps |
| 11 | Union-Report-2014 | Union (2014), The General Trade Union of Workers in Textile, Garment and Clothing Industries, Jordan: General Annual Report 2011–2013 |
| 12 | Union-Info | Union (unknown), You are not alone: Worker Leaflet |
| 13 | Union-2018 | Union (2018), Worker statistics 01/2016 -30/2016 |
| 14 | Chamber-2018 | Amman Chamber of Industry (2018), Directory of Jordanian Exporters 2017–2018 |
| 15 | BWJ-Evaluation-2016 | Tufts University (2016), The impact of Better Work: A joint program of the International Labour Organization and the International Finance Corporation |
| 16 | BWJ-Evaluation-2014 | Tufts University (2014), Better Work Jordan Impact Assessment: Manager Report |
| 17 | BWJ-Evaluation-2013 | Al Jidara (2013), Final Independent Evaluation of Better Work Jordan – Phase 1 |
| 18 | BWJ-Plan-2013 | Al Jidara (2013), Sustainability Plan for the Better Work Jordan |

(continued)

**Annex 3** (continued)

| | Code | Document |
|---|---|---|
| 19 | BWJ-Report-2009 | BWJ (2010), Better Work Jordan: Garment Industry, 1st Compliance Synthesis Report (Reporting Period: 03/2009–02/2010) |
| 20 | BWJ-Report-2010 | BWJ (2011), Better Work Jordan: Garment Industry, 2nd Compliance Synthesis Report (Reporting Period: 10/2009–12/2010) |
| 21 | BWJ-Report-2011 | BWJ (2012), Better Work Jordan: Garment Industry, 3rd Compliance Synthesis Report (Reporting Period: 02/2011–12/2011) |
| 22 | BWJ-Report-2012 | BWJ (2013a), Better Work Jordan: Garment Industry, 4th Compliance Synthesis Report (Reporting Period: 09/2011–09/2012) |
| 23 | BWJ-Report-2013 | BWJ (2013b), Better Work Jordan: Garment Industry, 5th Compliance Synthesis Report (Reporting Period: 10/2012–09/2013) |
| 24 | BWJ-Report-2014 | BWJ (2015), Better Work Jordan: Garment Industry, 6th Compliance Synthesis Report (Reporting Period: 10/2013–10/2014) |
| 25 | BWJ-Report-2015 | BWJ (2016), Better Work Jordan: Garment Industry, 7th Compliance Synthesis Report (Reporting Period: 11/2014–10/2015) |
| | BWJ-Report-2016 | BWJ (2017), Annual Report 2017: An Industry and Compliance Review, Better Work Jordan (Reporting Period: 11/2015–12/2016) |
| 26 | BWJ-Report-2017 | BWJ (2018), Annual Report 2018: An Industry and Compliance Review, Better Work Jordan (Reporting Period: 01/2017–12/2017) |
| | | BWJ (2018), Statement from the Project Advisory Committee of Better Work Jordan on the release of Better Work Jordan Annual Report 2018: An Industry and Compliance Overview |
| 27 | BWJ-Report-2018 | BWJ (2019), Annual Report 2019: An Industry and Compliance Review, Better Work Jordan (Reporting Period: 01/2018–12/2018) |
| 28 | BWJ-PAC-2018 | BWJ (2019), Statement from the Project Advisory Committee of Better Work Jordan on the release of the 10th Annual Report on working conditions in Jordan's garment sector |
| 29 | BWJ-Study-2012 | BWJ (2012), Employment of Jordanians in the Garment Industry: Challenges and Prospects |
| 30 | BWJ-Study-2013 | BW (2013), The Collective Bargaining Agreement in Jordan's Garment Industry. Case Study |
| 31 | BWJ-Study-2014 | BWJ (2014), Addressing Sexual Harassment in Jordan's Garment Industry: Case Study |
| 32 | BWJ-Study-2014 | BWJ (2014), Addressing Sexual Harassment in Jordan's Garment Industry: Case Study |
| 33 | BW-Study-2013 | BWJ (2013), Garment Industry Baseline Report: worker perspectives from the factory and beyond |
| 34 | BW-Study-2014 | BW (2014), Addressing Sexual Harassment in Jordan's garment industry: Case Study |
| 35 | BWJ-Survey-2014 | BWJ (2014), Better Work Jordan: Factory Survey 2014 |
| 36 | BWJ-Survey-2015 | BWJ (2016), Better Work Jordan: Factory Survey 2015 |
| 37 | BWJ-Survey-2019 | BWJ (2019), Worker Voice Survey Results |
| 38 | BWJ-Guide-2012 | BWJ (2012), Good Practices: Apparel – Better working conditions, better business |
| 39 | BWJ-Info-2016 | BWJ (2016), Better Work Jordan 8th Annual Buyers' Forum: Summary |
| 40 | BWJ-Info-2016 | BWJ (2016), Better Work Public Reporting: Jordan |
| 41 | BWJ-Info-2019 | BWJ (2019), Celebrating a ten-year journey in Jordan: Highlights and achievements |

(continued)

**Annex 3** (continued)

| | Code | Document |
|---|---|---|
| 42 | BW-Info-2016 | BW (2016), Public Reporting: Frequently Asked Questions |
| 43 | BHRRC-2018 | Business and Human Rights Resource Centre (2018), Jordan's Garment Sector: How are brands combatting worker exploitation and abuse? |
| 44 | CCC-2009 | Clean Clothes Campaign (2009), False Promises: Migrant workers in the Global Garment Industry, Discussion Paper |
| 45 | CSRWJ-2014 | CSR Watch Jordan (2014), Corporate Responsibility in the Manufacturing Sector: A comparative study on the practice of Corporate Responsibility among manufacturing companies in Jordan |
| 46 | EU-2014 | EU (2014), Strategic Plan of the Garment Manufacture Sector in Jordan |
| 47 | FIAS-2008 | FIAS (2008), Competitiveness and Corporate Social Responsibility in the Jordanian Apparel Industry |
| 48 | IGLHR-2011a | Institute for Global Labour and Human Rights (2011), Sexual predators and serial rapist run wild at Wal-Mart supplier in Jordan |
| 49 | IGLHR-2011b | Institute for Global Labour and Human Rights (2011), Another crisis in Jordan: Workers strike the IBGM Factory over illegal sweatshop conditions, strikers beaten and threatened with forcible deportation |
| 50 | IGLHR-2012 | Institute for Global Labour and Human Rights (2012), Ten years into the U.S.-Jordan Free Trade Agreement: Brutal sweatshop conditions endure at the Taiwanese-owned Rich Pine Factory with gruelling mandatory seven-day 96-hour work weeks |
| 51 | IGLHR-2013 | Institute for Global Labour and Human Rights (2013), President Obama may very well clean up the U.S.-Jordan Free Trade Agreement |
| 52 | RSS-FES-2013 | Royal Scientific Society of Jordan & Friedrich-Ebert-Stiftung Amman Office (2013), The Future of Jordan's Qualified Industrial Zones (QIZs) |
| 53 | NLC-2006 | National Labor Committee (2006), U.S.-Jordan Free Trade Agreement Descends into Human Trafficking & Involuntary Servitude |
| 54 | News-2014-10 | Jordan Times (23 October 2014), Culture of shame, low salaries driving Jordanians away from industry sector jobs |
| 55 | News-2015-04 | Jordan Times (05 April 2015), Jordan's garment sector offers more benefits to guest workers — report |
| 56 | News-2015-05 | Jordan Times (16 May 2015), Int'l survey highlights strong impact of refugee influx on employment among Jordanians |
| 57 | News-2015-07 | Jordan Times (13 July 2015), Garment factory workers strike to demand higher salaries |
| 58 | News-2016-08 | Jordan Times (18 August 2016), Violations at garment factories "minor" — Labour Ministry |
| 59 | News-2016-11a | Jordan Times (10 November 2016), Officials, activists act swiftly to rescue "abused foreign laborers" |
| 60 | News-2016-11b | Jordan Times (22 November 2016), Syrians' reluctance to work in industrial sector threatens future of "rules-of-origin" deal with EU |
| 61 | News-2017-02 | Jordan Times (25 February 2017), Minister orders factory closure after alleged abuse of guest workers |
| 62 | News-2017-07 | Jordan Times (24 July 2017), Only two Jordanian companies benefiting from EU relaxed rules of origin |
| 63 | News-2018-02 | Jordan Times (22 February 2018), Minimum wages need to be increased to cope with price, tax hikes – experts |
| 64 | News-2018-08 | Jordan Times (04 August 2018), Stakeholders join forces for effective partnerships in garment sector |

**Annex 4** Category system of the analysis

| Category system Part A. Enablers | | |
|---|---|---|
| Main category | Category | Subcategory |
| 1. Global-level developments | 1.1 Public actors | 1.1.1 Countries importing from Jordan: Pressure to improve decent work |
| | | 1.1.2 Countries importing from Jordan: Technical and financial support |
| | | 1.1.3 Int'l organizations: Pressure to implement global labor standards |
| | | 1.1.4 Int'l organizations: Technical and financial support |
| | 1.2 Business actors | 1.2.1 Global buyers: Pressure to implement private labor standards |
| | | 1.2.2 Global buyers: Private audits |
| | 1.3 Labor (social) actors | 1.3.1 Global unions: Technical and financial support of local actors |
| | | 1.3.2 Global NGOs: Reports on decent work realities in Jordan |
| 2. Local-industry-level developments | 2.1 Public actors | 2.1.1 Legal institutions: Development of national labor laws |
| | | 2.1.2 Labor inspectorate: Development of factory audit capabilities |
| | | 2.1.3 Labor initiatives: Technical support and training centres |
| | 2.2 Business actors | 2.2.1 Employer organizations: Decent work sensitivity and capabilities |
| | | 2.2.2 Employer organizations: Pressure to improve decent work |
| | 2.3 Labor (social) actors | 2.3.1 Union: Development of decent work sensitivity and capabilities |
| | | 2.3.2 Union: Increase of resources to lobby for decent work |
| | | 2.3.3 Union: Increase of resources to represent all workers, incl. migrants |
| | | 2.3.4 Local NGOs: Reports on decent work realities in Jordan |
| | | 2.3.5 Local communities: Pressure to provide decent work opportunities |
| | 2.4 Stakeholder collaboration | 2.4.1 Business-labor: Collective bargaining process |
| | | 2.4.2 Business-labor: Ad hoc dialog on conflicts and opportunities |
| | | 2.4.3 Public-business-labor: Dialog in BWJ PAC meetings |
| 3. Factory-level developments | 3.1 Employers | 3.1.1 Development of decent work sensitivity |
| | | 3.1.2 Development of decent work capabilities |
| | | 3.1.3 Development/existence of financial and organizational resources |

(continued)

**Annex 4** (continued)

| Category system Part A. Enablers | | |
|---|---|---|
| Main category | Category | Subcategory |
| | 3.2 Workers | 3.2.1 Development of worker bargaining power |
| | | 3.2.2 Development of decent work sensitivity |
| | 3.3 Dialog | 3.3.1 Development of formal dialog channels on factory level |
| | | 3.3.2 Improvement of day-to-day communication on factory floor |
| 4. Better Work Jordan | 4.1 Consulting and capacity building | 4.1.1 Business stakeholders |
| | | 4.1.2 Labor stakeholders |
| | | 4.1.3 Public stakeholders |
| | 4.2 Monitoring and factory assessment | 4.2.1 Business stakeholders |
| | | 4.2.2 Labor stakeholders |
| | | 4.2.3 Public stakeholders |
| | 4.3 Facilitation and coordination | 4.3.1 Transparency across stakeholders |
| | | 4.3.2 Dialog between stakeholders |
| | | 4.3.3 Problem-solving with stakeholders |
| | 4.4 Success factors | 4.4.1 (Legal) obligation to participate in BWJ program |
| | | 4.4.2 Acceptance and respect of BWJ across stakeholders |

| Category system Part B. Inhibitors | |
|---|---|
| Category | Subcategory |
| 1. Competitiveness of the Jordanian export industry | 1.1 Buyer price and procurement pressure |
| | 1.2 High labor costs in Jordan compared to global competition |
| | 1.3 High risks of investment due to frequent local institutional changes |
| | 1.4 Especially weak bargaining position of subcontractors |
| 2. Institutional environment in Jordan | 2.1 Intense bureaucracy and frequent changes in regulatory demands |
| | 2.2 Cooperation difficulties between public and business actors |
| | 2.3 Mismatch between national strategy for garment industry and business (exporters) needs |
| 3. Limits of social auditing | 3.1 Plurality of factory audits (BWJ audits, buyer audits, local audits) |
| | 3.2 Room for deceit and blind spots in factory audits |
| | 3.3 Limited scope of BWJ audits (focus on local laws, labor standards) |
| | 3.4 Limited reach of auditors (focus on factory floor) |
| 4. Worker vulnerability | 4.1 Vulnerability of migrant labor |
| | 4.2 Vulnerability of local labor |

**Annex 5** Excerpt of the coding frame

| Main category: 4. Enablers: Better Work Jordan | | | |
|---|---|---|---|
| Subcategory 1 | Subcategory 2 | Description | Sample quote |
| 4.1 Consulting and capacity building | 4.1.1 Business stakeholders | Statements highlighting that BWJ consults, trains, or coaches business stakeholders | Better Work teach us 90% from what auditors need. [J08F05] |
| | 4.1.2 Labor stakeholders | Statements highlighting that BWJ consults, trains, or coaches labor stakeholders | BWJ also expanded its training services through the piloting of innovative sexual harassment prevention training for managers, supervisors, and workers. [BWJ-Report-2013] |
| | 4.1.3 Public stakeholders | Statements highlighting that BWJ consults, trains, or coaches public stakeholders | BWJ is training labor inspectors via classroom and field training. Training topics include [...] the Better Work model and compliance tool and investigating and addressing sexual harassment. [BWJ-Report-2018] |
| 4.2 Monitoring and factory assessment | 4.2.1 Business stakeholders | Statements highlighting that BWJ monitors or assesses activities of business stakeholders | Better Work is watching you. That is good, Better Work, it is good that we have them. [...] One needs someone, an external buddy, who says, this is right, this is wrong. [J02E01] |
| | 4.2.2 Labor stakeholders | Statements highlighting that BWJ monitors or assesses activities of labor stakeholders | Given the single trade union structure prescribed by national law, all factories continue to be noncompliant under Freedom to Associate. [BWJ-Report-2018] |
| | 4.2.3 Public stakeholders | Statements highlighting that BWJ monitors or assesses activities of publics stakeholders | [Discrimination in pay increased to] 83% of factories assessed more than once. This is due to the fact that the minimum wage increase that went into effect February 2012 differed for migrant workers as compared to Jordanian workers. [BWJ-Report-2013] |
| 4.3 Facilitation and coordination | 4.3.1 Transparency across stakeholders | Statements highlighting that BWJ contributes to establishing transparency concerning labor issues across stakeholder groups | BWJ has 29 key assessment questions which they publish openly on the Internet: "Transparency Portal". [J04E03] |

(continued)

**Annex 5** (continued)

| Main category: 4. Enablers: Better Work Jordan | | | |
|---|---|---|---|
| Subcategory 1 | Subcategory 2 | Description | Sample quote |
| | 4.3.2 Dialog between stakeholders | Statements highlighting that BWJ contributes to establishing dialog on labor issues between stakeholder groups | [PICCs existed] since long time with the labor union. They've got more organized with BWJ, when BWJ started. BWJ started in 2009. That's when it got official, more documented [J07F04] |

# References


Aissi, J., Peels, R., & Samaan, D. (2018). Evaluating the effectiveness of labour provisions in trade agreements: An analytical and methodological framework. *International Labour Review, 157*(4), 671–698. https://doi.org/10.1111/ilr.12125

Anner, M. (2019). Predatory purchasing practices in global apparel supply chains and the employment relations squeeze in the Indian garment export industry. *International Labour Review, 158*(4), 705–727. https://doi.org/10.1111/ilr.12149

Azmeh, S. (2014). Labour in global production networks: Workers in the qualifying industrial zones (QIZs) of Egypt and Jordan. *Global Networks, 14*(4), 495–513. https://doi.org/10.1111/glob.12047

Bair, J. (2017). Contextualising compliance: hybrid governance in global value chains. *New Political Economy, 22*(2), 169–185. https://doi.org/10.1080/13563467.2016.1273340

Baumann-Pauly, D., Nolan, J., van Heerden, A., & Samway, M. (2017). Industry-specific multi-stakeholder initiatives that govern corporate human rights standards: Legitimacy assessments of the fair labor association and the global network initiative. *Journal of Business Ethics, 143*(4), 771–787. https://doi.org/10.1007/s10551-016-3076-z

Better Work Jordan. (2019a). *Annual Report 2019: An Industry and Compliance Review*. Geneva. Retrieved from ILO website: https://betterwork.org/portfolio/19446-2/

Better Work Jordan. (2019b). *Celebrating a ten year journey in Jordan: Highlights and achievements*. Retrieved from https://betterwork.org/wp-content/uploads/2020/01/10YearsJordan0412_rev3.pdf

Business and Human Rights Resource Center. (2018). *Jordan's Garment Industry: How are brands combatting worker exploitation and abuse?* Retrieved from https://www.business-humanrights.org/sites/default/files/Jordan%20Briefing%20FINAL.pdf

De Bel-Air, F., & Dergarabedian, A. (2006). Migrations de travail, globalisation et politique. Les Zones industrielles qualifiantes (QIZ) de Jordanie. In F. De Bel-Air & A. Dergarabedian (Eds.), *Migration et politique au Moyen-Orient* (pp. 37–60). IFPO.

Domat, G., Glass, B., & Brown, D. (2012). *The Apparel Industry: Jordan's comparative advantage in international trade* (Better Work Monitoring and Evaluation). Amman, Jordan. Retrieved from https://www.ilo.org/wcmsp5/groups/public/%2D%2D-arabstates/%2D%2D-ro-beirut/documents/genericdocument/wcms_209899.pdf

Enclude BV, & CMC. (2019). *Value Chain Analysis: Jordan Garments*. Zeist. Retrieved from https://www.cbi.eu/sites/default/files/2019_vca_jordan_apparel.pdf

Gereffi, G., & Lee, J. (2016). Economic and social upgrading in global value chains and industrial clusters: Why governance matters. *Journal of Business Ethics, 133*(1), 25–38. https://doi.org/10.1007/s10551-014-2373-7

Holzberg, B. (2019). Crossvergence of socially (ir)responsible employment practices in supplier firms. *Critical Perspectives on International Business, 16*(3), 305–324. https://doi.org/10.1108/cpoib-11-2017-0076

ILO (2019). *World Employment Social Outlook: Trends 2019*. Retrieved from https://www.ilo.org/wcmsp5/groups/public/%2D%2D-dgreports/%2D%2D-dcomm/%2D%2D-publ/documents/publication/wcms_670542.pdf

Khan, S. I., Bartram, T., Cavanagh, J., Hossain, M. S., & Akter, S. (2019). "Decent work" in the ready-made garment sector in Bangladesh. *Personnel Review, 48*(1), 40–55. https://doi.org/10.1108/PR-01-2018-0001

Kolben, K. (2013). Trade, Development, and Migrant Garment Workers in Jordan. *Middle East Law and Governance, 5*(1-2), 195–226. https://doi.org/10.1163/18763375-00501006

Kolben, K. (2015). Dialogic Labor Regulation in the Global Supply Chain. *Michigan Journal of International Law, 36*(3), 425–466.

Kolben, K. (November 2019). *The Jordanian Garment Industry and Better Work* (The Better Work Discussion Paper Series No. 34). Geneva. Retrieved from https://betterwork.org/wp-content/uploads/2020/01/DP34.pdf

Lund-Thomsen, P., Lindgreen, A., & Vanhamme, J. (2016). Industrial clusters and corporate social responsibility in developing countries: What we know, what we do not know, and what we need to know. *Journal of Business Ethics, 133*(1), 9–24. https://doi.org/10.1007/s10551-014-2372-8

Luque, A., & Herrero-García, N. (2019). How corporate social (ir)responsibility in the textile sector is defined, and its impact on ethical sustainability: An analysis of 133 concepts. *Corporate Social Responsibility and Environmental Management, 27*(42), 123. https://doi.org/10.1002/csr.1747

MacDonald, A., Clarke, A., & Huang, L. (2019). Multi-stakeholder partnerships for sustainability: Designing decision-making processes for partnership capacity. *Journal of Business Ethics, 160*(2), 409–426. https://doi.org/10.1007/s10551-018-3885-3

McKinsey & Company. (October 2018). *Is apparel manufacturing coming home? Nearshoring, automation, and sustainability – establishing a demand-focused apparel value chain*. Retrieved from https://www.mckinsey.com/industries/retail/our-insights/is-apparel-manufacturing-coming-home#

Nadvi, K., & Raj-Reichert, G. (2015). Governing health and safety at lower tiers of the computer industry global value chain. *Regulation & Governance, 9*(3), 243–258. https://doi.org/10.1111/rego.12079

National Labor Committee. (2006). *U.S.-Jordan Free Trade Agreement: Descents into human trafficking and involuntary servitude*. Retrieved from https://issuu.com/iglhr/docs/0605-iglhr-usjordanfreetradeagreeme

Niforou, C. (2015). Labour leverage in global value chains: The role of interdependencies and multi-level dynamics. *Journal of Business Ethics, 130*(2), 301–311. https://doi.org/10.1007/s10551-014-2222-8

Patton, M. Q. (2002). *Qualitative research and evaluation methods* (3rd ed.). SAGE Publications.

Robertson, R. (November 2019). *Working Conditions, Transparency, and Compliance in Global Value Chains: Evidence from Better Work Jordan* (The Better Work Discussion Paper Series No. 32). Geneva. Retrieved from https://betterwork.org/wp-content/uploads/2020/01/DP32.pdf

Ruwanpura, K. N., & Wrigley, N. (2011). The costs of compliance? Views of Sri Lankan apparel manufacturers in times of global economic crisis. *Journal of Economic Geography, 11*(6), 1031–1049. https://doi.org/10.1093/jeg/lbq036

Schreier, M. (2014). Qualitative content analysis. In U. Flick (Ed.), *The Sage handbook of qualitative data analysis* (pp. 170–183). SAGE Publications Ltd.. https://doi.org/10.4135/9781446282243.n12

Siegmann, K. A., Merk, J., & Knorringa, P. (2016). Positive class compromise in globalized production? The Freedom of Association Protocol in the Indonesian Sportswear Industry. *International Labour Review, 150*(3-4), 319. https://doi.org/10.1111/ilr.12036

Soundararajan, V., & Brown, J. A. (2016). Voluntary governance mechanisms in global supply chains: Beyond CSR to a stakeholder utility perspective. *Journal of Business Ethics, 134*, 83–102. https://doi.org/10.1007/s10551-014-2418-y

Van Roozendaal, G. (2017). Where symbolism prospers: An analysis of the impact on enabling rights of labour standards provisions in trade agreements with South Korea. *Politics and Governance, 5*(4), 19–29. https://doi.org/10.17645/pag.v5i4.1087

Yin, R. K. (2009). *Case study research: Design and methods*. Sage.

**Britta Holzberg** is a researcher and practitioner interested in the creation of decent and inclusive work environments worldwide. She has a master's degree in social and organizational psychology, a Diplom in business administration, and a PhD in international development studies. Her PhD project focused on HR practices and decent work in the textiles and garment supply chain, particularly in Egypt and Jordan. Britta has previously worked in both countries as a private sector consultant. In her consulting work, she supports public-private initiatives to promote decent and inclusive work across the MENA region and Europe.

# CSR Activities of Retail Chains Operating in Poland

**Anna Napiórkowska**

## 1 Introduction

Retail is an important sector of the economy whose functions extend beyond participation in the distribution of goods. The dynamic processes taking place in the modern economy cause significant changes in the functioning of retail companies. In particular, international retailers often act as supply chain coordinators and have a direct impact on all partners, both buyers of goods and services and suppliers (Stefańska & Śmigielska, 2015). Global sourcing allows for greater supplier variety and lower prices but also leads to public concern about working conditions, wages, child labor, health, and safety. In this case, managing and controlling the supply chain is difficult, mainly due to geographic distances.

As markets become more globalized and competitive, retailers are focusing their efforts on finding ways to differentiate themselves from their competitors. Price competitiveness, which used to be a strategy to distinguish from competitors, is no longer sufficient (Lee, 2021). Firms are under pressure from internal and external actors to meet rapidly changing expectations regarding their social responsibility. In recent years, legal regulations, trade unions, and increasing consumer awareness have led companies to take into account the economic, social, and environmental effects of their activities. As Lee (2021) indicates, pressure from the environment, company values, as well as personal values and beliefs of managers play a significant role in firms' involvement in corporate social responsibility (CSR) activities.

Over the years, there has been a noticeable upward trend in the number of research devoted to retailers' CSR. The subject of analysis in most studies is

A. Napiórkowska (✉)
SGH Warsaw School of Economics, Warszawa, Poland

Collegium of World Economy, Department of International Business, Warszawa, Poland
e-mail: anna.napiorkowska@sgh.waw.pl



I. G. Arraiano et al. (eds.), *Corporate Social Responsibility in a Dynamic Global Environment*, CSR, Sustainability, Ethics & Governance,
https://doi.org/10.1007/978-3-031-24647-0_9

retailers' sustainable and responsible activities and the perception of these by consumers (e.g., Dang et al., 2020; do Vale et al., 2020), as well as CSR impact on company's value (e.g., Lee, 2021; Nyame-Asiamah & Ghulam, 2020). Research was conducted mainly in developed countries (e.g., Schramm-Klein et al., 2015), while the specific characteristics of each country may have an impact on the intensity and type of CSR activities undertaken. Scholars and experts have shown a rising interest in the analysis of corporate social responsibility from a developing country perspective (e.g., Amos, 2018; Jamali & Karam, 2018). CSR in developing countries differs in characteristics from that practiced in developed countries and has been portrayed as less formalized, more philanthropic in nature, deeply based on cultural/religious values and primarily oriented on local communities (Vo & Arato, 2020).

This study examines retailers operating in Poland and provides an understanding of CSR practices for this sector. The study addresses a topic that has been barely covered by scientific literature (e.g., Śmigielska & Oczkowska, 2017; Stefańska, 2014) but is of great importance for the Central and Eastern Europe region. Among the research on the social responsibility of retailers in Poland, we can distinguish mainly qualitative studies that concern the impact of CSR activities on relationships in the supply chain (e.g., Bilińska-Reformat & Stefańska, 2013) or the identification of determinants of retailers' CSR activities directed at costumers (Śmigielska & Oczkowska, 2017). Quantitative research was carried out on consumers and concerned, inter alia, their perception of retail chains' CSR (e.g., Stefańska & Wanat, 2014).

The aim of this study is to analyze the areas of CSR activities undertaken by retail chains operating in Poland. Therefore, the following research questions were proposed:

Q1: With what intensity retail chains operating in Poland are involved in the chosen CSR areas?

Q2: Are there any differences between retail chains with only Polish capital vs. retail chains with foreign capital within the five chosen areas of CSR involvement?

## 2 Literature Review and Statement of the Problem

In Poland, the dynamic development of retail chains took place in the 1990s, when large-scale retail chains, including hypermarkets, supermarkets, and discount stores, began to appear on the Polish market. This was related to the expansion of international retail chains and foreign capital investments, as well as the development of retail chains managed by enterprises with Polish capital (Witek-Hajduk & Grudecka, 2018). With the development of the Polish economy, the strategies of retail chains have evolved. Gradually, new store formats began to emerge, and the growing competition was accompanied by practices and rules of conduct adopted by foreign retail chains toward consumers, as well as local communities and suppliers. Increasingly, importance was attached to creating the image of a retailer and communication of actions taken in the field of social responsibility (Witek-Hajduk & Grudecka, 2018). As Stefańska and Śmigielska (2015) indicate, currently retail chains, especially when they offer products under their own brands, very often control the entire value chain from producer to consumer. This is despite that sometimes they also have their own logistics centers and production plants. This often involves setting social responsibility standards for each link in the supply chain and monitoring compliance with them.

According to Statistics Poland (2019), in 2019 the share of trade in the creation of gross domestic product in Poland amounted to 15.5% and 17.7% in gross value added. Purchases of goods were made mainly directly from domestic producers and manufacturers (almost 53%). Retailers with foreign ownership most often made purchases directly from import (almost 45%) or directly from domestic producers and manufacturers (slightly over 43%).

Research shows that retailers' CSR activities focus mainly on environment, supply chain, workplace, and society (e.g., Bilińska-Reformat, 2013; Lee, 2021; Schramm-Klein et al., 2015). Moreover, Elg and Hultman (2011) draw attention to the internal aspect of CSR concerning the codification of standards applicable in the company. However, these activities occur with different intensity in each of the mentioned areas (Carrero & Valor, 2012). There is also variation in the categories of products sold. Vo and Arato (2020) identified eight main CSR themes in food retailers: fair operating practices, food quality and safety, transparency, procurement, labor rights at firms, environmental protection, philanthropy, and sustainable food offerings. On the basis of literature review on CSR in developing economies, Amos (2018) pointed out that social and environmental issues are the most widespread topics in CSR practices in developing countries, while ethics received the least attention.

Jones et al. (2005) note that the environment is the area of social responsibility most frequently indicated in the reports of large retailers. Retailers' activities related to environmental protection include reducing waste production, energy consumption and exhaust emissions, transport, consumption of raw materials and water, as well as the use of smaller packaging and recycling. According to Seuring and Müller (2008), retail companies communicate primarily about their CSR activities aimed at reducing energy and material consumption. Miles and Covin (2000) indicate low potential litigation expenses and lower insurance and energy costs as among the main drivers of retail businesses to take CSR actions that reduce their environmental impact.

Depending on the country and its regulations, retailers also differ in their employee-oriented CSR activities (Jany-Catrice & Lehndorff, 2002). The basic problems that retailers try to cope with in this regard include irregular working hours, quality (competences) and the number of employees (frequent rotation), and wages (Martinuzzi et al., 2011). Miles and Covin (2000) note that in CSR reports, retailers pay the least attention to the area related to the workplace, where the main activities indicated are employee training or occasional charity and volunteering.

The literature emphasizes the importance of the CSR activities in the supply chain (e.g., Liu et al., 2021; McGoldrick & Freestone, 2008), as retailers play a key role in this area. This makes them often responsible for the actions of other actors in the chain. Retailers are not only expected to behave responsibly in their own business practices but also to provide products manufactured sustainably by suppliers. Jones et al. (2005) indicate that sourcing is becoming often controversial issue in relation to large retailers due to the increasing level of international product search and the requirement of accountability for the actions of the company. In addition, while other industries within supply chains may address their own immediate environmental and community issues or workplace problems, CSR in the retail industry should focus on consumer pressures (Lee, 2021). Challenges faced by retailers include cooperation with suppliers from developing countries, in the case of which the control of their activities (often deviating from CSR principles) is difficult due to geographical distance. CSR refers here to a wide range of activities, from product and supplier selection, to promotional activities and the integration of the concept of sustainable development and ethical and social responsibility in all key activities of the retail chain (Musso & Risso, 2006).

Elg and Hultman (2011) draw attention to the relationship between the size of the company and the application of codes of conduct in relations with suppliers and sustainable development reports. International retailers are more likely to force suppliers to respect their own CSR principles, as suppliers are often more concerned with maintaining cooperation with them. On the other hand, as Musso and Risso (2006) point out, since retailers are closer to consumers, they know their needs and can use retail-level marketing tools that respond to local preferences for ethical and responsible business approaches.

Based on the literature review, this work covers five areas of corporate social responsibility of retail chains, i.e., (1) relations with employees; (2) responsibility for the natural environment; (3) relations with society; (4) relations in supply chain; and (5) codification and dissemination of CSR activities, which is the area related to the measure of the company's involvement in CSR.

## 3 Research Method and General Characteristics of the Studied Firms

To answer the research questions, a questionnaire survey was carried out on a sample of 259 medium and large retail chains operating in Poland. The respondents were retail chain managers responsible for activities related to corporate social responsibility. The data was obtained in November 2016 using the CATI technique – computer-assisted telephone interviews. The surveyed entities were retail chains defined as chains of retail stores or other retail outlets (at least two stationary points in Poland), which have a common administrator (e.g., owner or a franchisor) and operate under one brand. In the studied sample, medium-sized retail chains (from 50 to 249 employees) accounted for 49.4% of the respondents, while large (250 and more employees) for 50.6%. The vast majority of chains are retailers with only Polish capital (80.3%). The surveyed respondents are mainly nationwide retail chains (60.6%). Local range (activity only in one voivodeship) was indicated by 29.7% of the respondents; 9.7% are chains with international reach. 23.6% of the surveyed retail chains were established/started their operations in Poland before 1989, more than half (51%) in the 1990s, and 25.1% after 2011.

Retail chains' commitment to CSR has been measured through a series of statements across five areas of CSR. The first group of statements described the relations between retailers and employees, the second the retailers' attitude to the environment, the third the codification and dissemination of CSR activities, the fourth the retailer's relationship with the society, and the fifth the relationships in the respondent's supply chain (Table 1). In each of these areas, the respondents declared their commitment to CSR using a Likert scale, from 1 (I completely disagree) to 5 (I fully agree). Used statements were literature-derived. Retail chains were also asked about the adoption of international standards in the field of responsible management. Answering "yes" or "no," respondents declared the implementation of the appropriate standard in their company.

The analysis of results has two stages. First, using respondent's answers' frequencies, the intensity of CSR activities conducted by the surveyed companies was examined. In the second stage, as it was stated by Dancewicz and Struve (2019) "CSR in Poland is mainly driven by the influence of foreign investors," it was worthy to check if there are any significant differences in terms of capital share. To reach it, answers of two groups – (1) retail chains with only Polish capital and (2) retail chains with foreign capital – were compared. Statistical analysis was performed using the SPSS (Statistical Package for the Social Sciences) software.

## 4 CSR Activities Undertaken by Retail Chains Operating in Poland: Research Results Relationships with Employees

Responsibility toward employees and relations with them seem to be the most important dimension of CSR in which the surveyed retail chains operating in Poland are involved. Most, as much as 99.6% (answers 4 and 5 on Likert scale) of respondents, indicate that they support diversity in the workplace. The same number of respondents agrees with the statement that their employees know the values and rules of conduct in the company (including those related to CSR). Among the surveyed enterprises, 98.1% offer training and individual personal development programs for a network employee, as well as additional benefits in the form of, for example, food stamps, multisport cards, or employee pension programs.

Care for the balance of private and professional life of employees, e.g., through flexible working hours, is another important area indicated by retail chains (97.7% of respondents). Slightly less, 93.8% of retailers reward employees who report environmental and social initiatives beneficial for the enterprise, and 90.7% take into account the opinions of employees when formulating the company's strategy (Table 2).

**Table 1** Involvement of retail chains in CSR

| Statements | | | Source |
|---|---|---|---|
| **Relationships with employees** | **A1** | We support diversity and provide equal opportunities to our employees regardless of their age, gender, disability, and ethnic origin | Schramm-Klein et al. (2015) |
| | **B1** | We offer our employees a number of training courses and individual personal development programs | |
| | **C1** | We offer our employees valuable additional benefits | |
| | **D1** | We take into account the opinions of our employees when formulating the company's strategy | |
| | **E1** | We reward employees who submit environmental and social initiatives that are beneficial for the company | Witek-Hajduk and Zaborek (2016) |
| | **F1** | We care about work-life balance of our employees, including through flexible working hours | |
| | **G1** | Our employees know the values and rules of conduct in the company (including those related to CSR) | |
| **Natural environment** | **A2** | We recycle and reduce the amount of waste | |
| | **B2** | We reduce the consumption of energy, water, and paper and CO2 emissions | |
| | **C2** | We use alternative/renewable energy sources in the operation of our stores | Schramm-Klein et al. (2015) |
| | **D2** | In our stores, we actively promote reusable bags and/or biodegradable bags, instead of plastic bags | |
| | **E2** | We require suppliers to use environmentally friendly packaging in the transport of our goods and to limit secondary packaging | |
| | **F2** | We require our suppliers to use environmentally friendly raw materials and technologies | Witek-Hajduk and Zaborek (2016) |
| **Codification and CSR dissemination** | **A3** | We include CSR in the mission and values of the company | |
| | **B3** | We create the image of our retail chain and our own brands or products with reference to CSR values | |
| | **C3** | Our retail chain publishes and applies a code of good practice (voluntary set of rules of conduct – especially ethical and professional standards) | |
| | **D3** | We work with other companies and national or international organizations to disseminate CSR practices | |
| | **E3** | The company is a signatory to codes of self-limitation, e.g., the code of ethics in advertising | |

(continued)

**Table 1** (continued)

| Statements | | | Source |
|---|---|---|---|
| **Relations with society** | **A4** | We are involved in charity | |
| | **B4** | We are committed to sponsoring local or national sporting, educational, or cultural events | Schramm-Klein et al. (2015) |
| | **C4** | We actively support humanitarian projects in the countries of origin of our goods (e.g., poverty reduction, decent working conditions, reduction of child labor) | |
| | **D4** | When the activity of our company causes controversy among the local community, we quickly solve the problem | Witek-Hajduk and Zaborek (2016) |
| | **E4** | Currently, our company has no unresolved conflicts with local communities | |
| | **F4** | We encourage employees to participate in events organized by local communities, e.g., by allowing them to devote part of their working time to them and use the knowledge and/or material resources of the company | |
| | **G4** | Our company pays taxes where it operates | |
| **Supply chain relationships** | **A5** | We apply criteria referring to the CSR principles in the supplier selection procedure | Schramm-Klein et al. (2015) |
| | **B5** | We enforce international ethical and environmental standards from our suppliers | |
| | **C5** | We prefer suppliers who respect ethical working conditions | |
| | **D5** | We choose local suppliers first | |
| | **E5** | We always settle financial obligations toward our suppliers in a timely manner and in accordance with the terms of the contract | Witek-Hajduk and Zaborek (2016) |
| | **F5** | We apply solutions that ensure honesty and reliability of contacts in relations with suppliers and buyers | |
| | **G5** | We disclose general terms of sale and purchase in our publications | |
| | **H5** | We deal with customer complaints on an ongoing basis | Zaborek (2014) |
| | **I5** | We rarely have disputes with suppliers | Witek-Hajduk and Zaborek (2016) |
| | **J5** | We handle all disputes with suppliers efficiently | |

Source: Own elaboration

## *4.1 Natural Environment*

All surveyed retail chains declare recycling and reducing waste. In addition, 99.2% also reduce the consumption of energy, water, and paper and CO2 emissions, but only 24.3% of stores use alternative/ renewable energy sources. Among the surveyed

**Table 2** Answers' distribution in the area "relations with employees"

| Statements | | Answer on Likert scale | Frequencies (%) |
|---|---|---|---|
| A1 | We support diversity and provide equal opportunities to our employees regardless of their age, gender, disability, and ethnic origin | 3 | 0.4 |
| | | **4** | **47.1** |
| | | **5** | **52.5** |
| B1 | We offer our employees a number of training courses and individual personal development programs | 2 | 0.4 |
| | | 3 | 1.5 |
| | | **4** | **61.4** |
| | | **5** | **36.7** |
| C1 | We offer our employees valuable additional benefits | 2 | 0.4 |
| | | 3 | 1.5 |
| | | **4** | **68.3** |
| | | **5** | **29.7** |
| D1 | We take into account the opinions of our employees when formulating the company's strategy | 2 | 2.7 |
| | | 3 | 6.6 |
| | | **4** | **68.3** |
| | | **5** | **22.4** |
| E1 | We reward employees who submit environmental and social initiatives that are beneficial for the company | 2 | 1.2 |
| | | 3 | 5.0 |
| | | **4** | **68.7** |
| | | **5** | **25.1** |
| F1 | We care about work-life balance of our employees, including through flexible working hours | 2 | 1.2 |
| | | 3 | 1.2 |
| | | **4** | **66.0** |
| | | **5** | **31.7** |
| G1 | Our employees know the values and rules of conduct in the company (including those related to CSR) | 3 | 0.4 |
| | | **4** | **73.7** |
| | | **5** | **25.9** |

Source: Own elaboration
Bold values indicate the respondent's agreement with the statement, 4—I agree, 5—I fully agree

entities, 81.1% actively promote reusable bags (e.g., paper or canvas) and/or biodegradable bags in their stores. 67.2% of respondents require suppliers to use environmentally friendly raw materials and technologies, and slightly less, 66% also expect them to use environmentally friendly packaging in the transport of goods and to limit secondary packaging (Table 3).

## 4.2 *Codification and Dissemination of CSR Activities*

As many as 99.2% of the surveyed retail chains operating in Poland include CSR issues in the mission and values of their enterprise, and 96.9% create the image of the retail chain, private labels, or products with reference to CSR values. The situation is

**Table 3** Answers' distribution in the area "natural environment"

| Statement | | Answer on Likert scale | Frequencies (%) |
|---|---|---|---|
| A2 | We recycle and reduce the amount of waste | **4** | **34.4** |
| | | **5** | **65.6** |
| B2 | We reduce the consumption of energy, water, and paper and CO2 emissions | 3 | 0.8 |
| | | **4** | **27.4** |
| | | **5** | **71.8** |
| C2 | We use alternative/renewable energy sources in the operation of our stores | 1 | 2.3 |
| | | 2 | 40.2 |
| | | 3 | 33.2 |
| | | **4** | **19.7** |
| | | **5** | **4.6** |
| D2 | In our stores, we actively promote reusable bags and/or biodegradable bags, instead of plastic bags | 2 | 6.9 |
| | | 3 | 12.0 |
| | | **4** | **57.1** |
| | | **5** | **23.9** |
| E2 | We require suppliers to use environmentally friendly packaging in the transport of our goods and to limit secondary packaging | 2 | 9.7 |
| | | 3 | 24.3 |
| | | **4** | **50.6** |
| | | **5** | **15.4** |
| F2 | We require our suppliers to use environmentally friendly raw materials and technologies | 2 | 10.0 |
| | | 3 | 22.8 |
| | | **4** | **51.4** |
| | | **5** | **15.8** |

Source: Own elaboration
Bold values indicate the respondent's agreement with the statement, 4—I agree, 5—I fully agree

worse when it comes to codification – only 64.9% publish and apply a code of good practice, and 60.2% are signatories of codes of self-limitation, e.g., the code of ethics of the Direct Marketing Association, i.e., the code of ethics in advertising. Only 37.8% of the surveyed retail chains cooperate with other enterprises and national or international organizations in order to disseminate CSR practices (Table 4).

## 4.3 Relations with Society

Retail chains are very willing to engage in charity (98.8% of respondents). In the surveyed sample, 96.9% are involved in sponsoring local or national sports and educational or cultural events, and 96.5% encourage their employees to participate in events organized by local communities. Among the surveyed enterprises, 83.8% pay taxes where they operate (Table 5).

**Table 4** Answers' distribution in the area "codification and CSR dissemination"

| Statement | | Answer on Likert scale | Frequencies (%) |
|---|---|---|---|
| A3 | We include CSR in the mission and values of the company | 3 | 0.8 |
| | | **4** | **68.0** |
| | | **5** | **31.3** |
| B3 | We create the image of our retail chain and our own brands or products with reference to CSR values | 2 | 0.8 |
| | | 3 | 2.3 |
| | | **4** | **64.9** |
| | | **5** | **32.0** |
| C3 | Our retail chain publishes and applies a code of good practice (voluntary set of rules of conduct – especially ethical and professional standards) | 2 | 19.7 |
| | | 3 | 15.4 |
| | | **4** | **41.7** |
| | | **5** | **23.2** |
| D3 | We work with other companies and national or international organizations to disseminate CSR practices | 2 | 27.8 |
| | | 3 | 34.4 |
| | | **4** | **30.1** |
| | | **5** | **7.7** |
| E3 | The company is a signatory to codes of self-limitation, e.g., the code of ethics in advertising | 1 | 0.8 |
| | | 2 | 26.6 |
| | | 3 | 12.4 |
| | | **4** | **34.0** |
| | | **5** | **26.3** |

Source: Own elaboration
Bold values indicate the respondent's agreement with the statement, 4—I agree, 5—I fully agree

## 4.4 Relations in Supply Chain

Among retail chains operating in Poland, 90.7% prefer suppliers who comply with ethical working conditions, and 89.2% use criteria referring to CSR principles in the supplier selection procedure, but only 38.2% enforce international ethical and environmental standards (Table 6).

## 4.5 Retails Chains with Only Polish Capital vs. Retail Chains with Foreign Capital: Significant Differences

More than three-quarters of the surveyed companies were established or started their operations in Poland after 1990 and the economic transformation that took place after that time. Rapid expansion of foreign enterprises on the Polish market had a very important impact on the development of social responsibility in Poland. It is worth to analyze that there are any significant differences within the surveyed companies between retail chains with only Polish capital vs. retail chains with

**Table 5** Answers' distribution in the area "relations with society"

| Statement | | Answer on Likert scale | Frequencies (%) |
|---|---|---|---|
| A4 | We are involved in charity | 3 | 1.2 |
| | | **4** | **53.3** |
| | | **5** | **45.6** |
| B4 | We are committed to sponsoring local or national sporting and educational or cultural events | 2 | 0.8 |
| | | 3 | 2.3 |
| | | **4** | **71.8** |
| | | **5** | **25.1** |
| C4 | We actively support humanitarian projects in the countries of origin of our goods (e.g., poverty reduction, decent working conditions, reduction of child labor) | 1 | 0.4 |
| | | 2 | 23.9 |
| | | 3 | 15.8 |
| | | **4** | **33.6** |
| | | **5** | **26.3** |
| D4 | When the activity of our company causes controversy among the local community, we quickly solve the problem | 3 | 0.4 |
| | | **4** | **44.8** |
| | | **5** | **54.8** |
| E4 | Currently, our company has no unresolved conflicts with local communities | **4** | **41.3** |
| | | **5** | **58.7** |
| F4 | We encourage employees to participate in events organized by local communities, e.g., by allowing them to devote part of their working time to them and use the knowledge and/or material resources of the company | 2 | 0.8 |
| | | 3 | 2.7 |
| | | **4** | **63.7** |
| | | **5** | **32.8** |
| G4 | Our company pays taxes where it operates | 2 | 2.7 |
| | | 3 | 13.5 |
| | | **4** | **24.7** |
| | | **5** | **59.1** |

Source: Own elaboration

Bold values indicate the respondent's agreement with the statement, 4—I agree, 5—I fully agree

foreign capital. Table 7 shows areas were these two groups show statistically significant differences in engaging in particular CSR activities.

The most interesting differences between retailers with Polish capital and retailers with foreign capital are shown in the three areas:

(1) Codification and CSR dissemination – In the case of self-restriction codes, there is a statistically significant difference between retail chains with only Polish capital (54.8% of the surveyed retailers are signatories of such codes) and retail chains with foreign capital (82.4% signatories here).
(2) Relations with society – A significant difference appears in terms of activities aimed at active support of humanitarian projects in the countries of goods' origin (retail chains with only Polish capital: 54.3%; retail chains with foreign capital: 82.4%). Among the surveyed enterprises, 83.8% pay taxes where they operate.

**Table 6** Answers' distribution in the area "relations in supply chain"

| Relacje w łańcuchu dostaw | | | |
|---|---|---|---|
| Statement | | Answer on Likert scale | Frequencies (%) |
| A5 | We apply criteria referring to the CSR principles in the supplier selection procedure | 2 | 2.7 |
| | | 3 | 8.1 |
| | | **4** | **61.4** |
| | | **5** | **27.8** |
| B5 | We enforce international ethical and environmental standards from our suppliers | 1 | 0.8 |
| | | 2 | 20.5 |
| | | 3 | 40.5 |
| | | **4** | **28.2** |
| | | **5** | **10.0** |
| C5 | We prefer suppliers who respect ethical working conditions | 2 | 2.3 |
| | | 3 | 6.9 |
| | | **4** | **52.9** |
| | | **5** | **37.8** |
| D5 | We choose local suppliers first | 2 | 7.7 |
| | | 3 | 19.3 |
| | | **4** | **53.3** |
| | | **5** | **19.7** |
| E5 | We always settle financial obligations toward our suppliers in a timely manner and in accordance with the terms of the contract | 3 | 0.4 |
| | | **4** | **49.8** |
| | | **5** | **49.8** |
| F5 | We apply solutions that ensure honesty and reliability of contacts in relations with suppliers and buyers | 3 | 0.4 |
| | | **4** | **51.0** |
| | | **5** | **48.6** |
| G5 | We disclose general terms of sale and purchase in our publications | 2 | 0.4 |
| | | 3 | 2.3 |
| | | **4** | **56.0** |
| | | **5** | **41.3** |
| H5 | We deal with customer complaints on an ongoing basis | **4** | **46.3** |
| | | **5** | **53.7** |
| I5 | We rarely have disputes with suppliers | **4** | **59.1** |
| | | **5** | **40.9** |
| J5 | We handle all disputes with suppliers efficiently | **4** | **47.9** |
| | | **5** | **52.1** |

Source: Own elaboration
Bold values indicate the respondent's agreement with the statement, 4—I agree, 5—I fully agree

However, taking into account the capital structure of enterprises, 93.8% of retail chains with only Polish capital and only 43.1% of retail chains with foreign capital pay taxes in the place where they operate.

(3) Relations in supply chain – A statistically significant difference appears in the selection of local suppliers. 78.4% of retail chains with only Polish capital prefer

**Table 7** Significant differences in answers' distribution between retail chains with only Polish capital vs. retail chains with foreign capital

| Statements | Share of capital | Answer on Likert scale | Frequencies (%) | Dif. sig. (two-tailed) |
|---|---|---|---|---|
| **Relations with employees** | | | | |
| C1. We offer our employees valuable additional benefits | Polish | 2 | 0.5 | 0.002 ($\alpha = 1\%$) |
| | | 3 | 0.9 | |
| | | 4 | 65.3 | |
| | | 5 | 33.3 | |
| | Foreign | 3 | 4.3 | |
| | | 4 | 82.6 | |
| | | 5 | 13.0 | |
| G1. Our employees know the values and rules of conduct in the company (including those related to CSR) | Polish | 3 | 0.5 | 0.000 ($\alpha < 1\%$) |
| | | 4 | 69.0 | |
| | | 5 | 30.5 | |
| | Foreign | 4 | 95.7 | |
| | | 5 | 4.3 | |
| **Natural environment** | | | | |
| B2. We reduce the consumption of energy, water, and paper and CO2 emissions | Polish | 3 | 0.9 | 0.000 ($\alpha < 1\%$) |
| | | 4 | 32.4 | |
| | | 5 | 66.7 | |
| | Foreign | 4 | 4.3 | |
| | | 5 | 95.7 | |
| **Codification and dissemination of CSR** | | | | |
| A3. We include CSR in the mission and values of the company | Polish | 3 | 0.9 | 0.018 ($\alpha = 5\%$) |
| | | 4 | 64.8 | |
| | | 5 | 34.3 | |
| | Foreign | 4 | 82.6 | |
| | | 5 | 17.4 | |
| C3. Our retail chain publishes and applies a code of good practice (voluntary set of rules of conduct – especially ethical and professional standards) | Polish | 2 | 21.1 | 0.024 ($\alpha = 5\%$) |
| | | 3 | 17.8 | |
| | | 4 | 39.0 | |
| | | 5 | 22.1 | |
| | Foreign | 2 | 13.0 | |
| | | 3 | 4.3 | |
| | | 4 | 54.3 | |
| | | 5 | 28.3 | |
| E3. The company is a signatory to codes of self-limitation, e.g., the code of ethics in advertising | Polish | 1 | 0.9 | 0.000 ($\alpha < 1\%$) |
| | | 2 | 29.6 | |
| | | 3 | 14.1 | |
| | | 4 | 33.3 | |
| | | 5 | 22.1 | |
| | Foreign | 2 | 13.0 | |
| | | 3 | 4.3 | |
| | | 4 | 37.0 | |
| | | 5 | 45.7 | |

(continued)

**Table 7** (continued)

| Statements | Share of capital | Answer on Likert scale | Frequencies (%) | Dif. sig. (two-tailed) |
|---|---|---|---|---|
| **Relations with society** | | | | |
| B4. We are committed to sponsoring local or national sporting, educational, or cultural events | Polish | 2 | 0.9 | 0.000 ($\alpha < 1\%$) |
| | | 3 | 1.4 | |
| | | 4 | 68.5 | |
| | | 5 | 29.1 | |
| | Foreign | 3 | 6.5 | |
| | | 4 | 87.0 | |
| | | 5 | 6.5 | |
| C4. We actively support humanitarian projects in the countries of origin of our goods (e.g., poverty reduction, decent working conditions, reduction of child labor) | Polish | 1 | 0.5 | 0.000 ($\alpha < 1\%$) |
| | | 2 | 27.2 | |
| | | 3 | 17.8 | |
| | | 4 | 30.5 | |
| | | 5 | 23.9 | |
| | Foreign | 2 | 8.7 | |
| | | 3 | 6.5 | |
| | | 4 | 47.8 | |
| | | 5 | 37.0 | |
| E4. Currently, our company has no unresolved conflicts with local communities | Polish | 4 | 44.1 | 0.039 ($\alpha = 5\%$) |
| | | 5 | 55.9 | |
| | Foreign | 4 | 28.3 | |
| | | 5 | 71.7 | |
| G4. Our company pays taxes where it operates | Polish | 2 | 3.3 | 0.000 ($\alpha < 1\%$) |
| | | 3 | 2.8 | |
| | | 4 | 27.7 | |
| | | 5 | 66.2 | |
| | Foreign | 3 | 63.0 | |
| | | 4 | 10.9 | |
| | | 5 | 26.1 | |
| **Relations in supply chain** | | | | |
| B5. We enforce international ethical and environmental standards from our suppliers | Polish | 1 | 0.5 | 0.036 ($\alpha = 5\%$) |
| | | 2 | 23.0 | |
| | | 3 | 42.7 | |
| | | 4 | 23.0 | |
| | | 5 | 10.8 | |
| | Foreign | 1 | 2.2 | |
| | | 2 | 8.7 | |
| | | 3 | 30.4 | |
| | | 4 | 52.2 | |
| | | 5 | 6.5 | |

(continued)

**Table 7** (continued)

| Statements | Share of capital | Answer on Likert scale | Frequencies (%) | Dif. sig. (two-tailed) |
|---|---|---|---|---|
| D5. We choose local suppliers first | Polish | 2 | 7.0 | 0.000 ($\alpha < 1\%$) |
| | | 3 | 14.6 | |
| | | 4 | 54.5 | |
| | | 5 | 23.9 | |
| | Foreign | 2 | 10.9 | |
| | | 3 | 41.3 | |
| | | 4 | 47.8 | |
| G5. We disclose general terms of sale and purchase in our publications | Polish | 2 | 0.5 | 0.013 ($\alpha = 5\%$) |
| | | 3 | 2.8 | |
| | | 4 | 58.7 | |
| | | 5 | 38.0 | |
| | Foreign | 4 | 43.5 | |
| | | 5 | 56.5 | |
| I5. We rarely have disputes with suppliers | Polish | 4 | 56.3 | 0.044 ($\alpha = 5\%$) |
| | | 5 | 43.7 | |
| | Foreign | 4 | 71.7 | |
| | | 5 | 28.3 | |
| J5. We handle all disputes with suppliers efficiently | Polish | 4 | 50.7 | 0.047 ($\alpha = 5\%$) |
| | | 5 | 49.3 | |
| | Foreign | 4 | 34.8 | |
| | | 5 | 65.2 | |

Source: Own elaboration

goods from local producers in the first place; in the case of retail chains with foreign capital, it is only 51%.

More than a half (57.5%) of the surveyed retail chains have implemented the ISO 26000 sustainable management standard, 59% used the SA 8000 standard, and 64% of the surveyed retail chains declared fair trade procedures. Table 8 shows differences between retail chains with only Polish capital vs. retail chains with foreign capital in terms of international standard implementation. It can be noticed that there is a significant difference between these two groups as retail chains with foreign capital show much higher engagement in this area.

## 5 Conclusions, Limitations, and Recommendations for the Future Research

CSR is strategically important for consumer-oriented retail industry. Therefore, companies' activities directly affect their relationships with customers. Retailers are motivated to operate in a socially responsible manner to satisfy customers'

**Table 8** Implementation of responsible management standards

| Are the following standards implemented in your retail chain? | | ISO 26000 | SA 8000 | Fair trade |
|---|---|---|---|---|
| | | Frequencies (%) | | |
| All surveyed retail chains ($n = 259$) | **Yes** | **57.5** | **59** | **64** |
| | No | 42.5 | 41 | 36 |
| | Total | 100.0 | 100.0 | 100.0 |
| Retail chains with only Polish capital ($n = 208$) | **Yes** | **52.4** | **53.8** | **59.6** |
| | No | 47.6 | 46.2 | 40.4 |
| | Total | 100.0 | 100.0 | 100.0 |
| Retail chains with foreign capital ($n = 51$) | **Yes** | **78.4** | **80.4** | **84.3** |
| | No | 21.6 | 19.6 | 15.7 |
| | Total | 100.0 | 100.0 | 100.0 |

Source: Own elaboration
Bold values indicate the implementation of responsible management standard

demand or desire for ethical consumption (Lee, 2021). This research contributes to the corporate social responsibility literature by providing a multidimensional analysis of CSR activities in the retail sector in Poland. Key results show main areas dominating the CSR practices of retailers operating in the Polish market. The study addresses a topic that is of great importance for the Central and Eastern Europe region.

Responsibility towards employees seems to be the most important dimension of CSR in which the surveyed retail chains are involved. Retailers in Poland mainly declare supporting diversity and work-life balance. Moreover, they enable employees to improve their qualifications and education and offer additional benefits. Employees know the values and rules of conduct in the company (including those related to CSR) and are rewarded for reporting environmental and social initiatives beneficial for the company, and their opinions are taken into account when formulating the company's strategy. According to the natural environment protection, retail chains are mainly involved in recycling and reducing waste, energy, water, paper, and $CO_2$ emissions. Retail chains operating in Poland include CSR in the company's values and image. The codes of self-limitation are mostly used by retail chains with foreign capital.

Retail chains in Poland are also willing to engage in charity by sponsoring local or national sports and educational or cultural events. They also encourage their employees to participate in events organized by local communities. Active support for humanitarian projects in the countries of origin of goods occurs mainly in the case of retail chains with foreign capital. Less than half of the retail chains with foreign capital declare that they pay taxes locally, i.e., in the place where they operate. Retail chains operating in Poland prefer suppliers who comply with ethical working conditions and those who respect CSR values, but mostly do not enforce international standards on them. Retailers with only Polish capital prefer mostly local manufacturers, while in case of retail chains with foreign capital, it is only half of the respondents.

Conclusions from the conducted empirical study can be used as a recommendation on the strategy of CSR activities for retail chain managers. Taking into account the obtained results, it can be concluded that retailers, through their key role in the supply chain, can play a very important part in developing more responsible, ethical, and fair trade, both international and local. The study was limited to the Polish market, where corporate social responsibility is still developing. The study was also limited by the subjectivity of the respondents in assessing the results and activity. It is recommended to deepen this research by taking into account a larger number of enterprises with foreign capital, as well as extending the research to other countries to compare the results. Comparing CSR activities of retail chains from various industries also seems to be an interesting approach.

## Appendix

Descriptive statistics

| Std. Dev. | | | | | | Quest. Variance | N Skewness | Min Std. error | Max Kurtosis | Mean Std. error |
|---|---|---|---|---|---|---|---|---|---|---|
| A1 | 259 | 3 | 5 | 4.52 | 0.508 | 0.258 | −0.175 | 0.151 | −1.756 | 0.302 |
| B1 | 259 | 2 | 5 | 4.34 | 0.530 | 0.281 | −0.057 | 0.151 | 0.140 | 0.302 |
| C1 | 259 | 2 | 5 | 4.27 | 0.504 | 0.254 | 0.176 | 0.151 | 0.633 | 0.302 |
| D1 | 259 | 2 | 5 | 4.10 | 0.623 | 0.388 | −0.753 | 0.151 | 2.224 | 0.302 |
| E1 | 259 | 2 | 5 | 4.18 | 0.563 | 0.317 | −0.373 | 0.151 | 1.791 | 0.302 |
| F1 | 259 | 2 | 5 | 4.28 | 0.544 | 0.296 | −0.374 | 0.151 | 2.043 | 0.302 |
| G1 | 259 | 3 | 5 | 4.25 | 0.445 | 0.198 | 0.999 | 0.151 | −0.622 | 0.302 |
| A2 | 259 | 4 | 5 | 4.66 | 0.476 | 0.226 | −0.662 | 0.151 | −1.573 | 0.302 |
| B2 | 259 | 3 | 5 | 4.71 | 0.471 | 0.222 | −1.156 | 0.151 | −0.097 | 0.302 |
| C2 | 259 | 1 | 5 | 2.84 | 0.924 | 0.855 | 0.498 | 0.151 | −0.464 | 0.302 |
| D2 | 259 | 2 | 5 | 3.98 | 0.799 | 0.639 | −0.791 | 0.151 | 0.584 | 0.302 |
| E2 | 259 | 2 | 5 | 3.72 | 0.841 | 0.707 | −0.414 | 0.151 | −0.310 | 0.302 |
| F2 | 259 | 2 | 5 | 3.73 | 0.847 | 0.717 | −0.455 | 0.151 | −0.284 | 0.302 |
| A3 | 259 | 3 | 5 | 4.31 | 0.478 | 0.228 | 0.636 | 0.151 | −1.040 | 0.302 |
| B3 | 259 | 2 | 5 | 4.28 | 0.544 | 0.296 | −0.228 | 0.151 | 1.195 | 0.302 |
| C3 | 259 | 2 | 5 | 3.68 | 1.038 | 1.078 | −0.404 | 0.151 | −0.987 | 0.302 |
| D3 | 259 | 2 | 5 | 3.18 | 0.927 | 0.860 | 0.227 | 0.151 | −0.916 | 0.302 |
| E3 | 259 | 1 | 5 | 3.58 | 1.163 | 1.353 | −0.285 | 0.151 | −1.294 | 0.302 |
| A4 | 259 | 3 | 5 | 4.44 | 0.521 | 0.271 | −0.023 | 0.151 | −1.433 | 0.302 |
| B4 | 259 | 2 | 5 | 4.21 | 0.511 | 0.261 | −0.069 | 0.151 | 1.992 | 0.302 |
| C4 | 259 | 1 | 5 | 3.61 | 1.126 | 1.269 | −0.278 | 0.151 | −1.246 | 0.302 |
| D4 | 259 | 3 | 5 | 4.54 | 0.507 | 0.257 | −0.269 | 0.151 | −1.711 | 0.302 |
| E4 | 259 | 4 | 5 | 4.59 | 0.493 | 0.243 | −0.355 | 0.151 | −1.889 | 0.302 |
| F4 | 259 | 2 | 5 | 4.29 | 0.553 | 0.306 | −0.269 | 0.151 | 1.082 | 0.302 |

(continued)

| Std. Dev. | | | | | | Quest. Variance | N Skewness | Min Std. error | Max Kurtosis | Mean Std. error |
|---|---|---|---|---|---|---|---|---|---|---|
| G4 | 259 | 2 | 5 | 4.40 | 0.822 | 0.675 | −1.159 | 0.151 | 0.350 | 0.302 |
| A5 | 259 | 2 | 5 | 4.14 | 0.670 | 0.449 | −0.720 | 0.151 | 1.350 | 0.302 |
| B5 | 259 | 1 | 5 | 3.26 | 0.924 | 0.853 | 0.168 | 0.151 | −0.605 | 0.302 |
| C5 | 259 | 2 | 5 | 4.26 | 0.688 | 0.473 | −0.826 | 0.151 | 1.082 | 0.302 |
| D5 | 259 | 2 | 5 | 3.85 | 0.824 | 0.679 | −0.552 | 0.151 | −0.026 | 0.302 |
| E5 | 259 | 3 | 5 | 4.49 | 0.509 | 0.259 | −0.066 | 0.151 | −1.783 | 0.302 |
| F5 | 259 | 3 | 5 | 4.48 | 0.508 | 0.258 | −0.019 | 0.151 | −1.786 | 0.302 |
| G5 | 259 | 2 | 5 | 4.38 | 0.554 | 0.307 | −0.281 | 0.151 | 0.032 | 0.302 |
| H5 | 259 | 4 | 5 | 4.54 | 0.500 | 0.250 | −0.148 | 0.151 | −1.994 | 0.302 |
| I5 | 259 | 4 | 5 | 4.41 | 0.493 | 0.243 | 0.371 | 0.151 | −1.877 | 0.302 |
| J5 | 259 | 4 | 5 | 4.52 | 0.501 | 0.251 | −0.086 | 0.151 | −2.008 | 0.302 |

## References


Amos, G. J. (2018). Researching corporate social responsibility in developing-countries context: A systematic review of the literature. *International Journal of Law and Management, 60*(2), 284–310. https://doi.org/10.1108/IJLMA-04-2017-0093

Bilińska-Reformat, K. (2013). Wykorzystanie koncepcji CSR w działalności handlu detalicznego. *Handel Wewnętrzny, 3*, 72–79.

Bilińska-Reformat, K., & Stefańska, M. (2013). Cooperation of commercial chains with suppliers in the context of CSR concept, Visnik. *Logistika, 762*, 15–20.

Carrero, I., & Valor, C. (2012). CSR-labelled products in retailers' assortment. A comparative study of British and Spanish retailers. *International Journal of Retail & Distribution Management, 40*(8), 629–652. https://doi.org/10.1108/09590551211245425

Dancewicz, B., & Struve, F. (2019). *The Influence of MNEs on CSR in Poland*. In A. Długopolska-Mikonowicz, S. Przytuła, & C. Stehr (Eds.), *Corporate Social Responsibility in Poland. CSR, Sustainability, Ethics & Governance*. Springer. https://doi.org/10.1007/978-3-030-00440-8_11

Dang, V. T., Nguyen, N., & Pervan, S. (2020). Retailer corporate social responsibility and consumer citizenship behavior: The mediating roles of perceived consumer effectiveness and consumer trust. *Journal of Retailing and Consumer Services, 55*, 102082.

do Vale, R. C., Matos, P. V., & Herédia-Colaço, V. (2020). *Impact of CSR initiatives on consumer's perceptions and attitudes towards retailers*. In F. Martinez-Lopez, J. Gázquez-Abad, & E. Breugelmans (Eds.), *Advances in National Brand and Private Label Marketing. NB&PL 2020. Springer Proceedings in Business and Economics*. Springer. https://doi.org/10.1007/978-3-030-47764-6_15

Elg, U., & Hultman, J. (2011). Retailers' management of corporate social responsibility (CSR) in their supplier relationships – does practice follow best practice? *The International Review of Retail, Distribution and Consumer Research, 21*(5), 445–460. https://doi.org/10.1080/09593969.2011.618887

Jamali, D., & Karam, C. (2018). Corporate social responsibility in developing countries as an emerging field of study. *International Journal of Management Reviews, 20*(1), 32–61. https://doi.org/10.1111/ijmr.12112

Jany-Catrice, F., & Lehndorff, S. (2002). Who bears the burden of flexibility? Working conditions and labour markets in the European retail trade. *Transfer, 8*(3), 504–520. https://doi.org/10.1177/102425890200800313

Jones, P., Comfort, D., & Hillier, D. (2005). Corporate social responsibility and the UK's top ten retailers. *International Journal of Retail & Distribution Management, 33*(11/12), 882–893. https://doi.org/10.1108/09590550510634611

Lee, S. H. (2021). Effects of retailers' corporate social responsibility on retailer equity and consumer usage intention. *Sustainability, 13*, 3080. https://doi.org/10.3390/su13063080

Liu, S., Fengmin, Y., & Dongyan, C. (2021). CSR investment decision and coordination strategy for closed-loop supply chain with two competing retailers. *Journal of Cleaner Production, 310*, 127378. https://doi.org/10.1016/j.jclepro.2021.127378

Martinuzzi, A., Kudlak, R., Faber, C., & Wiman, A. (2011). CSR activities and impacts of the retail sector. *RIMAS Working Papers, 4*, 1–29. https://www.sustainability.eu/pdf/csr/impact/IMPACT_Sector_Profile_RETAIL.pdf

McGoldrick, P. J., & Freestone, O. M. (2008). Ethical product premiums: Antecedents and extent of consumers' willingness to pay. *The International Review of Retail, Distribution and Consumer Research, 18*(2), 185–201.

Miles, M. P., & Covin, J. G. (2000). Environmental marketing: A source of reputational, competitive, and financial advantage. *Journal of Business Ethics, 23*, 299–311. https://doi.org/10.1080/09593960701868431

Musso, F., & Risso, M. (2006). CSR within large retailers international supply chains. Symphonya. *Emerging Issues in Management, 1*, 79–92. https://doi.org/10.4468/2006.1.06musso.risso

Nyame-Asiamah, F., & Ghulam, S. (2020). The relationship between CSR activity and sales growth in the UK retailing sector. *Social Responsibility Journal, 16*(3), 387–401. https://doi.org/10.1108/SRJ-09-2018-0245

Schramm-Klein, H., Morschett, D., & Swoboda, B. (2015). Retailer corporate social responsibility: shedding light on CSR's impact on profit of intermediaries in marketing channels. *International Journal of Retail & Distribution Management, 43*(4/5), 403–431. https://doi.org/10.1108/IJRDM-04-2014-0041

Seuring, S., & Müller, M. (2008). From a literature review to a conceptual framework for sustainable supply chain management. *Journal of Cleaner Production, 16*, 1699–1710. https://doi.org/10.1016/j.jclepro.2008.04.020

Śmigielska, G., & Oczkowska, R. (2017). Retailers' responsibility towards consumers and key drivers of their development in Poland. *Administrative Sciences, 7*(3). https://doi.org/10.3390/admsci7010003

Statistics Poland (2019). *Rynek wewnętrzny w 2019 r. Główny Urząd Statystyczny (GUS), Warsaw, Poland.* https://stat.gov.pl/obszary-tematyczne/ceny-handel/handel/rynek-wewnetrzny-w-2019-roku,7,26.html, (access: 25.04.2021).

Stefańska, M. (2014). Competitive advantage based on CSR in supply chains–Suppliers about retailers strategy. *International Journal of Arts & Sciences, 7*(4), 35–44. http://www.universitypublications.net/ijas/0704/pdf/B4R318.pdf

Stefańska, M., & Śmigielska, G. (2015). *Theoretical foundations of CSR in retailing*. In M. Stefańska & R. Nestorowicz (Eds.), *Fair trade in CSR strategy of global retailers*. Palgrave Macmillan. https://doi.org/10.1057/9781137389046_1

Stefańska, M., & Wanat, T. (2014). Consumers' perception of corporate social responsibility (CSR) and their attitude toward retailers – Its significance for creating retailers' competitive advantage. *International Journal of Arts & Sciences, 7*(2), 47–58.

Vo, H. T. M., & Arato, M. (2020). Corporate social responsibility in a developing country context: a multi-dimensional analysis of modern food retail sector in Vietnam. *Agroecology and Sustainable Food Systems, 44*(3), 284–309. https://doi.org/10.1080/21683565.2019.1607797

Witek-Hajduk, M. K., & Grudecka, A. (2018). Positioning strategies of retailers' brands in the emerging market – A cluster analysis. *International Journal of Emerging Markets, 13*(5), 925–942. https://doi.org/10.1108/IJoEM-06-2017-0224

Witek-Hajduk, M. K., & Zaborek, P. (2016). Does business model affect CSR involvement? A survey of polish manufacturing and service companies. *Sustainability, 8*(93). https://doi.org/10.3390/su8020093

Zaborek, P. (2014). CSR and financial performance: The case of polish small and medium manufacturers. *International Journal of Management and Economics, 43*, 53–73. https://doi.org/10.1016/j.fbj.2017.12.002

**Anna Napiórkowska** is an assistant professor at the Department of International Business in the Collegium of World Economy SGH Warsaw School of Economics. In 2019, she obtained with honors the Ph.D. degree in the field of management and quality sciences at the Collegium of World Economy SGH. She landed second place in the Verba Veritatis competition for the best doctoral theses in the field of corporate social responsibility and business ethics and is the winner of the award of the Rector of SGH Warsaw School of Economics for the best doctoral dissertation in the discipline of management and quality sciences. She is the author or coauthor of publications in the field of corporate social responsibility, ethical leadership, CSR legalization, and retail chains' CSR strategies in peer-reviewed journals and scientific books.

# University Social Responsibility (USR): From the Perspective of Top 10 Universities in Japan

**Long Tam Pham**

## 1 Introduction

CSR has imposed itself in the economic area as a concept through which organizations consider the affairs and issues of their related communities or, in short, their surrounding stakeholders. According to Pinney (2001), CSR could most simply be defined as a set of management practices that ensure a business to minimize the negative impacts from its operations on society while maximizing its positive impacts. At the commencement of CSR, it was considered as a method to capture more profit, leading to the separate application and performance of the responsibilities. Those responsibilities were not related to the visions, missions, and objectives of a business (Baker, 2008). At that time, CSR activities were merely a tool for businesses to increase economic performances. Hence, corporations are often accused of emphasizing profits over other social responsibilities.

Initially, it was an unwritten consensus that CSR was voluntary practice and underestimated by most business. It results in the misleading perception of CSR focus—it should be a responsibility of a business, not just voluntary and charity (Van Marrewijk, 2003). That's why Porter and Kramer (2011) has challenged the consensus and developed a new concept called creating shared value (CSV). CSV could be considered as a transition and expansion from CSR, in which the main focus of CSV is to create value, which could be easily understood as before starting a business, launching a project, or applying an action that could affect the society, a business must take its positive impact toward the society into account (Porter & Kramer, 2011).

As the society is now concerned, social values have also been prioritized (Baker, 2014). In terms of values for society and community, the first thing that comes to

L. T. Pham (✉)
Graduate School of Economics, Osaka University, Osaka, Japan

 
I. G. Arraiano et al. (eds.), *Corporate Social Responsibility in a Dynamic Global Environment*, CSR, Sustainability, Ethics & Governance,
https://doi.org/10.1007/978-3-031-24647-0_10

mind is sustainability. Sustainable development has been defined in many ways, but the most common-shared definition belongs to the International Institute for Sustainable Development: "Sustainable development is development that meets the needs of the present without compromising the ability of future generations to meet their own needs" (United Nations 1987). Sustainability, therefore, is closely related to education, which sets the foundation for the development of the future generations, especially the role of higher educational institutions such as universities and academy. Hence, considering university as one of the society's organization, or in other words a corporate entity, the responsibilities toward society of corporates are also transferred to universities as well.

Acknowledging the importance of social responsibility in universities, a study concerned about the current situation of USR is necessary to be a groundbreaking step to further elucidate the research gap about USR, especially in Asian context in general and Japan in particular. Based on data analyses of a sample of top 10 Japanese universities using their official website and all sub-websites that contain the domain of the official one (all departments, schools, graduate schools, institutions, and others are also taken into account), the theoretical analysis are largely supported. This research makes two principal contributions to the literature:

- Understand the current situation of USR in Japan in particular.
- Define the characteristics, strengths, and weaknesses of USR in Japan.

## *1.1 Data Collection Methodology*

The research targets are top 10 universities in Japan based on QS World University Ranking (2020). There are two reasons for this choice. First, top 10 universities in Japan are of the former Imperial Universities, the most prestigious higher educational institutions in Japan and the best private universities. Second, their websites are both Japanese- and English-supported. In terms of data collection, data scraping is applied. Data scraping is also known as web extraction or harvesting, which is a technique to extract data from the World Wide Web (WWW) and save it to a file system or database for later retrieval or analysis (Zhao, 2017). Web scraping is widely accepted as a strong and effective method for gathering large amounts of data because of the massive volume of heterogeneous material that is continually created on the WWW (Mooney et al., 2015).

In this study, data crawling is employed to collect the data from ResearchGate, a database of over 135 million publications and 19 million researchers all over the world, to testify the accuracy and properness of the keywords that were used in data scraping—the process where the keywords were, respectively, searched in all websites that contain the domain of the official websites of proposed top 10 universities above.

## 2 Literature Review and Research Questions

### 2.1 *Social Responsibility*

The concept of social responsibility currently plays an increasingly significant role and has become part of the global argument on competitiveness and sustainability in the sense of globalization (Vasilescu et al., 2010). According to Investopedia, social responsibility means that individuals and companies have a duty to act in the best interests of their environment and society as a whole. A trade-off may exist between economic development, in the material sense, and the welfare of the society and environment, though this has been challenged by many reports over the past decades. Therefore, social responsibility means sustaining the equilibrium between the two aspects.

Historically, several alternative terms for social responsibility that carry the same meaning could be listed as "social responsiveness," "social performance," and "ethical responsibility" (Barrena Martínez et al., 2019). All these concepts vary in terms of words but share the same ideas about the "responsibility" toward the society in line with ethical activities.

Although these concepts and definitions vary, they all share the notion that social responsibility is not confined to social aspects but involved socioeconomic, environmental, and other relevant issues. These social problems, on the other hand, reflect directly to sustainability or in other words sustainable development. Hence, we can infer that social responsibility currently involves and impacts sustainability, despite the diversity of definitions and terms.

### 2.2 *From Corporate Social Responsibility to University Social Responsibility*

Social responsibility is applied not only to individuals but organization as well. In terms of organization and institution, the party that brought most impact to the society has been corporate. The first study that introduced corporate social responsibility (CSR) was counted for Howard Rothmann Bowen (1953) when he presented "Social Responsibility of the Businessman" in a comprehensive discussion. He pointed out that CSR expresses a fundamental morality in the way a company behaves toward society. It follows ethical behavior toward stakeholders and recognizes the spirit of the legal and regulatory environment. This study provides a framework for business leaders and scholars to view the subjects as part of strategic strategy and decision-making in management. Though published in another era, because of its importance to the present ethical concerns of business activities in the United States, it is frequently and increasingly cited by scholars and researchers.

Since then, CSR has evolved and transformed from pure philanthropy to regulated practices and instrumentality (Hamidu et al., 2015). In the context of globalization and integration, especially the foundation of the European Union, this event shed light on CSR, where they promoted CSR as a standard to preserve common values and increase the sense of solidarity and cohesion. With the presence of CSR, European companies should beware of their practices to keep them in accordance with the European values and international agreed norms and standards.

From the 1980s and during the new millennium, corporates are increasingly suffering more pressures on globalization problems, transparency, fierce competition, and market saturation (Husted & Allen, 2006). Therefore, it is important to emphasize that corporates are now in need of a tool which could simultaneously solve social problems while functioning as a competitive advantage in order to survive. The answer is clear; CSR could do both if it is not separated from business strategy and operations of the corporate. Based on this, Porter and Kramer (2011) has challenged the consensus and developed a new concept called creating shared value (CSV). CSV could be considered as a transition and expansion from CSR, in which its main focus swifts to create values, such as launching a project or applying an action that could affect the society. In other words, a business must take its positive impact toward the society into account (Porter & Kramer, 2011). CSR now belongs to the strategic management of a corporate, while maintaining competitive strategy and assuring sustainable development. Corporates are using CSR as a strategic tool to increase their competitive advantages, along with providing the society with applicable solutions to solve social problems.

Table 1 sums up the evolution of CSR in both specific areas and summaries throughout different periods.

As the evolution of CSR had surpassed the pure philanthropic and regulated characteristics, the core features of CSR in this era were said to be the management of externalities and the alignment of social and economic responsibilities (Hamidu et al., 2015).

1. Management of externalities

   CSR externalities refer to all kinds of causes that have an impact on the rights of various stakeholders that are not directly taken care of in a corporation or organization's decision-making process. Nowadays, most CSR activity deals with externalities involving workers' rights, and to minimize the impacts of rationalization, stakeholders must be managed well so that valid claims are not ignored and unsafe items are thrown out (Husted & Allen, 2006).
2. Alignment of social and economic responsibilities

   This function of CSR balances the interests of different parties, integrating the involvement of society and corporate. There has been ongoing debate about the focus on shareholders and profitability, but it should obviously be considered that CSR does not conflict with economic matters or profitability in other words. Many definitions of CSR from business and government stress that social and economic responsibilities must be aligned. This core characteristic brought up the case of M. Porter's CSV model—in which firms can benefit economically from being socially responsible by creating shared values before and during the operation processes (Carol, 2014).

**Table 1** The evolution of CSR through periods

| Periods and areas | Summaries of main focused areas |
|---|---|
| 1950s–1960s<br>• Religious and human philosophies<br>• Community development<br>• Unregulated philanthropy<br>• Poverty alleviation<br>• Obligation to the society | Philanthropy |
| 1970s–1980s<br>• Extension of CSR commitments<br>• CSR as symbol of corporate citizenship<br>• Stakeholder relationship management<br>• Corporate reputation<br>• Socioeconomic priorities<br>• Bridging governance gap<br>• Stakeholders' rights<br>• Legal and ethical responsibilities | Regulated CSR |
| 1990s–now<br>• Competitive strategy<br>• Environmental protection<br>• Sustainability<br>• Internationalization of CSR standards<br>• Transparency and accountability | Strategic CSR |

Source: Hamidu et al. (2015)

Both these above characteristics share the same idea: "Looking forward to maintaining sustainability." On the other hand, sustainability, to some extent, directly related to education, especially higher education and its institutions (Martin, 2015). The reason is simple, higher education or in short university plays a pivotal role in educating and nurturing the future generation, which impacts the sustainable development of a nation. Considering university as a corporate entity, the social responsibility of a corporate is also transformed into university social responsibility. In defining USR, Reiser (2007) firstly developed the concept, while focusing on the university management's impact on society, stating that "USR is a policy of ethical quality of the performance of the university community (students, faculty, and administrative employees) via the responsible management of the educational, cognitive, labor and environmental impacts produced by the University." In 2010, Vasilescu et al. defined USR from the citizenship perspective as "the need to strengthen civic commitment and active citizenship." In this study, he emphasized an ethical approach by developing a sense of civil citizenship by encouraging the students and the academic staff to provide social services to their local community or to promote ecological and environmental commitment for local and global sustainable development.

The definitions may vary. However, all previous studies agree that higher education institutions can cause "significant" impact of the society (Vallaeys, 2014; Kappo-Abidemi & Kanayo Kingsley, 2020). UNESCO underlines the current importance of such a contribution: "At no time in human history was the welfare of nations so closely linked to the quality and outreach of their higher education systems and institutions" (World Conference on Higher Education Partners, June 2003). Taking the global context into account, the author decided to define USR as the given definition by the United Nation: "The responsibilities of universities for the impacts of their decisions and activities on society and the environment through transparent and ethical strategies."

## 2.3 Recent Studies

In fact, USR has been far from developed compared to the development of CSR in private companies. From the first development of the concept, Jabbour (2010) stated that universities could bring a huge impact on environment. Another study by Wright (2010) examined how a group of university presidents and managers in Canada conceptualize sustainable development and its relationship to sustainable university. Nejati et al. (2010) investigated the issue of environmental sustainability of the world top 10 Universities and later on in 2011 elucidated the commitment of these universities toward social responsibility and their current concerns by exploring their website contents and annual reports. The former study showed that these universities are aware of their environmental impacts and have taken necessary steps toward sustainability while the latter stressed that these universities are also providing sufficient information on most of the core areas of CSR.

Giuffre and Ratto (2014) illustrated USR as the ability of university to disseminate and implement a set of general principles and specific values. They also stated four key processes of USR: management, teaching, research, and extension. Shek et al. (2017) concluded the social engagement of higher education institutions has become one of the cores and fundamental pillars of universities, with that being said social service is also considered as a main mission, which equals to teaching and researching.

Although there have been a few studies related to USR, up until 2017, the very first research that took Asian universities' context into account had been presented (Bokhari, 2017). In this paper, he aimed to present a framework that clarifies the role of social responsibility and its various aspects while strengthening the close relationship between it and sustainable development in Saudi Arabian universities' context.

As USR is a transformation from the pure CSR, it should contain most of the CSR's characteristics and functions. In order to measure the presence of CSR in universities, Nejati et al. (2011) have introduced seven core areas of CSR:

1. Organizational governance
2. Human right
3. Labor practices
4. Environment
5. Fair operating practices

6. Consumer issues
7. Community involvement and development

However, the concept of seven core areas was too narrow and, on the other hand, seems to be bias on the view of corporate rather than organization, as universities, although could be considered as a corporate entity which had been mentioned above, also carry different characteristics. Therefore, a wider concept that fits universities better while covering all the significant aspects as well must be deployed.

In order to find out the characteristics of an object, the best way is to define and analyze its definition. A. Dahlsrud (2008) studied 37 definitions of CSR (consisting of extensive literature review of both journal articles and web pages). Most definitions were published from 1998 onward, so the integrity of the definition is ensured. In this paper, Dahlsrud introduced the concept "the five dimensions of CSR" that he used to measure the presence of CSR's definition (Table 2).

As for the results of the study, stakeholder, social, economic, and voluntariness dimension appeared in more than 80% of the analyzed definitions. The environmental dimension received a significant lower dimension ratio than other dimensions with only 60% in terms of appearance. This could be explained by Carroll's literature review in 1999, in which she stated that the environmental dimensions were not included in the early definitions, mostly before 1990. However, when CSR is explained in more depth, and global issues have become more and more severe, the environmental dimensions are equally emphasized compared to others. Hence, it is safe to say that this concept is accurate and applicable for further research. In this study, the author will use the concept of five dimensions as the foundation to deeply exploit USR in Japanese universities as proposed.

**Table 2** The five dimensions of CSR

| Dimensions | The definition is coded to the dimension if it refers to | Example phrases |
|---|---|---|
| The environmental dimension | The natural environment | 'a cleaner environment'<br>'environmental stewardship'<br>'environmental concerns in business operations' |
| The social dimension | The relationship between business and society | 'contribute to a better society'<br>'integrate social concerns in their business operations'<br>'consider the full scope of their impact on communities' |
| The economic dimension | Socio-economic or financial aspects, including describing CSR in terms of a business operation | 'contribute to economic development'<br>'preserving the profitability'<br>'business operations' |
| The stakeholder dimension | Stakeholders or stakeholder groups | 'interaction with their stakeholders'<br>'how organizations interact with their employees, suppliers, customers and communities'<br>'treating the stakeholders of the firm' |
| The voluntariness dimension | Actions not prescribed by law | 'based on ethical values'<br>'beyond legal obligations'<br>'voluntary' |

Source: Dahlsrud (2008)

# 3 Empirical Analysis

## *3.1 Research Methodology*

This study uses a "content analysis methodology," which is considered as one of the most effective and suitable to analyze secondary data (Nejati, 2011). For example, the study on Turkish companies of Halici and Kucukaslan in 2005 evaluated ethical concepts of the companies' websites. Another fact is that content analysis on companies' annual report and CSR report has been a widely used method in the study of CSR since the 1970s (Milner, 1999). It is also argued that content analysis is a "distinctive approach to analysis," which seeks to quantify the content of a text in a "systematic and replicable manner" (Bryman, 2004).

Academically, content analysis can be considered as a type of qualitative approach (Low et al., 2015). According to Hennink et al. (2011), qualitative research allows in-depth examination of people's experiences through the use of a specific set of research methods, such as in-depth interviews, focus groups, direct observation, and content analysis, because interpretation and observation are important integral components of qualitative research in understanding the social world. As a result, qualitative methodology allows researchers to use documents as a primary source for obtaining a broad picture of an object's actions, and content analysis is a standard textual research technique that entails codifying qualitative and quantitative data into various categories based on predetermined criteria. Content analysis is a "research approach for deriving reproducible and accurate conclusions from data according to their context, " according to Krippendorff (1980). Furthermore, content analysis may generate repeatable and reliable conclusions from texts for a collection of conditions surrounding a scenario that is not the same as the original.

The process underlying the content analysis provides a text or texts as input and a theme as output. This procedure entails decomposing inputs into elementary units (analysis units) and categorizing these units inside categorical variables (categories of analysis). The units of analysis, which are classified within the categories of analysis, are next investigated using scientific quantitative techniques. The enumeration method can be as simple as binary coding, which shows if the category appears in documents, or it can be as complex as a computation of the frequency with which the category appears in the texts (frequency counts on specific keywords).

The research was divided into four phases: (1) selecting the sample of universities to be surveyed; (2) collecting text information from the selected universities, which were published on their respective institutional websites; (3) analyzing retrieved text using precisely defined reference parameters; and (4) visualizing the process and results using tables and figures.

## 3.2 University Selection

Top 10 Japanese universities (by QS Ranking, 2020, with Tokyo Institute of Technology excluded, due to the lack of schools and graduate schools in terms of numbers) include the following:

| | |
|---|---|
| • Tokyo University | • Hokkaido University |
| • Kyoto University | • Kyushu University |
| • Osaka University | • Waseda University |
| • Tohoku University | • Keio University |
| • Nagoya University | • Hitotsubashi University |

Regarding the population, although more than 75% of Japanese students are studying in private universities (Maruyama, 2008)—which is a huge proportion compared to other developed countries such as the United States (20%)—Huang (2017) stated that the difference between private and national is reflected in the academic backgrounds of university leaders: 87% of national university presidents—but only 70% of private university presidents—have doctoral degrees; 38.6% of national presidents—but only 5.9% of those at private universities—graduated from their current university. In addition, far more national than private university presidents received their degrees from national universities (46.5% compared with 25.8%) and conduct research abroad (33.7% compared with 7.9%). This directly illustrates the impact of universities' presidents on university strategic management, and it is also reflected through their websites, which are the main research objects of this study.

This paper studied the content of these above university official websites and its related sites that contain the official website's domain (e.g., osaka-u-ac.jp is considered as the main domain, so all sites that contain this domain such as econ.osaka-u.ac.jp is also taken into account). With this approach, not only the main official website but also all sites and pages of schools, graduate schools, and other academic institutions are also analyzed, which could provide the author with more sufficient results about USR's situation.

## 3.3 Category of Analysis Process

The following three phases can be used to carry out the content analysis technique: (1) Define the categories of analysis or the categories within which to frame the content of the documents, which, of course, depend on the investigation's goals; (2) define the unit of analysis or the unit through which to analyze the documents under consideration, which may consist, for example, of individual words or key symbols; and (3) construct the data quantification technique or specify the enumeration system (Bowen 1953).

**Table 3** Keywords for each dimension in both English and Japanese

| | English | 日本語 |
|---|---|---|
| Environment dimension | Environment, natural resources, nature, pollution | 環境、自然、汚染 |
| Economics dimension | Economic, development, profit, competitive | 経済、開発、発展、利益、競争 |
| Social dimension | society, community, fairness | 社会、協同、平等 |
| Stakeholder dimension | Stakeholder, shareholder, governance, policy | ステークホルダー、シェアホルダー、統治、ガバナンス、対策 |
| Voluntariness dimension | Voluntary, contribution, willingness, philanthropy | ボランティア、貢献やる気、慈善, |

Source: Author's elaboration

Concerning the categories, it was decided to use the concept "five dimensions of CSR" that is introduced before in the "Theoretical Framework" part. Several scholars stated that when examining paragraphs, it is easier and more precise to derive meaning from the keywords (Guthrie et al., 2004; Samkin & Schneider, 2008). Therefore, five "layer 1" keywords (environmental, economics, social, stakeholder, voluntariness) were chosen to draw the significant and represented the main categories, based on the five dimensions stated above.

The next step is to define the unit of analysis, which the author identified as the key symbols that represent the main idea of a text information sample. The key symbols, as mentioned, coincided either with a single word that had only one possible meaning, for example, natural resources, pollution, profit, competitive, society, stakeholder, contribution, willingness, and so on. The definition of key symbols was determined by analyzing the abstracts content of multiple journal articles via ResearchGate, in order to ensure the consistency in terms of meaning. For example, if an article contains "environment," "profit," and "community," we could infer that this article directly relates to environmental, economic, and social dimension (Table 3).

## *3.4 Measurement and Enumeration Process*

The measurement methodology is simple. For each appearance of a keyword in an article, input 1 points for its respective dimension (Fig. 1). The total points later were

| | | | | | |
|---|---|---|---|---|---|
| 結果、サイト:kyoto-u.ac.jp | | | | | |
| | Social Responsibility | | | | |
| | | Economics | | | |
| | | | development | | 790 |
| | | | Economic | | 2754 |
| | | | 経済 | | 2837 |
| | | | 発展 | | 132 |
| | | | competitive | | 53 |
| | | | 利益 | | 57 |
| | | | 競争 | | 34 |
| | | | profit | | 56 |
| | | | 開発 | | 398 |
| | | 総合 | | 7111 | |

**Fig. 1** Example results on the website of Kyoto University for articles that contain "social responsibility" by Economic Dimension Score and each keyword's points (both by English and Japanese). Source: Author's elaboration

summed up statistically to generate the "dimension score" for each university. The example of detailed points and dimensions' score is illustrated in Fig. 2.

# 4 Results and Discussions

## *4.1 Findings*

The dominance of Kyoto University is obviously noticeable. They scored the highest in the first four dimensions while ranking second in the last dimension—environmental. Tokyo University, often considered as the best university in Japan, although owned high scores in all dimensions, fell behind Kyoto University with four dimensions staying at second rank while ranking first in environmental dimensions. Osaka University, which is the third best Japanese university, surprisingly scored lower than a private university—Waseda. Despite being a private university, Waseda scored the third highest in the first four dimensions, losing the third highest environmental score to Hokkaido university. In general, universities with higher ranking (based on QS Ranking) tend to score higher than the rest, except for the astonishing performance of Waseda. Another worth mentioned result is the dominance of economic dimension.

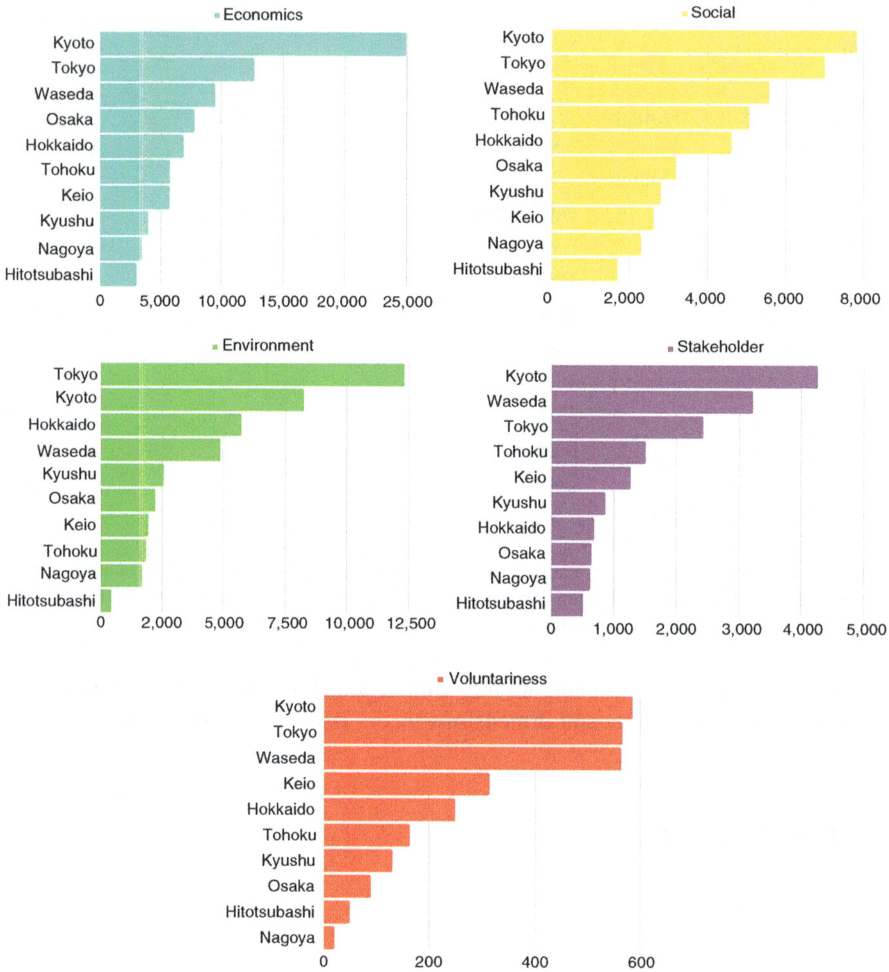


**Fig. 2** Scores by universities in each dimension. Source: Author's elaboration

## 4.2 *Dimension Ratio*

The dominance of economic dimension is clearly shown in Fig. 3. Nearly half of retrieved articles contain keywords related to economic, which proves that social and economic responsibilities are aligned together. Besides, universities are focusing on economic field of social responsibility that means USR's projects and activities at Japanese universities have been aiming at economic matters of society. Universities, with the special function that could impact both the input, future generation (the future labor force), and the output, public and private sectors, have the ability to connect these parties easily, thanks to its knowledge delivery function. For example, the Center for Global Initiatives of Osaka University currently conducts lots of USR

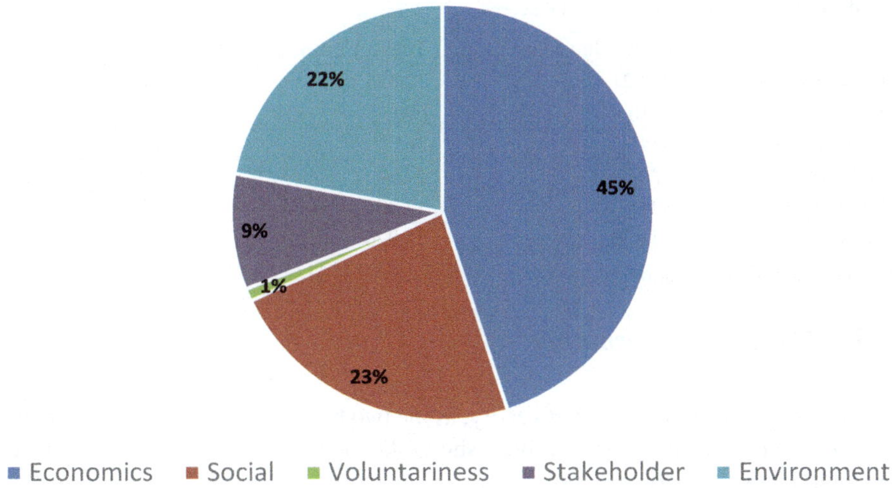


**Fig. 3** Total scores by dimension. Source: Author's elaboration

activities and projects. They are providing Sustainable Development Goals (SDGs) course as requirement credit for all their educational programs while organizing the relationship with Japan International Cooperation Agency (JICA) to create internship opportunities for students and provide Japanese SMEs with CSR knowledge. All these activities both directly and indirectly impact the socioeconomic rather than pure economic itself, so it reflected the high score of economic dimensions compared to the rests. Another reason could be the impact of Japanese government. Japan is going to host two global events—Olympic Tokyo 2021 and Osaka Expo 2025. In line with the direction of the United Nations, these events would be hosted in light of globalization and aimed to raise the awareness about global issues. SDGs are an effective tool to educate people; the UN in general and the Japanese government in particular are promoting the widespread propaganda of SDG to its citizen. Besides, most of the SDGs directly relate to economic, social, and environmental dimension; therefore, the dominance of these three dimensions is also illustrated in Fig. 3 with 45%, 23%, and 22%, respectively. Society and environment are currently considered as global issues that most developed countries have to deal with, in which the equal high score of social and environment indexes had emphasized the importance of the two and simultaneously shown the awareness of Japanese universities toward these dimensions.

**Table 4** Top 3 performances by total scores

| | Kyoto University | Tokyo University | Waseda University |
|---|---|---|---|
| Economic | 24,921 | 12,635 | 9455 |
| Social | 7816 | 6998 | 5565 |
| Voluntariness | 584 | 565 | 563 |
| Stakeholder | 4263 | 2424 | 3216 |
| Environment | 8246 | 12,304 | 4877 |
| Total | 45,830 | 34,926 | 23,676 |

Source: Author's elaboration

## 4.3 The Top Performance

Kyoto University and Tokyo University, as the two best universities in Japan, are the leaders in terms of total scores. Table 4 shows the total results of the top performance by total dimension scores.

In spite of being the second best university in Japan by QS Ranking, Kyoto University outperformed Tokyo University (top 1 by ranking) and the rest with better scores in most of the dimensions, especially in economic (2 times higher) and stakeholder (1.6 times higher). However, in general, the dimension scores of Tokyo University in terms of both dimension and total scores are by far higher than all of the rest. Tokyo ranked second in the first four dimensions and rank first in environmental. It is obvious that the top national universities in Japan have long traditional history and rich research resources that led to the good performance. Osaka University, which also ranked as one of the top national universities, although having low total dimension scores (13,832), has performed well in the main three dimensions: economic, social, and environmental (7711, 3178, and 2215, respectively).

Waseda University stood out as a dark horse in the total results. They ranked third in the first four dimensions while also receiving good scores in environmental. Waseda is often considered as the top private university in Japan, which has priority and reputation in politics and business education, which could be the explanation for the good results. Moreover, one of the most famous professors in CSR's fields (Professor Kanji Tanimoto) is also working simultaneously as a Professor at School of Commerce and as the Director of Waseda's Research Institute for Corporate Social Responsibility. Another noticeable thing is Waseda is the only one in the top 10 which owns a specified institute for CSR.

## 4.4 Level of Commitments

In general, the level of commitments of national university is worth mentioning, especially higher-ranking universities. This phenomenon reflects the vision of Japanese Ministry of Education, Culture, Sports, Science, and Technology (MEXT). At the moment, the orientation of MEXT toward higher education institutions is to aim at global education—or in other words to attract more international students and participate more in international academic activities. This vision is also transferred to each university as well through its vision, mission, or motto, such as Osaka University (Live locally, Grow globally), Tokyo University ("The University of Tokyo aims to be a world-class platform for research and education, contributing to human knowledge in partnership with other leading global universities"), and Kyoto University ("As an international institution, Kyoto University will promote foreign academic exchange and thereby strive to contribute to the well-being of the world").

The only two private universities (Waseda and Keio) also share good level of commitments toward USR. This could be explained by the enrollment and admission strategy of these universities. Because they have to financially support themselves, it is necessary to strive to link themselves with the community and social issue, in order to attract more potential students. At the moment, the more social knowledge the students have, the more effort these universities have to make.

However, there are still some universities showing less commitment than the rest such as Kyushu University, Nagoya University, and Hitotsubashi University. The situation of Kyushu and Nagoya is special. They are considered as the top university in their region; hence, there is barely no competition with others.

## 4.5 Orientation and Trend

The orientation of USR in top Japanese university is clear. Voluntariness and stakeholder account for very low proportion (1% and 9%, respectively). As mentioned before, this situation could be referred to the evolution of CSR. Nowadays, CSR has evolved and transformed into strategic CSR, where CSR has been integrated to the operation of the corporate and organization. Figure 4 illustrated the core characteristics of CSR during each evolution period.

Voluntariness dimension belongs to philanthropy, where voluntary was the core characteristics for all CSR's activities during the 1950s. After that, CSR had evolved

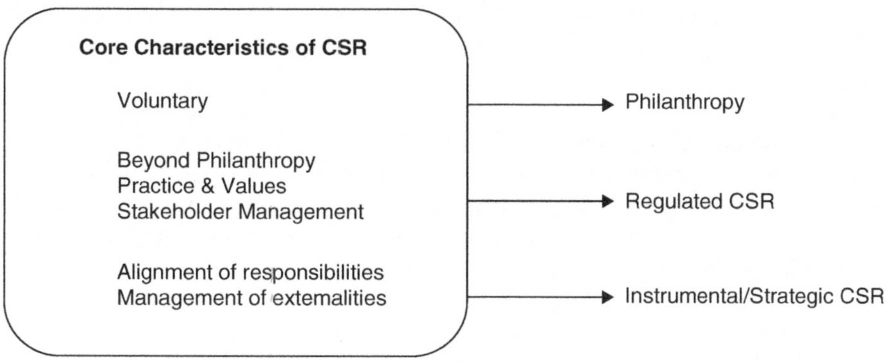


**Fig. 4** The core characteristics of CSR during each era. Source: Hamidu et al. (2015)

from barely philanthropic function to be stated in regulations and laws that corporate had to follow. In this stage, stakeholder management took part as a core function of CSR. However, these eras had passed. Current situation reflects the new era of CSR where companies and organizations strive to solve social problems while still maintaining economic indicators. The externalities (stakeholders that directly impact the operation of a corporate or an organization) are also concerned, to some extents. That is explained by the low accountability of voluntariness and stakeholder dimension.

## 5 Strengths and Weaknesses

**Strengths**

- In general, universities in Japan focus on social responsibilities in economic while maintaining their awareness toward social and environment problem. Variously, leading Japanese universities have still taken USR seriously and announced this in their social communication and reports in approaches.
- Being the third largest economy in the world at the moment, the results of Japanese universities in economic dimension are impressive. Although in recent years, due to several consequences, the economic growth of Japan economy has stalled, the government is trying to take back the previous position of Japanese economic performance. The dimension scores illustrated that the orientation and plan of the government is in line with research activities, social responsibilities, and strategy of the leading universities.
- Japan is assigned as the host for a few global events in the near future. These events provide Japan with lots of opportunities to attract global attention. By integrating and promoting SDGs, it is a good effort from both Japanese government and leading universities to educate people through these social activities.
- The top 3 universities in Japan, which are also considered as top universities in the world, shows good level of commitment, hence fulfilling their visions and missions of global attractiveness and international education. The results of these universities support their positions as the leaders of academic and education while also promoting and consolidating the image of Japanese education worldwide.

**Weaknesses**

- In the imbalance of dimensions' distribution, although currently it is the new era of social responsibility, it is necessary to mention the presence of voluntariness and stakeholder (especially voluntariness, as it accounted for only 1% of the scores). Japanese universities, at the moment, are bypassing the presence of these two dimensions.
- There is no diversity among dimensions and research interests of universities. Universities that own schools or graduate schools that are closely related to a specific dimension tend to focus only on their specialized fields. For example, universities such as Hokkaido, Kyushu, and Kyoto, which have graduate school

of agriculture and aquaculture, performed well in terms of environment dimension. Hokkaido University even ranked third in this dimension, outperforming Osaka University.

- Several national universities earning the title "best universities in their regions" might experience poor competitive enforcement motivation, reflecting in their low scores. This can be proven by the overview of higher education in Japan where the gap between the top 3 and the rest exists.
- Few universities, although considered to be strong at business-related fields, showed less commitment. That could infer that they are not considering USR as a strategic plan in their management.

## 6 Conclusions

This study shows that Japanese universities have engaged properly in social responsibility and sustainable development. Moreover, these activities, projects, research, and implementations have also been announced and reported in their website contents. By replacing the term corporate to university, this study also proves that the application and integration of social responsibility into universities is rational and feasible. Moreover, the findings of this study infer that top 10 Japanese universities are almost on the right track toward their social responsibility fulfillment, in which they are covering three important dimensions that are in line with the ultimate Sustainable Development Goals proposed by the United Nations in general and the orientation of the Japanese government.

There are few contributions that this research has brought. In terms of USR, this is one of the first studies that takes Asian context into account in general and could be considered as the groundbreaking for USR's research in Japanese higher education context. This work also contributes to the growing scholarly literature on the new issue of social responsibility in government entities (Dentchev et al., 2018; Dumay et al., 2010; Kappo-Abidemi & Ogujiuba, 2020).

In terms of limitations, firstly, the scope and scale of this study only covered the top 10 universities in Japan, mainly public universities. The reason the author mainly chose public Japanese universities is because they have social obligations beyond legal and economic ones (Atakan & Eker, 2007; Vasilescu et al., 2010), while private organizations typically aim to achieve economic benefits (Vázquez & Lanero, 2016). However, Japan is one of the most modernized educational systems with thousands of universities, from national to public and private, each owning different characteristics. Secondly, research methodology is also considered as lacking control object. It is derived from the qualitative methodology applied. There could be differences related to website designs or errors in data collection process that led to the inaccuracy of retrieved results. As a result, while the content analysis research method was able to determine whether universities incorporated actions regarding social responsibility issues into their websites and which social responsibility dimensions received more resources, the approach adopted had the limitation of just portraying a count of the key symbols sensitive to the issue under investigation, rather than displaying the real dimensions of what was accomplished. Several universities could have been conducting social responsibilities, but they do not announce or report the information in their websites, which could lead to the lack of information that this study's methodology cannot access.

While there are still several limitations as a starting point in research of this kind, future follow-up studies can take this into consideration and enhance them. More study is needed to better identify the degree of social responsibility integration within public sector organizations' medium-long-term planning processes, particularly in the university context, which expand the understanding of the current state of USR.

## References

Atakan, M. G., & Eker, T. (2007). Corporate identity of a socially responsible university—A case from the Turkish higher education sector. *Journal of Business Ethics, 76*, 55–68.

Baker, M. (2008). *Can you have social responsibility without ethics?* Mallen Baker—Change is a learnable process. Retrieved April 3, 2021, from http://mallenbaker.net/article/clear-reflection/can-you-have-social-responsibility-without-ethics

Baker, M. (2014). *Should CSR be made compulsory after all?* Mallen Baker—Change is a learnable process. Retrieved April 3, 2022, from http://mallenbaker.net/article/clear-reflection/should-csr-be-made-compulsory-after-all

Barrena Martínez, J., Llinàs Audet, X., & Ramos Monge, E. (2019). Drivers and barriers of University social responsibility: Integration into strategic plans. *World Review of Entrepreneurship, Management and Sustainable Development, 15*(1/2), 174.

Bokhari, A. (2017). Universities social responsibility (USR) and sustainable development: A conceptual framework. *International Journal of Economics and Management Studies, 4*(12), 8–16.

Bowen, H. R. (1953). *Social responsibilities of the businessman*. University of Iowa Press.

Bryman. (2004). *Social research methods* (2nd ed., p. 592). Oxford University Press.

Carol, M. (2014). *Corporate social responsibility and creating shared value: What's the difference?* Heifer International.

Dahlsrud, A. (2008). How corporate social responsibility is defined: An analysis of 37 definitions. *Corporate Social Responsibility and Environmental Management, 15*(1), 1–13. https://onlinelibrary.wiley.com/doi/epdf/10.1002/csr.132

Dentchev, N. A., Eiselein, P., & Kayaert, T. (2018). Social responsibility within Brussels municipalities: An exploratory study. In W. James & D. M. Wasieleski (Eds.), *Corporate social responsibility (business and society 360)* (Vol. 2, pp. 147–174). Emerald Publishing.

Dumay, J., Guthrie, J., & Farneti, F. (2010). GRI sustainability reporting guidelines for public and third sector organizations: A critical review. *Public Management Review, 12*, 531–548.

Giuffre, L., & Ratto, S. E. (2014). A new paradigm in higher education: University social responsibility (USR). *Journal of Education & Human Development, 3*(1), 231–238.

Guthrie, J., Petty, R., Yongvanich, K., & Ricceri, F. (2004). Using content analysis as a research method to inquire into intellectual capital reporting. *Journal of Intellectual Capital, 5*, 282–293.

Hamidu, A. A., Md Haron, H., & Amran, A. (2015). Corporate social responsibility: A review on definitions, Core characteristics and theoretical perspectives. *Mediterranean Journal of Social Sciences, 6*(4).

Hennink, M., Hutter, I., & Bailey, A. (2011). *Qualitative research methods*. Sage Publications.

Huang, F. (2017). Public versus private: university leadership in Japan. *Times Higher Education (THE)*. Retrieved April, 5 2022, from https://www.timeshighereducation.com/blog/public-versus-private-university-leadership-japan

Husted, B. W., & Allen, D. B. (2006). Corporate social responsibility in the multinational enterprise: Strategic and institutional approaches. *Journal of International Business Studies, 37*(6), 838–849.

Jabbour, C. J. C. (2010). Greening of business school: A systemic view, international journal of sustainability. *Higher Education, 11*(1), 49–60.

Kappo-Abidemi, C., & Kanayo Kingsley, O. (2020). Higher education institutions and corporate social responsibility: Triple bottomline as a conceptual framework for community development. *Entrepreneurship and Sustainability Issues, 8*, 1103–1119.

Krippendorff, K. (1980). *Content analysis: An introduction to its methodology*. Sage.

Low, M., Grant, S., & Li, Y. (2015). Voluntary reporting of intellectual capital: Comparing the quality of disclosures from New Zealand, Australian and United Kingdom universities. *Journal of Intellectual Capital, 16*, 779–808.

Martin, B. (2015). *University social responsibility: A common European reference framework final public report of the EU—USR project, 52709—LLP—2012—1—RO—ERASMUS—ESIN february 2015 co—funded by the EU lifelong learning Programme*. Retrieved April 4, 2021, from http://www.eu-usr.eu/wp-content/uploads/2015/04/D1.4-Final-Report-Public-Part-EN.pdf.

Maruyama, F. (2008). An overview of the higher education system in Japan. *The Journal of Finance and Management in Colleges and Universities*, 1–12.

Milner, M. J. (1999). Exploring the reliability of social and environmental disclosures content analysis. *Accounting, Auditing & Accountability Journal, 12*(2), 237–256.

Mooney, S. J., Westreich, D. J., & El-Sayed, A. M. (2015). Epidemiology in the era of big data. *Epidemiology, 26*(3), 390.

Nejati, M., Salamzadeh, Y., & Sharafi Farzad, F. (2010). Environmental sustainability: A study of world leading universities. *Proceedings of International Conference on Environmental Research and Technology* (ICERT 2010). Malaysia, pp. 175–178.

Nejati, M., Shafaei, A., Salamzadeh, Y., & Daraei, M. (2011). Corporate social responsibility and universities: A study of top 10 world universities' websites. *African Journal of Business and Management, 5*(2), 440–447.

Pinney, C. (2001). *Imagine speaks out*. How to manage corporate social responsibility and reputation in a global marketplace: The challenge for Canadian business. http://www.imagine.ca/content/media/teamcanada.china.paper.asp?section=media

Porter, M. E., & Kramer, M. R. (2011). Creating shared value. *Harvard Business Review*. https://hbr.org/2011/01/the-big-idea-creating-shared-value

QS Ranking. (2020). *QS World University Ranking 2020*. https://www.topuniversities.com/university-rankings/world-university-rankings/2020

Reiser, J. (2007). Managing university social responsibility (USR). In *International sustainable campus network: Best practices-future challenges*.

Samkin, G., & Schneider, A. (2008). Adding scientific rigor to qualitative data analysis: An illustrative example. *Qualitative Research in Accounting & Management, 5*, 207–238.

Shek, D. T., Yuen-Tsang, A. W., & Ng, E. C. (2017). USR network: A platform to promote university social responsibility. In *University social responsibility and quality of life* (pp. 11–21). Springer.

United Nations. (1987). Report of the World Commission on Environment and Development: Resolution / adopted by the General Assembly. https://digitallibrary.un.org/record/153026?ln=en

Vallaeys, F (2014). University social responsibility: A mature and responsible definition. *GUNI Report 2014, Higher Education in The World*.

Van Marrewijk, M. (2003). Concepts and definitions of CSR and corporate sustainability: Between agency and communion. *Journal of Business Ethics, 44*, 95–105.

Vasilescu, R., Barna, C., Epure, M., & Baicu, C. (2010). Developing university social responsibility: A model for the challenges of the new civil society. *Procedia—Social and Behavioral Sciences, 2*(2), 4177–4182.

Vázquez, J. L., & Lanero, A. (2016). Citizens' perceptions on social responsibility in public administration organizations: A case study on Spain. *Transylvanian Review of Administrative Sciences, 12*, 166–183.

Wright, T. (2010). University presidents' conceptualization of sustainability in higher education. *International Journal of Sustainability in Higher Education, 11*(1), 61–73.

Zhao, B. (2017). Web scraping. In L. Schintler & C. McNeely (Eds.), *Encyclopedia of big data*. Springer.

**Long Tam Pham** is a Doctoral Student at Graduate School of Economics, Faculty of Management Science Research, Osaka University, Japan. He also earned his MBA in Business and Management at the same institution. Prior to coming to Japan, he earned the Japanese Government Scholarship (MEXT) for the duration of 2018—2024. Long has been working as Teaching Assistant for Centre for Global Initiatives, Osaka University since 2019.

His research interests lies in Social Responsibility and Sustainability. He is the author of multiples academic articles mainly focused on Education for Sustainable Development (ESD) and Business-Society connection. Currently he is working on the facilitation of ESD—Towards the achievement of Sustainable Development Goals 2030 at the Institute for Global Environment Strategies (IGES) as Visiting Researcher Intern.

# Institutionalization of CSR Practices in Nutrition

**Carmona-Garces Isabel Cristina and Londono-Correa Diana**

## 1 Introduction

In Colombia and worldwide, overweight, obesity, and chronic noncommunicable diseases (NCDs) are a public health concern not only because of their consequences for health and life but also because of the economic cost that these problems represent for the health systems of states. The NCDs Risk Factor Collaboration concluded that if the trend that began in 2000 continues, the global goal of good health and well-being will not be met (NCD, 2016). Colombia is no stranger to this problem; according to the latest National Survey on the Nutritional Situation (ICBF, 2019), 56% of the Colombian population is overweight.

This problem has become an issue of interest for the food producing, processing, and marketing industry. The food industry is under increasing pressure from its stakeholders who require of them to be jointly responsible for the prevention and control of this public health problem. The food industry actions have begun to be institutionalized as Corporate Social Responsibility Practices in Nutrition (CSRPN). This practices include the production and processing of healthier foods, the provision of accurate and appropriate information to consumers through nutrition labeling and advertising, and the promotion of healthy workplaces, among others.

C.-G. I. Cristina (✉)
Escuela de Nutrición y Dietética, Universidad de Antioquia, Medellín, Colombia

Grupo de Investigación Interdisciplinaria en Educación para la Salud y Educación Nutricional GIIESEN, Universidad de Antioquia, Medellín, Colombia

Universidad EAFIT, Medellín, Colombia
e-mail: iccarmonag@eafit.edu.co

L.-C. Diana
Universidad EAFIT, Medellín, Colombia
e-mail: dlondono@eafit.edu.co



I. G. Arraiano et al. (eds.), *Corporate Social Responsibility in a Dynamic Global Environment*, CSR, Sustainability, Ethics & Governance,
https://doi.org/10.1007/978-3-031-24647-0_11

To date, CSRPN have not been approached from the perspective of organization studies, and for this reason, the interest of this paper is to embark on a path in this direction. It has been identified that a possible lens for this analysis could be the New Sociological Institutionalism (NIS), which allows us to approach practices of this type, which are the product of a social construction. For this reason, the interest of this article is to explore how CSRPN could be approached from the NIS theory, to undertake empirical research paths in this regard. To achieve the above, this article addresses the CSRPN as an object of study and considers the NIS as a theoretical lens, based on which assumptions are made about the problem and conclusions are drawn in this regard.

## 2 Corporate Social Responsibility Practices in Nutrition as an Object of Analysis

CSRPN have been institutionalized over time as a strategy to achieve the sustainable development goals (SDGs) and as part of the food industry's CSR.

From 2000 onward, important actors began to emerge with claims to mobilize action around CSR, one of the main ones being *the Global Compact* (UN, 1999), which invites corporations to develop, implement, and disseminate CSR policies and practices. In it, the UN calls for the incorporation of ten principles in business activities around the world, including respecting fundamental human rights and not violating them, favoring the environment, and promoting environmental responsibility. These principles are closely related to the fundamental right to life, health, and adequate food, in which the food production and marketing industry play a key role on several fronts.

The World Summit on Sustainable Development (WSSD) held in Johannesburg also confirmed the need for the business world to be part of the solution to the sustainable development challenges facing the planet (UN, 2002); among them, it is worth mentioning four sustainable development goals related to the food industry: the end of poverty, zero hunger, health and wellness, and responsible production and consumption, just to mention the most relevant and which will be related to the problem to be addressed in this text.

Traditionally, the efforts of food producing and marketing corporations have been focused on the second sustainable development goal "to eliminate hunger" (UN, 2002), this is why it is common for them to contribute to the problem through philanthropic programs; however, due to the excessive growth of the overweight pandemic, obesity, and NCDs there is now a clear pressure from health organizations, academia, and consumers for them to contribute to other objectives such as reducing hunger" (UN, 2002, obesity, NCDs as well as achieving "health and wellness" and "responsible production and consumption" (UN, 2002). This work addresses specifically responsible production, commercialization, marketing, and consumption of processed and ultra-processed foods.

Processed and ultra-processed foods have been linked to the problem of overweight, obesity, and NCDs by the World Health Organization (WHO) and the Pan American Health Organization (PAHO) (PAHO & WHO, 2015), as these products are usually high in calories and low in nutritional value; high in fat, salt, sugar, and trans fats; and low in fiber, protein, vitamins, minerals, and other bioactive compounds. In addition, research shows that as the consumption of this type of food increases, the prevalence of overweight and obesity increases worldwide (PAHO & WHO, 2015).

To address NCDs and the global pandemic of overweight and obesity, different strategies have been proposed from a food systems perspective that brings together the value chain in the food and beverage sector, the environment, and consumer behavior (HLPE, 2017), so that all elements of the system contribute to the prevention and control of this problem. Among them, the WHO has proposed global plans and recommendations, such as the *Global Strategy on Diet, Physical Activity and Health* (WHO, 2004); the Global Action Plan for the Prevention and Control of NCDs 2013–2020 (WHO, 2013); and the recommendations of the Expert Commission on Ending Childhood Obesity (WHO, 2016). All recommendations propuse cross-sectoral actions, including actors in the food and beverage value chain (producers, processors, marketers, among others). The recommended actions focus on promoting healthy environments; producing, processing, and marketing healthy food; and providing adequate information about healthy food to consumers.

Colombia as a member of the WHO has begun to implement some of the recommended strategies, including the enactment in 2009 of Law 1355 of 2009 (Congress of the Republic of Colombia, 2009), which established obesity and NCDs as a public health priority and in its article XVIII urges food producing, importing, and marketing companies to establish CSR programs. These programs should include actions that contribute to the prevention and control of overweight, obesity, and NCDs.

In addition to the recommendations of health agencies and country legal guidelines, there are other initiatives that recommend and evaluate PRSEN. For example, in 1999 the Global Reporting Initiative (GRI) was created, a nonprofit organization whose orientation is the generation of sustainability reports (Raufflet et al., 2012) that include performance in the elements of the TBL (Raufflet et al., 2012). Many companies have adopted this report which includes health, nutrition, and welfare criteria for food producing companies within its reporting model. In the same year, the Dow Jones Sustainability Index (DJSI) was created. This index measures the development, construction, implementation, and management of resources to quantify the companies´ and contribution of the food production and marketing industry to the prevention and control of overweight, obesity, and NCDs.

Both reports, GRI and DJSI, include the evaluation of aspects such as having a defined nutritional policy, demonstrating that this policy exists and is implemented through different strategies aimed at promoting the health and adequate nutrition of its stakeholders. They also evaluate the plans to ensure that its product portfolio is increasingly healthier based on a program of innovation and reformulation for health and nutrition reasons. In adition, these reports consider if the food company is providing adequate information to consumers about their products (voluntary labeling beyond mandatory labeling). These reports also assess the commitment and evidence of responsible marketing (self-regulation of advertising aimed at children, responsible advertising, among others); and check the actions that allow the company to be a healthy workplace. These are the CSRPN referred to in this article.

## 3 Theoretical Possibilities for the Study of CSRPN from an Organizational Studies Perspective

Systematic reviews have identified different authors that have focused their efforts on analyzing the approaches and theories from which CSR has been studied.

For example Garriga and Melé (2004) classified the theories for analyzing CSR into four groups: (1) Instrumental theories, which understand CSR as another way of achieving the company's economic results; (2) political theories, which address the power of companies over society and the responsible use of this power; (3) integrative theories, which focus on the satisfaction of social demands; and (4) ethical theories, which focus on the responsibilities of companies toward the society in which they operate. The authors concluded that it was necessary to develop a theory that integrates these four perspectives.

In other review focusing on Political CSR, Frynas and Stephens (2015) found that the theories that dominated the field were stakeholder theory and institutional theory. Later Frynas and Yamahaki (2016), conducted a systematic review with the aim of identifying the theories to which CSR has been related, and identified some that respond to internal drivers such as resource-based view and agency theory, and others that respond to external drivers such as Stakeholder, legitimacy, resource dependency theory and institutional theory from its three approaches: economic, sociological and political. These authors recommended the development of multi-theoretical empirical studies, in order to obtain new conceptual developments and transcend to studies that integrate the different levels of analysis. This recommendation had already been made by Aguinis and Glavas (2012) who suggested multilevel studies in order to understand underlying CSR linking mechanisms.

On the other hand, Gond (2012) proposed four approaches to the study of CSR anchored in the paradigms of Burrell and Morgan (2001). (1) The functionalist approach sees CSR as a mechanism of social regulation, in which the company is privileged as the object of study and the interest is in demonstrating a positive link between CSR and financial performance, and CSR as a competitive advantage; thus, authors such as Porter and Kramer (Porter & Kramer, 2002) consider CSR as a tool for integrating business and society. (2) The sociological approach conceives CSR as a power relation; this approach focuses on power and the logic of conflict between business and society. According to the author, this is an objectivist approach that goes beyond the positioning of CSR; contributions to this approach are highlighted in Klein (2009). (3) The culturalist approach sees CSR as a cultural product in which representations, norms, and values are disseminated and exchanged. CSR is therefore studied as a narrative (Vigneau et al., 2015), as a generator of meaning (Basu & Palazzo, 2008), and consider the relationship with macro-level stakeholders, the state, among others. This perspective makes it possible to understand differences between companies and countries. (4) Constructivist approach understands CSR as a socio-cognitive construction of the sphere of business and society through a complex and reciprocal framing of identities, value systems, and social issues. This perspective allows to understand how discourses, practices, and policies are being formed over time.

The instrumental or functionalist approach has responded to administrative purposes and is therefore the most widely used for the study of CSR; however, the remaining approaches proposed by Garriga and Melé (2004) and Gond (2012) with some differences, allow to cover relevant aspects for organizational studies. These views recognize CSR beyond administrative practice, while providing a broader view of the phenomenon of interest.

By understanding CSR as a socially constructed practice (Prieto, 2017), it becomes a phenomenon of interest for organisational studies, since, paraphrasing Prieto, this practice implies continuous processes of social construction in the organization Understanding CSRPN as a socially constructed practice and as an organizational phenomenon, the NIS perspective could be consider a promising lens for its analysis.

## 4 The New Sociological Institutionalism to Address CSRPN

As clarified in the previous section, the NIS has been used as one of the theoretical perspectives to address CSR, and consequently it is seen as a possibility to address CSRPN. There are several benefits for addressing CSRPN from the NIS perspective. First, the NIS directs its attention to the institutional environments. Second, because it is a perspective interested more in legitimacy than in efficiency, which is largely related to the aims of CSRPN. And third, because the NIS approaches explain in different ways "how environmental pressures shape organizations" through environmental pressures and by institutional isomorphism (Meyer & Rowan, 1977). The NIS allows for the possibility to address CSRPN through a variety of approaches (Rowan, 1977), and furthermore it allows explaining that "organisations in the same environment will become structurally similar to the extent that they respond to similar pressures; that is, they will exhibit isomorphism" (DiMaggio & Powell, 1991).

From this perspective, this article proposes to review the following assumptions:

S1 *There is an institutional environment that is causing CSRPN to become institutionalized as a product of institutional isomorphism.*

S2 *This institutionalized environment is putting pressure on food producing and marketing organizations, and consequently, they adopt the CSRPN as a "rationalized myth."*

**S1: There Is an Institutional Environment That Is Causing CSRNP to Become Institutionalized as a Product of Institutional Isomorphism**

Institutional environments are those characterized by rules and requirements to which individual organizations must conform if they are to receive support and legitimacy. The requirements may originate from nation-state sanctioned regulatory agencies, from professional or trade associations, from generalized belief systems that determine how specific types of organizations should behave, and from other similar sources (Meyer & Rowan, 1977; Powell & DiMaggio, 2012); whatever the source, organizations are favored when they conform to the rules or beliefs (Scott & Meyer, 1994).

It is therefore important to argue whether an institutionalized environment is taking shape, which is consequently leading to the incorporation of CSRPN. An institutionalized environment is characterized by the configuration of institutionalized rules, which may be taken for granted or may be supported by public opinion or the force of law. Institutions inevitably entail normative obligations, but they often enter social life mainly as facts that the actors must consider; therefore, it is important to explain some facts that today's food producing and trading organizations must consider. In some cases institutions already entail legal, normative obligations or are simply supported by public opinion. Based on the above, it is seen that an institutional environment around CSRPN is taking shape, for example:

- *Setting rules supported by law:* WHO member countries have begun to implement various regulatory strategies based on WHO recommendations to address the obesity and NCDs pandemic, including the following:
  - Tax regulations: An example of such regulations is the consumption tax on sugary drinks implemented in Mexico in 2014, which has had a positive effect in terms of an average 12% reduction in purchases of sugary drinks and a 4% increase in purchases of non-taxed drinks (National Institute of Public Health, 2015). Such taxes have already been approved in more than 30 countries. In Colombia this proposal was made during 2016 to be implemented within the tax reform; however, due to lobbying and action by sugar and soft drink companies, it was not even brought to discussion for approval (Díaz-García et al., 2020).
  - Regulation of food production: These aim to encourage companies to innovate and reformulate existing products in order to reduce saturated fat, sodium, and sugar and eliminate trans fats. In Colombia, various bills and resolutions are being proposed in this area; the sodium reduction bill is in the process of being approved, as is the trans fat elimination bill.
  - Regulation of nutrition labeling and the marketing and advertising of unhealthy foods and beverages: Several bills have been discussed in Colombia in order for the country to adopt a nutrient profiling model for processed and ultra-processed food products. In addition, control the advertising of foods and beverages aimed specifically at children and adolescents, establish mandatory front-of-pack nutrition labeling. Many of these bills have been modified during the discussion process or shelved, in several cases due to lobbying

paid for by food and beverage companies (Díaz-García et al., 2020; Sandoval, 2018).

- *Setting up rules based on standards:* Although they are not mandatory, pressures derived from the need to legitimize companies and their responsible actions are generating more and more types of reports and accreditations, such as those mentioned in the theoretical section (GRI and DJSI) that include sections or indicators on CSRPN.
- *Public opinion-supported rule-setting*: Societal awareness of the overweight and obesity pandemic is increasing. It can be observed in the pressures from various interest groups, including the following:
  - Pressure from consumers: Who are increasingly affected by the problem of overweight and obesity and who also have more information available on health and nutrition and are more interested or aware of the issue.
  - Market pressure: The market for healthy foods continues to grow (The Nielsen Company, 2018), and the food industry is aware of this opportunity and is trying to respond to this demand by launching new "healthy" products that help to improve its RSEN indicators in the process.
  - Pressure from civil organizations: For the example of Colombia, two civil organizations—Educar Consumidores and Red Papáz—are focusing their efforts on the defense of the right to life, health, adequate food, and transparent consumer information, especially for children.

After reviewing the environmental pressures that exist today in the Colombian environment, it could be concluded that there is evidence of an institutionalized environment around the CSRPN that favors the interest of their ceremonial incorporation by food producing and marketing organizations to legitimize themselves.

Consequently, the institutionalized environment generates institutional isomorphisms which according to Meyer and Rowan (1977) have some decisive consequences for organizations: (a) they incorporate elements that are externally legitimized, rather than in terms of efficiency; (b) they use external or ceremonial evaluation criteria to define the value of structural elements; and (c) dependence on externally fixed institutions reduces turbulence and maintains stability and legitimacy.

**S2: This Institutionalized Environment Is Putting Pressure on Food Producing and Marketing Organizations and Consequently They Adopt the CSRPN as a "Rationalized Myth"**

The notion of rationalized myth is explained in new institutionalism by Meyer and Rowan (1977):

> In institutionalised environments organisations incorporate practices and procedures defined by the prevailing rationalised concepts of organisational and institutionalised work in society, to enhance their legitimacy and prospects for survival ... regardless of the immediate effectiveness of the acquired practices and procedures. Institutionalised products, services, techniques, policies and programmes function as powerful myths that organisations adopt ceremonially (p. 79).

Organizations' adoption of these myths can generate, among other things, conflict with efficiency criteria, so in order to maintain ceremonial conformity, organizations tend to protect their formal structures from the uncertainties of technical activities by loosely integrating and differentiating between their formal structures and actual work activities, with the result that organizations strongly reflect the myths of their institutional environments, even more than the demands of their work activities.

As CSR is a socially constructed practice that is incorporated for the purposes of legitimacy and sustainability in institutionalized environments, there is an excessive eagerness for its application as a management model, measured through sustainability and CSR indicators, communicated through reports, and publicized by the organizations themselves, but with little reflection, and even incongruent with the organization's real work activity. In this regard, Prieto (2017) argues that rationalized myths, being rationalized prescriptions, are not questioned because they create meaning and provide confidence, and in that same logic they continue to be reproduced.

Meyer and Rowan (1977) posit three theses regarding rationalized myths (1) that environments and environmental domains that have institutionalized a large number of rational myths generate more formal organizations; (2) that organizations that incorporate myths are more legitimate and successful and are more likely to survive; and (3) that organizational control efforts, particularly in highly institutionalized environments, are devoted to ritual conformity, both internally and externally. These three theses have implications for the investigation of an organizational phenomenon from this perspective. These could be validated in the empirical field, and could be applied to specific CSR practices such as nutrition.

The analysis of the CSRPN as a rationalized myth is then considered a possibility, since these are incorporated as corporate or business nutritional policies that are managed, communicated, and publicized through corporate websites, media, and reports, among others. These practices are a way to show that the company is acting responsibly and consequently gain legitimacy. The lens of NIS can enable the study of CSRPN not only because of its explanatory power of the phenomenon as such but also because through it the problematic generated by this rationalized myth, the consequences of its ill-considered incorporation, and its incongruency with the actual working activity of the organization, can be evidenced. According to Romero (1999, p. 23) "it can enable explicit theorising about the reciprocal influence between institutional constraints and political strategies and an understanding of the way ideas, interests and institutions interact." p. 23).

## 5 Conclusion

As the CSRPN are understood as ways of doing and saying that arise from competences, meanings, and materialities, their analysis as an organizational phenomenon from the NIS and the notion of rationalized myth become relevant. For a rationalized myth to be incorporated by the organizations, an institutionalized environment is required, and it is evident that in Colombia, regulatory, normative, and cultural rules are beginning to take shape, which must be taken into account by the organizations that produce and commercialize food.

The institutional environment generates environmental isomorphism, so that organizations of interest adopt PRSENs as rationalized myths to increase their legitimacy and prospects of survival. The adoption of CSRPN as rationalized myths is problematic, as it is a practice that is incorporated and managed in a quick and reflected manner, even with a lack of congruence between the company's goals and social goals. From the perspective of the NIS, there are ways to analyze the phenomenon of interest, the CSRPN; however, empirical studies are needed to unveil the rationalities of this myth and the logics and aims of its incorporation by food producing organizations.

It is therefore relevant and necessary to study how this myth is incorporated in organizations and its social consequences, its impact on public health, and the effects on the organization itself, for which other NIS concepts such as isomorphisms, logics, and institutional mechanisms can be very useful.

It is important to observe whether CSRPN go beyond the ceremony and isomorphic practices that are reported, communicated, and publicized, in other words, whether they go beyond the symbolic and really permeate the organization's aims to match a need and a public health issue that merits it. There is an interesting opportunity to analyze CSRPN from the perspective of the NIS to highlight ambiguous, complex practices and to unveil dynamics that underlie organizations and their structures.

## References


Aguinis, H., & Glavas, A. (2012). What we know and don't know about corporate social responsibility: A review and research agenda. *Journal of Management, 38*(4), 932–968.

Basu, K., & Palazzo, G. (2008). Corporate social responsibility: A process model of sensemaking. *Academy of Management Review, 33*(1), 122–136.

Burrell, G., & Morgan, G. (2001). *Sociological paradigms and organisational analysis: Elements of the sociology of corporate life*. Routledge.

Congress of the Republic of Colombia. (2009). *Law 1355 of 2009. whereby obesity and the chronic non-communicable diseases associated with it are defined as a public health priority and measures are adopted for their control, care and prevention. 14 October 2009.200947.502.*

Díaz-García, J., Valencia-Agudelo, G. D., Carmona-Garcés, I. C., & González-Zapata, L. I. (2020). Interest groups and consumption tax on sugar-sweetened beverages in Colombia. *Lecturas de Economía, 93*, 155–187.

DiMaggio, P., & Powell, W. (1991). Introduction: The new institutionalism in organizational analysis. In *The new institutionalism in organizational analysis* (pp. 1–38). University of Chicago Press.

Frynas, J. G., & Stephens, S. (2015). Political corporate social responsibility: Reviewing theories and setting new agendas. *International Journal of Management Reviews, 17*(4), 483–509.

Frynas, J. G., & Yamahaki, C. (2016). Corporate social responsibility: Review and roadmap of theoretical perspectives. *Business Ethics: A European Review, 25*(3), 258–285.

Garriga, E., & Melé, D. (2004). Corporate social responsibility theories: Mapping the territory. *Journal of Business Ethics, 53*(1), 51–71.

Gond, J. P. (2012). Four perspectives on CSR. In E. Raufflet, B. Félix, & C. G. de la Torre (Eds.), *Responsabilidad Social Empresarial* (pp. 29–41). *Pearson.*

HLPE. (2017). *Nutrition and food systems. A report by the high level panel of experts on food security and nutrition of the committee on world food security*. Rome.

ICBF. (2019). *National survey of the nutritional situation ENSIN 2015*. Bogotá.

National Institute of Public Health. (2015). *Reduction in consumption of taxed beverages after the implementation of the tax in Mexico*. Retrieved February 20, 2017, from http://www.insp.mx/epppo/blog/3666-reduccion-consumo-bebidas.html

Klein, N. (2009). *No logo: No space, no choice, no jobs* (Vol. 46). Picador.

Meyer, J., & Rowan, B. (1977). Institutionalized organizations: Formal structure as myth and ceremony. *American Journal of Sociology, 83*(2), 340–363.

NCD Risk Factor Collaboration, Di Cesare, M., Bentham, J., Stevens, G., Zhou, B., Danaei, G., et al. (2016). Trends in adult body-mass index in 200 countries from 1975 to 2014: A pooled analysis of 1698 population-based measurement studies with 19-2 million participants. *The Lancet, 387*(10026), 1377–1396. https://doi.org/10.1016/S0140-6736(16)30054-X

WHO. (2004). *Global strategy on diet, physical activity and health*. WHO.

WHO. (2013). *Global action plan for the prevention and control of noncommunicable diseases 2013–2020*.

WHO. (2016). *Report of the commission on ending childhood obesity*. Geneva.

UN. (1999). *Global compact*. Retrieved January 1, 2017, from http://pactoglobal-colombia.org/index.php/sobre-pacto-global/que-es-pacto-global

PAHO & WHO. (2015). *Ultraprocessed foods and beverages in Latin America: trends, effect on obesity and implications for public policy*. http://iris.paho.org/xmlui/bitstream/handle/123456789/7698/9789275318645_esp.pdf?sequence=5

PAHO, & WHO. (2015). *Ultra-processed foods are driving the obesity epidemic in Latin America*. Retrieved February 20, 2017, from http://www.paho.org/bol/index.php?option=com_content&view=article&id=1774:nota2sept

Porter, M. E., & Kramer, M. R. (2002). The competitive advantage of corporate philanthropy. *Harvard Business Review, 80*(12), 56–68.

Powell, W., & DiMaggio, P. (2012). *The new institutionalism in organizational analysis*. University of Chicago.

Prieto, M. (2017). Corporate social responsibility: An object of organisational analysis. In A. De la Rosa & M. Perez (Eds.), *Documentos de Posgrado, Estudios Organizacionales* (Vol. 1, pp. 32–50). Remineo.

Raufflet, E., Lozano, J. F., Barrera, E., & Garcia de la Torre, C. (2012). *Corporate social responsibility* (1st ed.). Pearson Educación.

Romero, J. J. (1999). Introductory Study: "The new institutionalisms: their differences, their proximities". In W. Powell & P. Di Maggio (Eds.), *El Nuevo Institucionalismo En El Análisis Organizacional* (pp. 33–78). Colegio Nacional de Ciencias Políticas y Administración Pública, AC, Universidad Autónoma Del Estado de México, Fondo de Cultura Económica.

Sandoval, M. (2018). *Taxing sugary drinks: lies and truths*. https://lasillavacia.com/silla-llena/red-social/historia/impuesto-las-bebidas-azucaradas-mentiras-y-verdades-69374

Scott, W. R., & Meyer, J. W. (1994). *Institutional environments and organizations: Structural complexity and individualism*. Sage.

The Nielsen Company. (2018). *Healthy segment continues to outperform in sales vs. indulgent segment*. https://www.nielsen.com/co/es/insights/news/2018/el-segmento-saludable-continua-con-un-mejor-desempeno-en-ventas-vs-el-segmento-indulgente.html

UN. (2002). *Report of the world summit on sustainable development (WSSD)*. UN.

Vigneau, L., Humphreys, M., & Moon, J. (2015). How do firms comply with international sustainability standards? Processes and consequences of adopting the global reporting initiative. *Journal of Business Ethics, 131*(2), 469–486.

**Carmona-Garces Isabel Cristina** is Nutritionist Dietitian; Mg. in Education; PhD (c) in Administration. Associate Professor, and member of Interdisciplinary Research Group in Health Education and Nutrition Education, Center for Research in Nutrition and Food, Escuela de Nutrición y Dietética, Universidad de Antioquia, Medellín-Colombia. Member of the Research Group in Administration and Organizations, Universidad EAFIT. Her research interests are related to the role of the food and beverage industry in food environments to achieve health and wellness, one of the global objectives. Among my topics of interest are the process of institutionalization of CSR practices in nutrition, nutrition labeling and food and beverage advertising.

**Londono-Correa Diana** is PhD Professor in Administration, Universidad EAFIT, Medellín, Colombia. Business Administrator, Master of Science and PhD in Administration, Universidad EAFIT, Medellín, Colombia. Member of the research group in Business History Universidad EAFIT, which deals with studies in the field of economics and business history. Her topics of interest are related to corporate and business strategy, corporate political action, frontier games and corporate synergies.

# Do the ESG Factors Affect Bank Insolvency? A Study Applied to Spanish Credit Cooperatives (Cajas Rurales) Between 2010 and 2015

**Antonio Madera del Pozo and Natalia Cassinello Plaza**

## 1 Introduction

Although more and more individuals (investors, authorities, media, and the general public) are worried about the need to promote inclusive and sustainable economic growth to respect future generations' rights, the analysis of environmental, social, and governance (ESG) risks in business management is not a new issue, as there is evidence that shows that in the Victorian era, those entrepreneurs who kept their workers happy, for example, by providing accommodation near the factories, observed higher productivity than those others who did not take into account these criteria in the day-to-day management (Plinke, 2016).

However, it is from the Rio Summit in 1992, which clarifies the concept of sustainable development, and the Paris Summit in 2015, in which they established some measures to reduce greenhouse gases emissions, when the concept of sustainable economy began to be relevant.

After that, we highlight the approval of the SDG principles (Sustainable Development Goals) and the establishment of the PRI (Principles for Responsible Investment), both by the United Nations, for their positive contribution to sustainable finance, in addition to the recent work developed within the European Commission launching the first taxonomy of sustainable finance in Europe.

A. M. del Pozo
Ethifinance Ratings, Paris, France

Universidad Europea de Madrid, Madrid, Spain

Universidad Pontificia Comillas - ICADE, Comillas, Spain
e-mail: amdelpozo@comillas.edu

N. C. Plaza (✉)
Universidad Pontificia Comillas - ICADE, Comillas, Spain
e-mail: ncassinello@comillas.edu

 
I. G. Arraiano et al. (eds.), *Corporate Social Responsibility in a Dynamic Global Environment*, CSR, Sustainability, Ethics & Governance,
https://doi.org/10.1007/978-3-031-24647-0_12

At the same time, and at the academic level, the volume of research related to sustainable finance and ESG criteria has been growing in recent years, highlighting works carried out by Orlitzky et al. (2003), Margolis and Walsh (2003), Margolis et al. (2009), Goss and Roberts (2011), and Stellner et al. (2015), where an in-depth literature review is carried out. They confirmed the relationship between ESG and the financial sustainability of companies (the greater the awareness, the lower the risk).

However, most of the academic research has been focused on corporates, obviating the financial sector, and this despite its importance as credit channeler and the necessity to control its risks because of its systemic status.

In this sense, we have only found researches carried out by Batten et al. (2016) and Dietz et al. (2016), in which they analyze the impact of environmental risks on the stability of the financial system, in addition to Campiglio et al. (2017) who analyze the impact of ESG in central banks. No research has been found about ESG and the probability of default of financial institutions, only investigations that propose incorporating these criteria into the risk policy of the financial institution, as indicated by Ahmed and Rahman (2014). No research has been found ESG criteria and the probability of default of a financial institution.

We believe that in the future, capital requirements of the financial sector will consider these risks, although, previously, it becomes necessary to quantify the relationship between ESG risks and probability of default. To fill this lack of knowledge, we have analyzed the relation between ESG risks and the probability of default of financial institutions, specifically the Spanish credit cooperatives sector (*Cajas Rurales*) during 2010 and 2015.

To do this, we have used a logit model with a dependent variable that takes the value 1 in the case of the default of the entity (as it is defined in Madera (2017)) and the value 0 otherwise and a series of explanatory variables that measure environmental, social, and governance aspects extracted exclusively from public financial reports.

As far as our knowledge reaches, this is the first empirical research that directly matches ESG criteria with the probability of default of a financial institution in Spain.

Our research is organized as follows: In the second section, we define the methodology to be used. In the third section, we emphasize, in the definition of the dependent variable, the explanatory variables and the sample. In the fourth section, we present the results of this investigation. Finally, the fifth section includes the conclusions and defines the main lines of future research.

## 2 Methodology

To achieve the objective of this research, that is, to quantify the influence of the ESG risks on the probability of default of a credit cooperative, we will use a logistic regression because this technique has showed a strong predictive power in all insolvency investigations of credit unions analyzed (see Table 1).

**Table 1** The state of art

| Technique | Researches |
|---|---|
| Discriminant | Simon (1980) |
| Logit/probit | Dabós (1996), Gama et al. (2004), Porath (2006), Gama et al. (2011) and Lima (2012) |
| Survivor analysis | Maggiolini and Mistrulli (2005), Braga et al. (2006), Cabo and Rebelo (2010), Fiordelisi and Mare (2013) and Carvalho et al. (2015) |

Source: own

A binary logistic regression matches a dichotomous variable with a set of independent variables, being used in those cases in which a simple or multiple regression cannot be used, linear or not (Silva and Barroso 2004), because the basic assumptions are not met, that is, the non-normality of the errors, heteroscedasticity, values of the probability outside the interval [0,1], and the presence of a coefficient of determination that is not adequate as a measure of adjustment of the model (Ramajo et al. 2002).

Its general expression being the following (Eq. 1):

$$P = \frac{1}{1 + e^{-\alpha - \beta_1 X_1 - \beta_2 X_2 - \ldots - \beta_k X_k}} \tag{1}$$

According to Silva and Barroso (2004), the process of finding the parameters that maximize the likelihood function is complicated and requires the application of a numerical procedure, such as the Newton-Raphson algorithm that requires the use of iterations, and although it is not the object of this research to define the steps to find the parameters, to explain the procedure starts by calculating the value of the constant term that maximizes that value of the likelihood, without including any of the explanatory variables (Eq. 2):

$$a = ln \frac{\sum_{i=1}^{n} Y_i}{n - \sum_{i=1}^{n} Y_i} \text{ y } b_1 = b_2 = \ldots = b_k = 0 \tag{2}$$

being a, $b_1, b_2, \ldots, b_k$ the estimation of the following parameters: $\alpha$, $\beta_1, \beta_2, \ldots, \beta_k$.

Subsequently, iterations of the model are calculated including one to one each explanatory variable, observing the values taken by the verisimilitude in order to find the answer that maximizes that likelihood.

For the interpretation of the parameters, the first derivative is calculated, arriving at the expression collected in Eq. (3):

$$\frac{dP}{dX_1} = \frac{-e^{-\alpha-\beta_1 X_1} \times (-\beta_1)}{(-e^{-\alpha-\beta_1 X_1})^2} = \beta_1 P(1-P) \quad (3)$$

If β1 is positive, dP/dx will also be positive, obtaining the highest value of dP/dx when $P = 0.5$.

Following Ramajo et al. (2002), in the general case of the model with K variables, an alternative interpretation of the coefficients can be made considering that from the expression dP/dx, we obtain that β_j/β_k = (dP/dx_j)/(dP/〖dx〗_k) and, therefore, the ratio of estimated components provides a measure of the relative change in probabilities.

## 3 Sample, Dependent, and Independent Variables

*Cajas Rurales* are a type of Spanish credit cooperative that combines the characteristics of a credit institution with those of a cooperative society. For this reason, they must to comply both the regulations applicable to credit institutions and the regulations governing the activity cooperative.

Their business is very territorialized, offering traditional banking services to a range of customers closely linked to agricultural and livestock activity.

At present, the segment of Spanish credit cooperatives is made up of a total of 62 cooperatives, of which 59 correspond to *Cajas Rurales*, grouped around the National Union of Credit Cooperatives (UNACC), the Spanish Cooperative Bank, and the Cajamar Cooperative Group. We have discarded from our study the three credit cooperatives that are not Caja Rural.

A sample of panel data is constructed with information related to each ESG factors between 2010 and 2016. This information has been obtained from the Bank of Spain, the National Union of Credit Cooperatives, and the Cajamar Cooperative Group.

To estimate the model, as is usual within this line of research (Madera, 2017), we will use the information from the year prior to the insolvency event.

As a dependent variable, we define a dichotomous variable that takes the value 1 if the credit cooperative has been declared in an insolvency process, and 0 nonthewise.

Madera (2017) explains that the insolvency of Spanish credit cooperatives is not common, similar to what was observed in Argentina or Portugal. Conversely, they observe a greater number of merger processes that, as they demonstrate in their investigation, were entirely due to insolvency problems.

For the determination of the dependent variable, we use the Registry of the Bank of Spain. In Graph 1 we show the distribution of insolvencies (dependent variable equal to 1) by year, observing a greater concentration between 2011 and 2012:

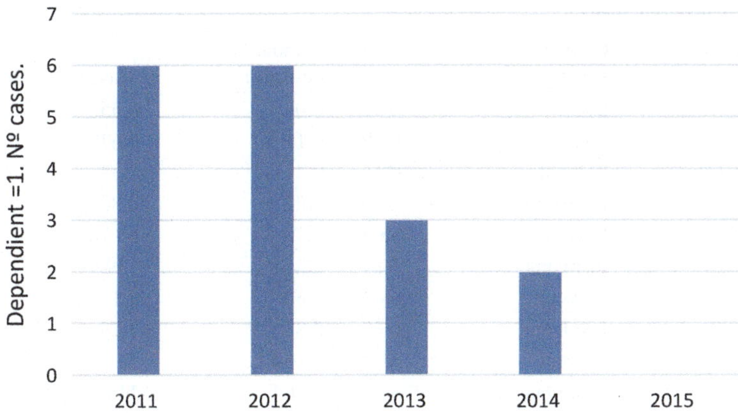


**Graph 1** Temporal distribution of dependent variable. Source: own

**Table 2** Explanatory variables

| Type | Denomination | Ratios |
|---|---|---|
| Environmental | *Suministros 2* | $\frac{\text{Suppliers}}{\text{Employees}}$ |
| Social | *Equidad* | $\frac{\text{Women}}{\text{Men}}$ |
| | *Formacion2* | $\frac{\text{Traning exp.}}{\text{Employees}}$ |
| | *Salarios2* | $\frac{\text{Wages}}{\text{Employees}}$ |
| | *Red* | $\frac{\text{ATM}}{\text{Customers}}$ |
| Governance | *Directivos* | $\frac{\text{Director}}{\text{Employees}}$ |
| | *Equidad_Consejo* | $\frac{\text{Women} - \text{Director}}{\text{Men} - \text{Director}}$ |

Source: own

Regarding the explanatory variables, we have selected seven variables (see Table 2) grouped by categories, one for the environmental pillar, four for the social pillar, and two for the governance pillar, all of them selected according to their respective use in other ESG risk studies or of own contribution.

## 4 Results

At first, we perform a univariate analysis through a *t*-test of equality of means. We define the null hypothesis as "there are no differences between the two groups." The alternative hypothesis is defined as "there are differences between both groups."

In Table 3 we show the main figures of each explanatory. The preliminary analysis shows that the insolvent group presents the lowest values in equality (*Equidad*), equality within the board of directors (*Equidad_Consejo*), efficiency in the use of their supplies (*Suministros2*), and training expenses per employee

**Table 3** Univariate analysis

| Explanatory variable | Cooperative situation | Mean | Standard deviation | Maximum | Minimum |
|---|---|---|---|---|---|
| Equidad | Solvent | 0.42 | 0.07 | 0.55 | 0.30 |
| | Insolvent | 0.40 | 0.11 | 0.67 | 0.25 |
| Equidad_Consejo | Solvent | 0.16 | 0.21 | 0.83 | – |
| | Insolvent | 0.15 | 0.22 | 0.86 | – |
| Salarios2 | Solvent | 45.48 | 5.78 | 56.70 | 35.37 |
| | Insolvent | 50.43 | 28.11 | 149.44 | 1.09 |
| Directivos | Solvent | 0.44 | 0.60 | 2.17 | – |
| | Insolvent | 0.38 | 0.83 | 3.25 | 0.01 |
| Suministros2 | Solvent | 31.77 | 10.54 | 53.17 | 14.04 |
| | Insolvent | 30.52 | 23.64 | 116.56 | 1.78 |
| Red | Solvent | – | – | 0.01 | – |
| | Insolvent | 0.08 | 0.30 | 1.23 | – |
| Formacion2 | Solvent | 3.34 | 3.90 | 14.00 | – |
| | Insolvent | 1.16 | 1.42 | 4.20 | – |

(*Formacion2*). However, they presented the highest salary levels per employee (*Salarios2*) and the largest network of ATMs per customer (*Red*).

In Table 4, we show the results of the *t*-test for equality of means. In it, we can observe that there are differences between both groups defined by the dependent variable (solvent and insolvent), except in the variable *Equidad*, because the standard error is below the confidence level of 0.05%, not rejecting the null hypothesis. Next step of our research is the estimation of the logit regression.

Table 5 shows the results of the model, in which we can see that all the explanatory variables related to the ESG criteria considered have been significant, except the equality within the members of the board (*Equidad_Consejo*); for this reason, this variable has been eliminated in the final version of our model:

In fact, we can observe that explanatory variables like equality between women and men (*Equidad*), training expenses per employee (*Formacion2*), and efficiency per employee (*Suministros2*) exert a negative influence on the probability of default (the more value, the less probability) as evidenced in the column (Exp(B)) where everyone takes values above zero. On the other hand, wages (*Salarios 2*), number of members of the board of directors (*Directivos*), and number of ATM per customer influence positively over the probability of default (the more value, the more probability) because of their values (Exp(B)) below zero.

The estimated model shows a good predictive power, with a Nagelkerke R squared of 0.64, as shown in Table 6, showing a predictive power of 82% in the operative group and 76% in the group of the merged ones, as shown in Table 7.

**Table 4** Independent samples test

| Explanatory variable | | Levene's test for equality of variances | | t-test for equality of means | | | | | 95% Confidence Interval of the difference | |
|---|---|---|---|---|---|---|---|---|---|---|
| | | F | Sig. | t | df | Sig. (2-tailed) | Mean difference | Std. Error Difference | Lower | Upper |
| Equidad | Equal variances assumed | 3.27 | 0.08 | 0.62 | 32.00 | 0.54 | 0.02 | 0.03 | −0.05 | 0.09 |
| | Equal variances not assumed | | | 0.62 | 27.68 | 0.54 | 0.02 | 0.03 | −0.05 | 0.09 |
| Equidad_Consejo | Equal variances assumed | 0.01 | 0.95 | 0.08 | 32.00 | 0.94 | 0.01 | 0.07 | −0.14 | 0.16 |
| | Equal variances not assumed | | | 0.08 | 31.98 | 0.94 | 0.01 | 0.07 | −0.14 | 0.16 |
| Salarios2 | Equal variances assumed | 1.79 | 0.19 | −0.71 | 32.00 | 0.48 | −4.95 | 6.96 | −19.13 | 9.23 |
| | Equal variances not assumed | | | −0.71 | 17.35 | 0.49 | −4.95 | 6.96 | −19.61 | 9.71 |
| Directivos | Equal variances assumed | 8.71 | 0.01 | 2.16 | 32.00 | 0.04 | 2.18 | 1.01 | 0.13 | 4.23 |
| | Equal variances not assumed | | | 2.16 | 20.18 | 0.04 | 2.18 | 1.01 | 0.08 | 4.28 |
| Suministros2 | Equal variances assumed | 0.34 | 0.57 | 0.20 | 32.00 | 0.84 | 1.25 | 6.28 | −11.54 | 14.04 |
| | Equal variances not assumed | | | 0.20 | 22.12 | 0.84 | 1.25 | 6.28 | −11.77 | 14.26 |
| Red | Equal variances assumed | 4.47 | 0.04 | 1.03 | 32.00 | 0.31 | −0.07 | 0.07 | −0.22 | 0.07 |
| | Equal variances not assumed | | | −1.03 | 16.00 | 0.32 | −0.07 | 0.07 | −0.23 | 0.08 |
| Formacion2 | Equal variances assumed | 0.29 | 0.59 | 0.25 | 32.00 | 0.80 | 0.06 | 0.25 | −0.44 | 0.57 |
| | Equal variances not assumed | | | 0.25 | 29.15 | 0.80 | 0.06 | 0.25 | −0.45 | 0.57 |

**Table 5** Model's output

| | | B | S.E. | Wald | df | Sig. | Exp(B) |
|---|---|---|---|---|---|---|---|
| Step 1[a] | Equidad | −30.80 | 14.80 | 4.33 | 1 | 0.04 | – |
| | Salarios2 | 0.20 | 0.08 | 5.55 | 1 | 0.02 | 1.22 |
| | Formacion2 | −1.10 | 0.59 | 3.51 | 1 | 0.06 | 0.33 |
| | Directivos | 2.23 | 1.99 | 1.26 | 1 | 0.26 | 9.31 |
| | Suministros2 | −0.15 | 0.08 | 3.85 | 1 | 0.05 | 0.86 |
| | Red | 441.48 | 195.13 | 5.12 | 1 | 0.02 | 5.37E+191 |
| | Constant | 7.49 | 5.32 | 1.99 | 1 | 0.16 | 1,787.58 |

[a]Variables entered on step 1: Equidad, Salarios2, Formacion2, Directivos, Suministros2, Red

**Table 6** Model summary

| Step | -2 Log likelihood | Cox & Snell R Square | Nagelkerke R Square |
|---|---|---|---|
| 1 | 24,824[a] | 0.481 | 0.642 |

[a]Estimation terminated at iteration number 11 because parameter estimates changed by less than 0.001

**Table 7** Classification table (cut value is 0.5)

| | | Predicted | | |
|---|---|---|---|---|
| | | Solvent | Insolvent | Percentage correct |
| Step 1 | Solvent | 14 | 3 | 82.40 |
| | Insolvent | 4 | 13 | 76.50 |
| | Overall percentage | | | 79.40 |

Graph 2 represents the distribution of default probabilities assigned by the model to each group (solvents and insolvent). In it, we can observe that those Cajas Rurales that disappeared show higher probabilities based on a worse management of the ESG factors than the rural savings banks that remained operative during the study period (Graph 3).

With this work, we have tried to quantify the relation between ESG factors and the probability of insolvency of a financial institution, since we have detected a lack of scientific knowledge in the analyzed bibliography.

For this, we have resorted to the specific case of the Spanish rural savings banks, on which we have built a panel data sample with information regarding ESG factors between 2010 and 2016. We have used public information.

We have defined a dependent variable following Madera (2017) definition, that is, insolvency by merger, and a series of explanatory variables.

Through a univariate analysis, a *t*-test of equality of means, and a logit model, we have been able to determine that ESG variables related to equality between men and women, efficiency in the use of resources, salaries per employee, ATM network in function of its clients, and training expenses for employees have turned out to be

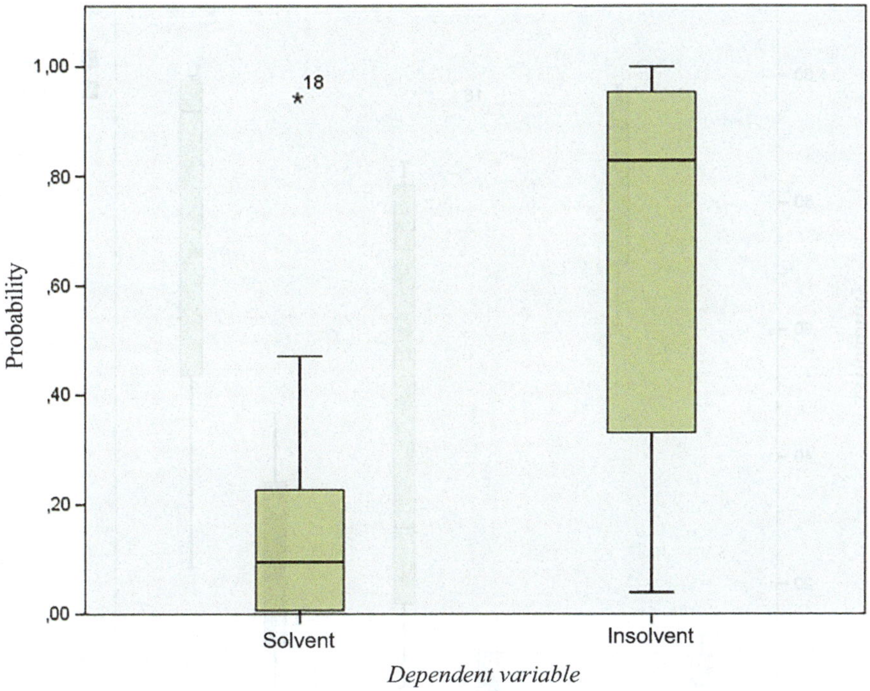


**Graph 2** Distribution of probabilities. Source: own

significant. On the other hand, equality within the board of directors is not significant.

In fact, our model assigns greater probabilities of insolvency to the credit cooperatives that really disappeared, with a high explanatory power.

The limitations of our model are based on the availability of information, since we have only had access to public reports that limit the information. In fact, we think that our study could be reinforced with information about the credit portfolio, the environmental policy, the carbon footprint, or the remuneration to the board of directors; however, we believe that our research constitutes a first empirical demonstration that those companies, in this case credit cooperatives, which actively manage ESG risks, present a lower probability of insolvency isolating financial aspects.

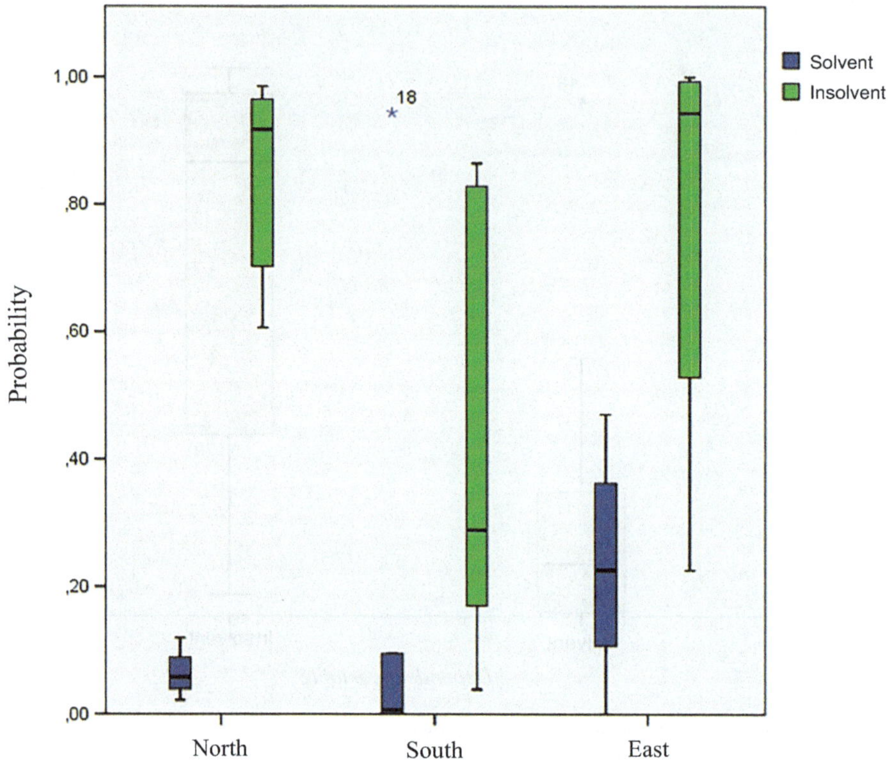


**Graph 3** Distribution of probabilities by territory. Source: own

## References


Ahmed, S. U., & Rahman, M. (2014). Incorporating ESG risk in Bank-lending in Bangladesh. *International Research Journal of Finance and Economics, 120.*

Batten, S., Sowerbutts, R., & Tanaka, M. (2016). *Let's tal about the weather: The impact of climate change on central banks*. Bank of England.

Braga, M., Fully, V., Colosimo, E., & Bressan, A. (2006). Investigating the solvency of Brazilian credit unions using a proportional hazard model. *Annals of Public and Cooperative Economics, 77*(1), 83–106.

Cabo, P., & Rebelo, J. (2010). The survival of Portuguese credit co-operatives: An econometric approach. *In Co-operatives contributions to a plural economy* ICA European Research Conference.

Campiglio, E., Dafermos, Y., Monnin, P., Ryan-Collins, J., Schotten, G., & Tanaka, M. (2017). Finance and climate change: What role for central banks and financial regulators. *In CEP-DNB workshop on Central Banking and Green Finance (pp.28-29)*

Carvalho, F., Diaz, M., Bialoskorski Neto, S., & Kalatzis, A. (2015). Exit and failure of credit unions in Brazil: A risk analysis. *Revista Contabilidade y Finanças, 26*(67), 70–84.

Dabós, M. (1996). Crisis Bancarias y medición del riesgo de default: métodos y el caso de los Bancos Cooperativos en Argentina. *Desarrollo económico*, 215–230.

Dietz, S., Bowen, A., Dixon, C., & Gradwell, P. (2016). Climate value at risk of global financial assets. *Nature Climate Change, 6*(7), 676–679.

Fiordelisi, F., & Mare, D. (2013). Probability of default and efficiency in cooperative banking. *Journal Of International Financial Markets, Institutions And Money, 26*, 30–45.

Gama, V., Braga, M., & de Lima, J. (2004). Análise de Insolvência das Cooperativas de Crédito Rural do Estado de Minas Gerais. *Estudos Economicos, 34*(3), 553–585.

Gama, V., Braga, M., Bressan, A., & Resende, M. (2011). Avaliaçao de insolvência em cooperativas de crédito: uma aplicaçao do sistema PEARLS. *Revista De Administração Mackenzie, 12*(2), 113–144.

Goss, A., & Roberts, G. S. (2011). The impact of corporate social responsibility on the cost of banks loans. *Journal of Banking & Finance, 35*(7), 1794–1810.

Lima, D. (2012). Forecasting insolvencies of Portuguese co-operative banks. *Third prize in the first EACB award Young Researchers on Co-operative banks.*

Madera, A. (2017). *Análisis de la sostenibilidad financiera de las cajas rurales a través de modelos logity regresión de cox. Propuesta de un indicador sintético de salud financiera (Doctoral dissertation)*. Universidad Pontificia Comillas.

Maggiolini, P., & Mistrulli, P. (2005). A survival analysis of de novo co-operative credit banks. *Empirical Economics, 30*(2), 359–378.

Margolis, J. D., & Walsh, J. P. (2003). Misery loves companies: Rethinking social initiatives by business. *Administrative Science Quarterly, 48*(2), 268–305.

Margolis, J. D., Elfenbein, H. A., & Walsh, J. P. (2009). Does it pay to be good... and does it matter? A meta-analysis of the relationship between corporate social and financial performance. *SSRN Electronic Journal*, 1–68.

Orlitzky, M., Schmidt, F. L., & Rynes, S. L. (2003). Corporate social and financial performance: A meta-analysis. *Organization Studies, 24*(3), 403–441.

Plinke, E (2016). *Explorar las inversiones ESG. Vontobel.* https://www.rankiapro.com/wp-content/uploads/2019/03/Gu%C3%ADa-encontrar-inversi%C3%B3n-ESG-adecuada-vontobel-am-rankiapro.pdf

Porath, D. (2006). Estimating probabilities of default for german saving banks and credit cooperatives. *Schmalenbach Business Review, 58*, 214–233.

Ramajo, J., Márquez, M. A., & Nogales, L. (2002). *Econometría aplicada: técnicas y modelos básicos*. ICE Universitas Ed.

Silva, L. C., & Barroso, I. M. (2004). *Regresión Logística*. La Muralla Ed.

Simon, C. J. (1980). *Predicting the failure of credit unions: An application of multivariate logit analysis*. Massachusetts Institute of Technology.

Stellner, C., Klein, C., & Zwergel, B. (2015). Corporate social responsibility and Eurozone corporate bonds: The moderating role of country sustainability. *Journal of Banking & Finance, 59*, 538–549.

**Antonio Madera del Pozo** , economist, PhD in Financial Economics from the Universidad Pontificia Comillas of Madrid and winner of the Ramón de San Pedro Prize for the best doctoral thesis defended in the faculties of Economics and Law at the Universidad Pontificia Comillas. He also holds a Master's Degree in Research in Economics and a Master's Degree in Finance, both from ICADE, and a Master's Degree in International Trade from ESIC. He currently works as Senior Director at Ethifinance ratings and is a lecturer at both the Universidad Pontificia Comillas—ICADE—and the Universidad Europea de Madrid. Previously he worked in the risk department of Banco Cooperativo Español and Caja Madrid. He is a regular contribiutor to the economic press, as well as to academic journals and conferences.

**Natalia Cassinello Plaza** , PhD got a degree in Business Administration (1990) from the Universidad Pontificia Comillas as well as a PhD in business with a thesis on Derivatives IFRS in the banking sector. Outside of the Universidad Pontificia Comillas, she has also completed a Master's degree in Taxation from the Instituto de Empresa (Madrid) in 1995.

She worked as a strategic consultant at McKinsey, as Tax advisor in a Spanish Law firm (Ramón y Cajal Abogados) and as Recruiting Manager at BCG- Boston Consulting Group.

Nowadays, she is working at Universidad Pontificia Comillas where she acts both as a professor (lecturer) in finance and accounting since 1994 as well as a program director of two undergraduate programs since 2019: A Degree in Business Administration and International Affaires (E6) and a degree in Business Analytics and International Affaires (E6A). She is also collaborating with Bank of Spain as a member since 2012 of the official supervision recruiting process. Besides all this, she is also a member of the Board of the NPO (non-profit organisation) Professional Women's Network Madrid since June 2020 as secretary of the board.

Outside of Spain, Natalia Cassinello has completed international executive programs. These include the international teaching & researchers' program (ITP) at HEC, Women Leadership Program at INSEAD and a Behavioural Economics course at London School of Economics. Beyond educational international experiences, she also has international experience as a professor at Bocconi in Milan and in research at Neoma in Paris/Reims.

Her current areas of interest for research include: sustainability ESG (environmental social government); gender equality; the digitalization of the banking sector; credit cooperatives; fintech innovation; mutual funds; behavioural finance or economics; housing sector and economic well being (more specifically in relation to vaccines).

# Relation Between Internal Social Responsibility and Job Satisfaction in Serbian Service Companies

**Milena Damnjanović**

## 1 Introduction

Employee's well-being is essential for each company, as it is well-known that employees are one of the most important stakeholders. Internal social responsibility activities are activities directly related to the employees' physical and psychological working environment. Research on perception of social responsibility practices that influence employees as internal stakeholders and their work attitude and behavior need more exploration in service industry (Nazir et al., 2021; Wang et al., 2020; Kim et al., 2020; Hu et al., 2019; Su & Swanson, 2019), particularly in the context of developing countries (Wang et al., 2020; Chan & Hasan, 2019).

Service industry has a growing significance in world economy, as it creates large revenues, has important role in income creation, and contributes to employment increase. Service sector employs more than 70% of people in the world's most developed countries. In the third quarter of 2019, Serbian service sector employed 57.4% of all employees (RS RSZ, 2020), and the similar results were in the year before.

Employees are considered to be the most valuable asset in any company. Employees can be influenced by, but also they can influence on different organizational activities, so they have a key role in achieving organizational effectiveness. Company's success or failure in the market completely depends on them. It is expected for employees to have opinions about companies' social responsibility activities, and those opinions could influence their attitudes and behavior. Service

M. Damnjanović (✉)
Department of Entrepreneurship Management at the Faculty of Technical Sciences in Čačak, University of Kragujevac, Kragujevac, Serbia
e-mail: milena.damnjanovic@ftn.kg.ac.rs



I. G. Arraiano et al. (eds.), *Corporate Social Responsibility in a Dynamic Global Environment*, CSR, Sustainability, Ethics & Governance,
https://doi.org/10.1007/978-3-031-24647-0_13

company's social responsibility activities lead to a positive impact on a firm' performance higher than a manufacturing company's (Jeon et al., 2020, p. 3).

In general, prior research in the service industry indicates that a company's social responsibility activities can increase customers' trust and satisfaction, repurchase and revisit intentions, financial performance, employees' green behavior, and positive work outcomes by service employees (Jeon et al., 2020, p. 3; Su & Swanson, 2019, p. 438).

This research focused on understanding about the impact of internal social responsibility on employees' job satisfaction in service industry in Serbia. Several studies found positive relation between internal social responsibility and job satisfaction (Koirala & Charoensukmongkol, 2020; Story & Castanheira, 2019; Chan & Hasan, 2019; Low & Ramayah, 2018; Bogan et al., 2018; Obeidat et al., 2018; Alkayed, 2017; Low & Ong, 2015a, 2015b), so this study begins from these findings. Research was done in service industry companies because of specific nature of service "production" and the significance of employees for that "production." This research will contribute to existing literature in this field, as there is a lack of similar research, especially in developing countries and service industry. Practical contributions are in suggestions for managers to develop internal social responsibility strategy and measure its impact on employees as they are key stakeholders for companies' success.

## 2 Literature Review

### 2.1 *Internal Social Responsibility*

Many scholars in their research made distinction between internal and external social responsibility (Story & Castanheira, 2019; Chan & Hasan, 2019; Cek & Eyupoglu, 2019; Perdana & Mardiana, 2018; Low et al., 2018; Obeidat et al., 2018; Stojanović-Aleksić et al., 2016; Low & Ong, 2015b). European Commission (2001) made difference between internal and external dimension of corporate social responsibility in the green paper "Promoting European Framework for Corporate Social Responsibility." External stakeholders have benefits of external corporate social responsibility activities, and employees have benefits of internal corporate social responsibility activities.

External social responsibility refers to socially responsible activities which are directed toward external stakeholders. Social responsibility spreads outside of the company into the local community and involves a variety of stakeholders: business partners and suppliers, customers, public authorities and NGOs representing local communities, and environment (European Commission, 2001).

Internal social responsibility refers to socially responsible activities oriented toward internal stakeholders. Within the company, socially responsible practices primarily involve employees and relate to issues such as investing in human capital, health and safety, and managing change (European Commission, 2001, p. 8). It can be anything done inside the organization to improve the life of the employees which affect their productivity and directly affect the profitability (Obeidat et al., 2018, p. 108). Internal social responsibility practices refer as employee oriented social responsibility activities which are directly related to the physical and psychological working environment of the employees (Low & Ramayah, 2018, p. 47; Low et al., 2018, p. 708; Low & Ong, 2015b, p. 115; Low, 2014, p. 23; Turker, 2009, p. 192) concerning their health and well-being, their training and participation in the organizational activities, equal opportunities, and also the work-life balance (Low, 2014, p. 23). Activities like employee training, continuing education programs, safe working environments, diversity policies, daycare programs, and ethical labor practices are parts of internal social responsibility (Story & Castanheira, 2019; Chan & Hasan, 2019; Low & Ramayah, 2018). Internal social responsibility includes respect for individual differences and problems that employees face and mutual respect and understanding, promoting humanity and care for people that is not limited to work hours but includes an attempt to make balance between work, family, and leisure also (Stojanović-Aleksić et al., 2016).

Companies have limited resources, so it may occur that investment in external socially responsible activities is done by compromising internal socially responsible activities, such as employee welfare, training, and improvement of working conditions. Employees expect and demand that if their companies are socially responsible toward external stakeholders, then they should be socially responsible toward internal stakeholders at the similar way.

When employees perceive that their company is investing in them through socially responsible activities, these practices are rewarded by employees as they show positive attitude and behavior toward the company. Internal social responsibility provides a way to build a relationship between an organization and its employees (Chan & Hasan, 2019, p. 2). It has become a significant aspect of attracting and retaining employees as better working environment and good internal relations help an organization to enhance employee job satisfaction and increase their commitment. As internal social responsibility practices are related to betterment of working conditions, these practices could be related to the level of satisfaction one may have about their job (Story & Castanheira, 2019). Internal social responsibility practices create shared value between employees and organization, through fulfilling the needs of their employees through a competitive salary, reasonable benefit packages, and job responsibilities (Low, 2014, p. 25). Managers need to comprehend that considering internal social responsibility means companies manage their internal stakeholders with responsible strategy (Perdana & Mardiana, 2018, p. 180). Establishment of good relation between employees and company needs using an internal social responsibility as strategic tool (Damnjanović, 2019, p. 446).

Frontline service employees' perceptions of social responsibility increase the belief that they are performing important work, resulting in greater job satisfaction (Su & Swanson, 2019, p. 439). A survey among the employees at Booking.com showed that 84% of employees desire to be employed in organizations that contribute actively to social responsibility activities (Khaskheli et al., 2020, p. 2).

The majority of research about social responsibility effects deals with its impact on customers or investors, while the impact on employees is often neglected (Low, 2014). That is surprising when it is well-known how much employees' attitudes and behavior are critical for overall success of organizations. Employees are looking for inspiration and leadership. They want to be satisfied with their choice of employer.

## 2.2 Job Satisfaction

Job satisfaction may be one of the most researched employee outcomes due to the increased complexity and competitiveness of workplace. Also, job satisfaction has central role in many theories and models of individual behaviors and attitudes, and it has practical applications for improving organizational effectiveness (Judge & Klinger, 2008).

The most often cited definition of job satisfaction was given by Locke in 1976, which explains job satisfaction as a pleasurable or positive emotional state resulting from appraisal of one's job or job experiences (Khaskheli et al., 2020, p. 5; Judge & Klinger, 2008, p. 394; Story & Castanheira, 2019, p. 1362; Lu et al., 2020, p. 3; Bogan et al., 2018, p. 54; Low & Ong, 2015b, p. 116). Spector (1985, p. 695) explained job satisfaction as a global feeling about the job or as a related constellation of attitudes about various aspects or facets of the job.

Some researchers consider job satisfaction as a concept which represents satisfaction regarding different aspects or facets of work. The facets approach is used to find out which parts of the job produce satisfaction or dissatisfaction (Spector, 1985). Spector (1985, p. 700) used in his Job Satisfaction Survey nine facets: pay, promotion, supervision, benefits, contingent rewards, operating procedures, coworkers, nature of work, and communication. This scale is one of the most often used in research, when job satisfaction is measured using facets. Story and Castanheira (2019, p. 1364) proposed measuring job satisfaction using Job Descriptive Index, where employees should express their satisfaction with five different job areas: pay, relationship with coworkers, supervision, opportunity for promotion, and work itself. On the other hand, job satisfaction can be considered also as a global concept which represents overall job satisfaction. Cammann et al. (1983) gave their overall job satisfaction scale which measures global job satisfaction using three items. Also, Brayfield and Rothe's job satisfaction scale is commonly used to measure overall job satisfaction (Judge & Klinger, 2008, p. 397). Sometimes both approaches can be used to get a complete picture of employee job satisfaction. For the purpose of this research, Cammann et al. (1983) approach was used.

Employee satisfaction could be also explained using intrinsic job satisfaction and extrinsic job satisfaction (Khaskheli et al., 2020; Obeidat et al., 2018; Abdallah et al., 2017). Intrinsic satisfaction refers to how people feel about the nature of the job tasks themselves. Key factors relevant for intrinsic job satisfaction refer to creativity, achievement, moral values, independence, and authority. Extrinsic job satisfaction refers to how employees feel about aspects of the work situation that are external to job tasks or the work itself (Abdallah et al., 2017, p. 32). Key factors relevant for extrinsic job satisfaction are advancement, company policy, compensation, and recognition. Using this approach, general job satisfaction is a combination of the both intrinsic and extrinsic dimensions.

Importance of job satisfaction comes from several reasons. Firstly, employees deserve to be treated fairly. Second, employees may affect organizational functioning through behavior which is a result of job satisfaction. And job satisfaction can also be a reflection of organizational functioning and sign of potential trouble spots in organization. Job satisfaction is considered a valuable, effective, and cogent mediator of employee attitude and behavior (Lu et al., 2020, p. 3). It is one of the most important dimensions of individual's happiness at work, as most people spend large amounts of time at their work (Abdallah et al., 2017, p. 32).

Job satisfaction could result from job environment and job itself but also from some factors that are not necessarily directly linked to job tasks. Social responsibility perceptions may influence job satisfaction (Cek & Eyupoglu, 2019; Bauman & Skitka, 2012), as responsible activities of company can impact the degree to which employees are satisfied with their job (Story & Castanheira, 2019, p. 1362). Social responsibility activities of organizations can lead to job satisfaction of employees in terms of security, safety, self-esteem, feeling of belongingness, fulfillment of social values, and promotion of organizational purposes (Bauman & Skitka, 2012). Employees having a perception that the organization shows fair and honest behaviors in the outside environment bears an important impact on the employees' job satisfaction and organizational commitment (Bogan et al., 2018; Im et al., 2016). Some research showed that hotel employees' perception of social responsibility can influence their quality of working life, job satisfaction, and overall quality of life (Hu et al., 2019, p. 130). Job satisfaction could be examined as one of the consequences of internal social responsibility (Low & Ramayah, 2018, p. 49; Kowalczyk & Kucharska, 2019, p. 599).

Relation between social responsibility, particularly internal social responsibility, and job satisfaction needs attention and further investigation, especially in service industry, where research about this relation are modest.

## 3 Methodology

The subject of this research was the relation between internal social responsibility and employees' job satisfaction in service companies in Serbia. Employees should be the main beneficiaries of internal social responsibility practices, so the main assumption was that those practices positively affect employees' job satisfaction. The main goal was to show if there was a positive correlation between internal social responsibility and employees' job satisfaction. Additional goals were to show if there was statistically significant difference in the perceptions of internal social responsibility between respondents using different criteria and to show if there was statistically significant difference between respondents in the level of job

**Table 1** The structure of respondents

| Characteristics | Criterion | % of respondents |
|---|---|---|
| Sex | Male | 30 |
| | Female | 70 |
| Age | 20–30 | 20 |
| | 31–40 | 42 |
| | 41–50 | 20 |
| | 51–60 | 15 |
| | >61 | 3 |
| Degree of vocational education | Secondary school qualifications | 16 |
| | 2-year postsecondary school qualifications | 15 |
| | Bachelor | 31 |
| | Master of science | 33 |
| | Magister | 3 |
| | Ph.D. | 2 |
| Work experience | 1–5 years | 34 |
| | 6–10 years | 16 |
| | 11–15 years | 27 |
| | >16 years | 23 |
| Position in company | Managerial position | 55 |
| | Nonmanagerial position | 45 |

Source: Authors' work

satisfaction. The data were collected using questionnaire. It had three parts. First part was dedicated to general questions about respondents. The second part was dedicated to internal social responsibility scale, which was adapted from Turker (2009), Akremi et al. (2018), and Stojanović-Aleksić et al. (2016). It consisted of eight items. The third part was dedicated to job satisfaction scale, developed by Cammann et al. (1983), and it consisted of three items. All scales were frequently used in similar research. All items were rated on a five-point Likert-type scale ranging from strongly agree to strongly disagree. The analysis was done through SPSS. Descriptive statistics, correlation analysis, one-way ANOVA, and independent samples t-tests were done. Factor analysis was done to test the appropriateness of items for selected scales and Cronbach's alpha coefficient for reliability of variables.

The sample counted 100 respondents, from various service companies in Serbia. The structure of respondents participated in this research using different criteria is given in Table 1. The majority of respondents were women (70%), while male respondents were 30% of all participants in the research. The most respondents were aged 31–40 years (42%). Considering education background, the most respondents had bachelor (31%) degree and master degree (33%). When work experience was considered, 34% of respondents had less than 5 years of working experience, but 27% of respondents had 11–15 and 23% had more than 16 years of working experience. Fifty-five percent of respondents were at managerial position.

# 4 Results

All items of internal social responsibility scale and job satisfaction scale were subject of principal component analysis (PCA) in SPSS, version 20. Before PCA, data adequacy for factor analysis was estimated. The number of cases for each item was appropriate, and there had been many correlation coefficients in correlation matrix with value above 0.3. The value of Kaiser-Meyer-Olkin Measure of Sampling Adequacy was 0.897, which was higher from recommended value of 0.6, and the Bartlett's Test of Sphericity was statistically significant at $p = 0.000$. All conditions for factor analysis were fulfilled.

Principal component analysis found two components with characteristic values above 1, which explained 58.837% and 10.689% of variance. Scree plot showed obvious breaking point after second component. So decision was made to keep two factors. This two-factor solution explained 69.526% of total variance, while the contribution of the first factor was 58.837% and the contribution of the second factor was 10.689%. For better understanding of these factors, direct Oblimin rotation was done. Both components had a lot of high factor loadings, and all items gave high values to only one of two components. All items of the first component (internal social responsibility) gave high factor loadings only to component 1, and all items of the second component (job satisfaction) gave high factor loadings only to component 2. These results were presented in Table 2.

**Table 2** Factor loadings matrix and correlation coefficients matrix for PCA with two-factor direct Oblimin rotation for internal social responsibility and job satisfaction scale (the main factor loadings and correlation coefficients for each item are given in bold font)

| Items | Factor loadings | | Correlation coefficients | |
|---|---|---|---|---|
| | Component | | Component | Component |
| | 1 | 2 | 1 | 2 |
| Internal social responsibility 8 | **0.825** | −0.008 | **0.829** | −0.412 |
| Internal social responsibility 7 | **0.821** | 0.150 | **0.747** | −0.252 |
| Internal social responsibility 5 | **0.800** | −0.160 | **0.878** | −0.551 |
| Internal social responsibility 3 | **0.794** | −0.191 | **0.888** | −0.580 |
| Internal social responsibility 2 | **0.711** | −0.157 | **0.788** | −0.505 |
| Internal social responsibility 6 | **0.708** | −0.226 | **0.819** | −0.573 |
| Internal social responsibility 4 | **0.647** | −0.246 | **0.767** | −0.563 |
| Internal social responsibility 1 | **0.611** | 0.136 | **0.544** | −0.163 |
| Job satisfaction 3 | −0.086 | **−0.920** | 0.364 | **−0.878** |
| Job satisfaction 1 | 0.137 | **−0.841** | 0.549 | **−0.908** |
| Job satisfaction 2 | 0.318 | **−0.735** | 0.678 | **−0.891** |
| Extraction method: Principal component analysis<br>Rotation method: Oblimin with Kaiser normalization | | | | |
| a. Rotation converged in six iterations | | | | |

Source: Authors' work

**Table 3** Descriptive statistics

| | Mean (1–5) | Std. deviation | Variance | *N* |
|---|---|---|---|---|
| Internal social responsibility | 3.8213 | 0.95565 | 0.913 | 100 |
| Job satisfaction | 4.0267 | 0.94836 | 0.899 | 100 |

Source: Authors' work

**Table 4** Results of correlation analysis

| | | Job satisfaction |
|---|---|---|
| Internal social responsibility | Pearson correlation | 0.658** |
| | Sig. (two-tailed) | 0.000 |
| | N | 100 |

Source: Authors' work
**Correlation is significant at the 0.01 level (two-tailed)

Between these two factors, medium negative correlation was found ($r = -0.49$). The results of this analysis supported the use of these items for internal social responsibility scale and for job satisfaction scale, as it was suggested.

Reliability analysis for internal social responsibility scale was measured using Cronbach's alpha coefficient, and it was 0.907. Cronbach's alpha coefficient for job satisfaction scale was 0.887. This reliability analysis had proven good reliability and internal consent of scales for this sample.

Descriptive statistics for internal social responsibility and job satisfaction were presented in Table 3. The mean value for internal social responsibility suggests that employees valued companies' efforts in this field. Results also suggest that employees were satisfied with their jobs, as the mean value was 4.0267, on the scale from 1 to 5.

Relation between employees' perceptions about internal social responsibility and job satisfaction was explained through Pearson correlation coefficient. Preliminary analysis was done in order to prove that all assumptions about normality, linearity, and homogeneity of variance were satisfied. The results showed strong positive correlation between internal social responsibility and employees' job satisfaction, $r = 0.658$, $n = 100$, $p < 0.005$, where high perceptions about internal social responsibility were followed with high levels of job satisfaction (Table 4). Coefficient of determination, $R^2$, was 0.4329, which means that employees' perceptions about internal social responsibility explained 43.29% of variance in employees' responses on the scale of job satisfaction.

The results also showed strong positive correlation between internal social responsibility and employees' job satisfaction for employees with 6–10 years of work experience, $r = 0.862$, $n = 16$, $p < 0.005$, and medium positive correlation between internal social responsibility and employees' job satisfaction for employees with 11–15 years of work experience, $r = 0.484$, $n = 27$, $p < 0.005$, and this difference between coefficients was statistically significant, as $z_{obs} = 2.2156$.

The independent samples t-tests were done in order to show if there was statistically significant difference in the perceptions of internal social responsibility

**Table 5** Descriptive statistics for job satisfaction for employees with different amounts of work experience

| Work experience | *N* | Mean | Std. deviation |
|---|---|---|---|
| Group 1 (1–5 years) | 34 | 3.6569 | 1.17632 |
| Group 2 (6–10 years) | 16 | 4.3958 | 0.66909 |
| Group 3 (11–15 years) | 27 | 4.0864 | 0.76568 |
| Group 4 (>16 years) | 23 | 4.2464 | 0.77991 |
| Total | 100 | 4.0267 | 0.94836 |

Source: Authors' work

between employees at different positions within the company. Levene's statistic was 0.012, $p = 0.914 > 0.05$, so equal variances could be assumed.

The results of independent samples t-tests showed that there was statistically significant difference in the perceptions of internal social responsibility between employees at managerial position ($M = 4.0318$, SD = 0.95265) and nonmanagerial employees ($M = 3.5639$, SD = 0.90439), t (98) = 2.5, $p = 0.014$. Mean difference between groups (0.46793, 95% CI: 0.09645 to 0.83941) was medium (eta square = 0.06).

One-way ANOVA was done in order to show if there were statistically significant differences in the level of job satisfaction between employees with different amount of work experience.

The results of one-way ANOVA showed that there was statistically significant difference in the level of job satisfaction between employees with different amount of work experience. The results of descriptive statistics for all groups were given in Table 5.

Levene's statistic was 3.043, at confidence level $p = 0.033$, so the assumption of homogeneity of variances was not fulfilled, so the results of Welch and Brown-Forsythe tests were used, as they are resistant to breach of presumption of equality of means.

The results of Welch test showed that there was statistically significant differences in the level of job satisfaction in four groups of employees (F (3; 49.446) = 2.8, $p = 0.05$). The Brown-Forsythe test also showed that there was statistically significant difference in the level of job satisfaction in four groups of employees (F (3; 92.368) = 3.677, $p = 0.015$). That difference, calculated through eta-squared indicator, was medium, as eta-squared was 0.09. Additional comparisons using Tukey HSD test showed that there was statistically significant difference in the average levels of job satisfaction between employees with 1–5 years of work experience ($M = 3.6569$, SD = 1.17632) and employees with 6–10 years of work experience ($M = 4.3958$, SD = 0.66909). There were not statistically significant differences between other groups of employees.

## 5 Discussion

Employees are the most important stakeholders in each company, especially in service companies, due to their irreplaceable role in service "production." They directly determine if customers will repurchase the services and increase their spending in the company. Employees value beneficial treatment through internal CSR practices, while employers seek loyalty and dedication in their work performance (Low & Ong, 2015b, p. 121). As some research identified job satisfaction as one of the consequences of internal social responsibility (Kowalczyk & Kucharska, 2019; Low & Ramayah, 2018), the motivation for this research could be understood.

The results showed strong positive correlation between internal social responsibility and employees' job satisfaction. Similar studies revealed existence of positive relationship between internal social responsibility and job satisfaction also (Koirala & Charoensukmongkol, 2020; Chan & Hasan, 2019; Story & Castanheira, 2019; Low & Ramayah, 2018; Low & Ong, 2015a, 2015b; Bogan et al., 2018; Obeidat et al., 2018; Alkayed, 2017). Employees' perceptions about internal social responsibility explained 43.29% of variance in employees' responses on the scale of job satisfaction. Obeidat et al. (2018) and Alkayed (2017) had even better results (67% and 64.8%), while Low and Ramayah (2018) found similar result (41%) and Low and Ong found 33.7% (Low & Ong, 2015a) and 34.9% (Low & Ong, 2015b). Chan and Hasan (2019) had a little lower result (28%), but they considered only legal and ethical dimension of internal social responsibility. The results of similar studies were consistent with the results presented in this paper.

The results also found that working experience was relevant for these analyses, as there had been found statistically significant differences in correlation coefficients for internal social responsibility and employees' job satisfaction relation for different groups of employees, considering working experience. Also, one-way ANOVA proved the existence of statistically significant difference in the level of job satisfaction between employees with different amount of work experience.

Managers and nonmanagerial employees had different perceptions about internal social responsibility, and this difference had statistical significance, which was proven using independent t-tests. The existence of differences in perceptions of internal social responsibility between employees at different working positions within company was also confirmed in similar research (Perdana & Mardiana, 2018).

## 6 Conclusion

The results clearly showed the positive impact of the perceptions about internal social responsibility practices that have enhanced employee's job satisfaction. That was the main goal of this research. These results were consistent with results in this field.

The contribution of this research to the social responsibility knowledge is in highlighting the importance of implementation of internal social responsibility practices in service companies, as it had significant influence on job satisfaction of employees. Satisfied employees provide better service and enhance customer loyalty, which is extremely important for service companies. Other contribution refers to the fact that there is a lack of similar research, especially in developing countries and service industry. Practical contribution was in suggestions for managers to develop internal social responsibility strategy and measure its impact on employees as they are the key stakeholders for companies' success. This research also suggests that it is important to include employees from all positions in company in socially responsible activities, as there were differences in perceptions of internal social responsibility by managerial and nonmanagerial employees.

There were several limitations of this research. First refers to the sample size. Bigger sample would provide more reliable results. Other limitation refers focus to service industry only. It could be interesting to compare results of the same research in different industry than give some conclusion about similarities and differences. Third, job satisfaction scale could be expanded with different facets of job satisfaction, in order to determine how internal social responsibility influences different facets. Future research will try to overcome these limitations.

## References

Abdallah, A. B., Obeidat, B. Y., Aqqad, N. O., Al Janini, M. N. K., & Dahiyat, S. E. (2017). An integrated model of job involvement, job satisfaction and organizational commitment: A structural analysis in Jordan's Banking sector. *Communications and Network, 9*, 28–53. https://doi.org/10.4236/cn.2017.91002

Akremi, A. E., Gond, J. P., Swaen, V., De Roeck, K., & Igalens, J. (2018). How do employees perceive corporate responsibility? Development and validation of a multidimensional corporate stakeholder responsibility scale. *Journal of Management, 44*(2), 619–657.

Alkayed, W. (2017). The effect of corporate social responsibility on job satisfaction (Empirical Study on Jordanian Private Hospital). *European Journal of Business and Management, 9*(12), 1–5. ISSN 2222–2839.

Bauman, C., & Skitka, L. (2012). Corporate social responsibility as a source of employee satisfaction. *Research in Organizational Behavior, 32*, 63–86.

Bogan, E., Turkay, O., & Dedeoglu, B. B. (2018). Perceived corporate social responsibility and job satisfaction: The mediator role of organizational identification. *International Journal of Business and Management Studies, 10*(2), 51–67.

Cammann, C., Fichman, M., Jenkins, G. D., & Klesh, J. R. (1983). Assessing the attitudes and perceptions of organizational members. In S. E. Seashore, E. E. Lawler, P. H. Mirvis, & C. Cammann (Eds.), *Assessing organizational change: A guide to methods, measures, and practices* (pp. 71–138). John Wiley & Sons.

Cek, K., & Eyupoglu, S. Z. (2019). Does teachers' perceived corporate social responsibility lead to organisational citizenship behaviour? The mediating roles of job satisfaction and organisational identification. *South African Journal of Business Management, 50*(1), a1481. https://doi.org/10.4102/sajbm.v50i1.1481

Chan, T. J., & Hasan, N. A. M. (2019). Internal corporate social responsibility practices and employees' job satisfaction in a Malaysian Banking Company. *Journal Pengurusan, 55*, 97–109. https://doi.org/10.17576/pengurusan-2019-55-08

Damnjanović, M. (2019). Investing in Employees in Serbian Companies—Internal Dimension of CSR. In M. Milković, S. Seljan, M. Pejić Bach, S. Peković, & Đ. Perovic (Eds.), *ENTRENOVA—ENTerprise REsearch InNOVAtion*, 5, 1 (pp. 441–448). IRENET.

European Commission. (2001). *Green paper: Promoting a European framework for corporate social responsibility*. Retrieved April 15, 2021, from europa.eu/rapid/press-release_DOC-01-9_en.pdf

Hu, B., Liu, J., & Qu, H. (2019). The employee-focused outcomes of CSR participation: The mediating role of psychological needs satisfaction. *Journal of Hospitality and Tourism Management, 41*, 129–137.

Im, S., Chung, Y. W., & Yang, J. Y. (2016). Employees' Participation in CSR and Organizational Outcomes. *Sustainability, 9*(28), 1–14.

Jeon, M. M., Lee, S., & Jeong, M. (2020). Perceived corporate social responsibility and customers' behaviors in the ridesharing service industry. *International Journal of Hospitality Management, 84*, 10234.

Judge, T. A., & Klinger, R. (2008). Job satisfaction: Subjective well-being at work. In M. Eid & R. J. Larsen (Eds.), *The science of subjective well-being* (pp. 393–413). Guilford Press.

Khaskheli, A., Jiang, Y., Raza, S., Qureshi, M., Khan, K., & Salam, J. (2020). Do CSR increase organization citizenship behavior among employees? Mediating affective commitment and job satsfaction. *Corporate Social Responsibility and Environmental Management, 27*(6), 1–15. https://doi.org/10.1002/csr.2013

Kim, H., Rhou, Y., Topcuoglu, E., & Kim, Y. G. (2020). Why hotel employees care about corporate social responsibility (CSR): Using need satisfaction theory. *International Journal of Hospitality Management, 87*, 102505.

Koirala, M., & Charoensukmongkol, P. (2020). Contributions of CSR perception to employees commitment and job satisfaction: Does personal income matter? *Kasetsart Journal of Social Sciences, 41*, 303–307.

Kowalczyk, R., & Kucharska, W. (2019). Corporate social responsibility practices incomes and outcomes: Stakeholders' pressure, culture, employee commitment, corporate reputation, and brand performance. A Polish–German cross-country study. *Corporate Social Responsibility and Environmental Management, 27*, 595–615.

Low, M. P. (2014). Internal corporate social responsibility: An overview. *Australian Journal of Basic and Applied Sciences, 8*(16), 18–29.

Low, M. P., Choe, K. L., & Lau, T. C. (2018). Perceived roles of ethics and social responsibility, internal corporate social responsibility and employee engagement of academicians. *The Journal of Social Sciences Research, Special Issue, 2*, 706–717.

Low, M. P., & Ong, S. F. (2015a). The manifestation of internal corporate social responsibility on employee's behavior in small medium sized enterprises. *Journal of Social Science Studies, 2*(2), 259–277.

Low, M. P., & Ong, S. F. (2015b). The role of internal corporate social responsibility in professional service sector: An empirical study from Klang Valley, Malaysia. *Asia Pacific Journal of Advanced Business and Social Studies, 1*(1), 113–123.

Low, M. P., & Ramayah, T. (2018). Adopting consistent PLS in assessing the influence of internal corporate social responsibility among small medium sized enterprises employees. *Management Research Journal, 8*(1), 46–62.

Lu, J., Ren, L., Zhang, C., Wang, C., Ahmed, R., & Streimikis, J. (2020). Corporate social responsibility and employee behavior: Evidence from mediation and moderation analysis. *Corporate Social Responsibility and Environmental Management, 27*(4), 1719–1728. https://doi.org/10.1002/csr.1919

Nazir, O., Ul Islam, J., & Rahman, Z. (2021). Effect of CSR participation on employee sense of purpose and experienced meaningfulness: A self-determination theory perspective. *Journal of Hospitality and Tourism Management, 46*, 123–133.

Obeidat, B., Altheeb, S., & Masa'deh, R. (2018). The Impact of internal corporate social responsibility on job satisfaction in Jordanian pharmaceutical companies. *Modern Applied Science, 12*(11), 105–120.

Perdana, K., & Mardiana, N. (2018). How managers perceive internal corporate social responsibility: An empirical study of Indonesian women's employment. *Redefining Corporate Social Responsibility Developments in Corporate Governance and Responsibility, 13*, 177–192.

Republika Srbija Republički zavod za statistiku (RS RSZ). (2020). *Trendovi, IV kvartal 2019*. Retrieved March 12, 2021, from https://publikacije.stat.gov.rs/G2020/Pdf/G20208001.pdf

Spector, P. (1985). Measurement of Human Service Staff Satisfaction: Development of the Job Satisfaction Survey. *American Journal of Community Psychology, 13*(6), 693–713.

Stojanović-Aleksić, V., Erić-Nielsen, J., & Bošković, A. (2016). Društvena odgovornost u bankarskom sektoru: iskustva iz Srbije. *Bankarstvo, 45*(2), 34–55.

Story, J., & Castanheira, F. (2019). Corporate social responsibility and employee performance: Mediation role of job satisfaction and affective commitment. *Corporate Social Responsibility and Environmental Management, 26*, 1361–1370.

Su, L., & Swanson, S. (2019). Perceived corporate social responsibility's impact on the well-being and supportive green behaviors of hotel employees: The mediating role of the employee-corporate relationship. *Tourism Management, 72*, 437–450.

Turker, D. (2009). How CSR influence organizational commitment. *Journal of Business Ethics, 89*, 189–204.

Wang, C., Hu, R., & Zhang, T. (2020). Corporate social responsibility in international hotel chains and its effects on local employees: Scale development and empirical testing in China. *International Journal of Hospitality Management, 90*. https://doi.org/10.1016/j.ijhm.2020.102598

**Milena Damnjanović** (born Stanisavljević), is a Teacher Assistant at the Department of Entrepreneurship Management at the Faculty of Technical Sciences in Čačak, University of Kragujevac, Serbia. She is a PhD student at the Faculty of Economics, University of Kragujevac, Serbia. She is interested in the field of social responsibility and entrepreneurial economics. She published several scientific papers in international and national journals and participated in many scientific international and national conferences.

# Part III
# Corporate Governance

# The Evolution of Chinese Corporate Governance

**Shy Lih Wong**

## 1 Introduction

In 1978, China opened its economy to the outside world in 1978. Since then, the transformation of modern CG in China has been intimately associated the unanticipated rapid growth of its capital markets. The Chinese state-owned enterprises (SOEs) began share issuance in the 1980s, and two stock exchanges were established in Shanghai and Shenzhen in the early 1990s. In 1992, the China Securities Regulatory Commission (CSRC) launched the issuance criteria and listing procedures for SOEs. The Asian financial crisis in 1997 has intensified the major reforms of CG system in China in the late 1990s, from board independence and board structure to audit committees and quarterly reporting.

Since 2000, the Chinese CG mechanism has transformed drastically in spite of the history of its institutional and regulatory settings being relatively young compared to the United States and Germany. Chinese regulator initiated the adoption of the laws and regulation in relation to universally approved CG standard in the early twenty-first century. At that time, the leading CG mechanism was the Organisation for Economic Co-operation and Development (OECD) Principles of Corporate Governance, based primarily on the shareholder value CG model. Since then, the international standard of Anglo-American or shareholder primacy CG code has been further extended to company and securities legislation in China.

Over the past three decades, China has remained remains one of the fastest growing economies in the world (Cumming et al., 2021). Both the Shanghai and Shenzhen Stock Exchanges eased the rules for Qualified Foreign Institutional Investors (QFII), and the inclusion of more than 200 leading A shares in the MSCI

S. L. Wong (✉)
Graduate Business School, UCSI University, Kuala Lumpur, Malaysia
e-mail: 1001955457@ucsiuniversity.edu.my

 
I. G. Arraiano et al. (eds.), *Corporate Social Responsibility in a Dynamic Global Environment*, CSR, Sustainability, Ethics & Governance,
https://doi.org/10.1007/978-3-031-24647-0_14

Emerging Markets Index in June 2018 were landmark achievement of China's financial market in the past decade. Despite these accomplishments, the majority of foreign investor respondents are far from fully understanding the Chinese characteristics of China's CG regime according to a survey conducted by the Asian Corporate Governance Association in 2018 (Allen & Li, 2017).

The recent financial irregularities of TAL Education Group and Luckin Coffee, Inc. (Costa, 2021) again highlight the corporate misconduct of Chinese corporations listed in the United States. This has led to an intense concern among the investor community over the efficacy of CG reforms of Chinese firms and is likely to inhibit some investors from investing into Chinese firms. Thus, the current standard of CG system in Chinese companies is evidently not satisfactory and far from ideal.

Besides the rising consideration of Chinese CG by foreign investors and foreign corporate executives, research interest in the CG of China is also escalating due to its present status as the world's second largest economy. Furthermore, the underlying CG mechanisms of Chinese firms and the distinctive characteristics of the Chinese financial system, legal and institutional frameworks, as well as business practices provide a unique context for researchers to further investigate their efficacy even though the Anglo-American CG code is generally adopted by China. In addition, it is vital to keep pace with the latest empirical findings on Chinese CG because the establishment of new codes and regulations implies that prior empirical evidence is no longer applicable in current times.

Based on the above premises, this study reviews the Chinese CG mechanism. Its aim is to overview the current CG issues in China, synthesize empirical evidence of the effects of selected Chinese CG themes, and provide an alternative point of view to help tackle the problems arising from the Chinese CG model to enhance firms' governance performance as well as protect the interests of relevant stakeholders. Also, suggestions for future research also presented.

The method that this study utilized is narrative review approach (Baumeister & Leary, 1997; Rousseau, 2012; Rousseau et al., 2008; Wong et al., 2013) which is a more popular review technique as compared to meta-analyzes and systematic reviews (Hodgkinson & Ford, 2014). The narrative review technique is considered to be appropriate if the purpose of the review is to summarize or appraise the results of several aspects of a large area of research (Snyder, 2019). Thus, this approach is applicable to this study given the broad nature of the Chinese CG literature. In addition, the scope of this study is limited to synthesizing only empirical findings of selected themes that are directly relevant to the major concerns of foreign institutional investors and foreign corporate executives, i.e., the protection of their firms' current and future investment in China as well as the level of transparency and dependability of China's CG policy. The selected themes derived from the major concerns include board independence, state ownership, concentrated ownership, supervisory board, and institutional investors.

Firstly, the terms of each theme of Chinese CG, "Chinese Capital Market" and "China Securities Regulatory Commission," were individually searched in the online databases of Emerald Publishing, Springer, SAGE, and EBSCOhost. Then, the selection of peer-reviewed articles after the initial search was based on their suitability, significance, and the extent of the journal's impact factor (Latapí Agudelo et al., 2019). Some of the references in this review were outside of the initial search in the online academic research database, which included reports from international organizations and business news from online news agencies. They were selected and used based on their applicability and importance to support the arguments and discussions in this review.

The literature review offered in this study will not only contribute to the expanding literature in connection with the Chinese CG but will also enhance the understanding of foreign institutional investors in relation to the current CG framework, institutional environment, and background of the Chinese capital market so as to lessen their portfolio risk before investing in China. This review also offers foreign corporate executives a contemporary overview of the unique characteristics of Chinese CG and capital market regulatory authority which allows them to gain a more detailed insight into Chinese CG and financial markets before evaluating the potential direct investment of their firms in China. Furthermore, this study sheds light on whether the present Chinese CG is converging toward the Anglo-American model or the other way around which may be useful for regulators and policymakers in determining their decision to formulate new CG codes, regulations, and policies.

This study is organized as follows: Section 1 of the study presents an overview of the Chinese capital market and CSRC. Given that China is rapidly developing, and that new rules and regulations are constantly changing China's corporate governance landscape; this study updates the latest overview. Section 2 of the study synthesizes the selected themes of the Chinese CG. The characteristics and issues of the chosen themes that are, for the most part, unique to China are then compared and contrasted with Anglo-American as well as German mechanisms. Section 3 reviews the convergence forces, convergence possibilities, and its underlying challenges. Finally, concluding remarks are presented in relation to the implications for foreign institutional investors, foreign corporate practitioners, Chinese policymakers, and regulatory authorities. The future research directions are also proposed.

## 1.1 Chinese Capital Market

In the early 1990s, two exchanges were established in the China's equity market. The Shanghai Stock Exchange and the Shenzhen Stock Exchange were launched in December 1990 and July 1991, respectively, and they remain the major stock exchanges as of today. China's stock market has undergone a significant expansion since its establishment in the 1990s. At the end of December 2019, China's stock market capitalization reached USD 8.5 trillion (World Bank, 2020), with a global ranking second only to the United States.

The distinctive background of the Chinese stock market was seen as closely related to state-owned enterprises (SOEs) reform. SOEs controlled China's economy, in particular its industrial sector before the 1978 economic reform (Clarke, 2003). The Chinese government adopted several approaches to enhance SOEs' performance but mostly failed (Rajagopalan & Zhang, 2008). At the end of 1994, more than half of the SOEs operated at a loss (Kaye, 1995) with the state industry's losses reaching a record high of approximately RMB 80 billion in 1996 (Lardy, 1998). Also, China had the greatest number of nonperforming loans among Asian countries between 2000 and 2002 (The Asian Banker Journal, 2002).

During the mid-1990s, the whole Chinese banking sector was technically insolvent if the international standard accounting requirements were applied (Jiang et al., 2008; Lardy, 1998; Wang et al., 2004; Xu & Wang, 1999). Hence, the Chinese government required to recourse to another fundraising mechanism for SOEs, and the newly launched stock exchanges were employed for this enormous task.

The capital market is expected to be dominated by SOEs (directly or indirectly owned) in view of the fundamental mission of the Chinese capital market which is to finance SOEs (Jiang et al., 2008). For the purpose of maintaining the control over the listed companies, two classes of shares were created by the Chinese government, i.e., tradable and non-tradable (Jiang et al., 2008). Tradable shares are offered to public investors, while non-tradable shares are owned by either the state or legal persons representing state. About 84% of listed companies were under state control in a 2003 study (Liu & Sun, 2005). Listed SOEs comprised of more than half of the total market capitalization at the beginning of 2013.

## *1.2 China Securities Regulatory Commission*

Previous empirical findings found that strict laws and effective enforcement are critical aspects of CG (La Porta et al., 1998, 2002). These vital features can help address firms' agency problems and safeguard the interests of minority shareholders concurrently. Their evidence reveals that countries like the United States and the United Kingdom which practice common law generally provide stronger legal protection than countries like Germany and China which practice civil law. Also, their evidence discovered that common law countries usually outperform civil law countries in terms of economic and financial outcomes at the national and firm levels.

Chinese policymakers and regulators in China were aware of the significance of legal protection which induced them to develop a contemporary regulatory mechanism for legal protection in the mid-1980s. Since then, China has introduced and implemented several reforms of laws and regulations aim at strengthening CG and increasing legal protection for minority shareholders and investors. However, the legal protection of shareholders and investors in China was poor and ranked much lower in comparison with other emerging economies during the 1990s and the early 2000s.

In empirical research conducted by Allen et al. (2005) in comparing Chinese shareholder, investor and creditor rights as well as law enforcement to other nations were explored by La Porta et al. (1998). The overall findings demonstrated China ranked lower in all aspects as compared to other English-origin nations. Also, weak investor protection has negative impact on share price and its liquidity. La Porta et al. (1998) argue that shareholders are reluctant to invest when investor protection is weak which indicates low share liquidity. On the other hand, strong protection for investor rights can encourage investment. Johnson et al. (2002) assert that when the risk of expropriation is low, entrepreneurs are more confident and willing to begin and reinvest in their businesses which implies strong legal protection for investors and minority shareholders is an essential requirement for economic growth. This is particularly important for China which is undergoing transition from planned to market-oriented economy.

In brief, Allen et al. (2005) claim that there is a lack of appropriate regulation and adequate supervision in governing the Chinese capital market as faced by many other emerging economies. The CSRC was established in 1992 as a specialized regulatory authority to safeguard the interests of shareholders. However, it only officially attained legal status as a regulator in 1998 when the securities law was enacted (Jiang & Kim, 2015).

The CSRC is accorded to regulate listed firms and is authorized to investigate fraud and impose anti-fraud measures. Apart from in charge of monitoring the market activities of exchanges, the CSRC is allowed to intervene in the authorization of seasoned equity offerings and initial public offerings (IPO) on the stock markets to ensure market liquidity, the quality of listed companies, and other economic and financial objectives of the government. For instance, the CSRC has temporarily suspended IPO approvals nine times for extended periods, and all IPO applications were rejected in 2013 (Jiang et al., 2017). The market intervention by regulators is uncommon and not welcome in many matured capital markets such as the United States and the United Kingdom. The CSRC is granted more authority than capital market regulators in advanced nations.

The first Listed Company Governance Code was introduced in 2002 and revised recently in 2018. Some key changes under the revised code include setting up a division for representatives of the Communist Party to implement state objectives and policies; encouraging board diversity; enhancing the functions of audit committee; reducing the power and influence of majority shareholders; and establishing environmental, social, and governance requirements.

## 2 Themes of Chinese CG

There are several themes of CG in the context of China. However, the focus of this study is to review the importance and effectiveness of board independence, SOEs, ownership concentration, supervisory board, and the role played by institutional investors on CG in the Chinese setting.

## 2.1 Board Independence

CG is regarded as a system in which a board of directors is an important monitoring tool to mitigate the problems arising from the principal-agent relationship. The majority of the prior and current research on CG is based upon agency theory. However, agency theory which emphasizes the separation of ownership and management has resulted in principal-agent problems in modern firms (Berle & Means, 1932).

Agency theory argues that the board will oversee the behavior of management to safeguard the interests of owners, and hence the directors must be independent (Fama & Jensen, 1983; Jensen & Meckling, 1976). Baldenius et al. (2014) reveal that the effectiveness of a board's monitoring is significantly be affected by the balance of power between the CEO and shareholders. On the other hand, a study showed that higher board gender diversity reduces the monitoring capability of the board (Usman et al., 2018).

From another perspective, some researchers argue that over-disciplining the managers may ruin firm performance. For example, U-shaped curvilinear association between the function of board independence and firm's research and development is found which indicates excessive overseeing may demotivate the firm's intention to innovate (Guldiken & Darendeli, 2016). Additionally, Al Dah (2018) questions the validity of CEO being monitored by board.

In 2001, the CSRC (CSRC, 2001) began to emulate the US Sarbanes-Oxley Act in embracing a new CG approach to safeguard the interests of minority shareholders and investors as well as strengthen the economic reform. The very first Chinese Code of Corporate Governance for Listed Companies was drawn up in 2002 which outlined the requirement for the compulsory appointment of independent directors on the board of each listed firm. Listed firms were required to have one-third of their total directors on the board as independent directors by June 2003. The implementation of the new CG code has largely strengthened the protection of minority shareholders' and investors' interest. The legal status of the independent directors was not authorized until the company law of China was revised in 2005. However, empirical evidence is still conflicting in relation to the impacts of board independence (Hanson, 2019; Li & Naughton, 2007; Liu et al., 2015; Meng et al., 2018). The general consensus and consistent empirical results on the effectiveness of board independence remain inconclusive.

According to Liu et al. (2015), there is a positive relationship between independent directors and firm performance in China. Also, the effect of board independence on firm performance is associated with ownership concentration in China, i.e., board independence is positively related to firm performance if the concentration of ownership is reduced. However, empirical work by Bai et al. (2004) and Li and Naughton (2007) provides contradictory evidence that the independent directors have no significant influence on company performance. Jiang and Kim (2015) opine that the Chinese independent directors are generally inactive and ineffective at oversight role but merely fulfilling the requirements of the CSRC regulation even though the number of independent directors in listed firms has grown over the years. Jiang and Kim (2015) further highlight that the nominations and appointments of independent directors in China are mainly made by majority shareholders. They further contend that the independent directors only represent the minority on boards even if they are active.

In a recent research, Meng et al. (2018) find that there is a negative impact on firm performance with the presence of independent directors, specifically firms with higher information costs. In spite of the majority of negative empirical evidence on the efficacy of the Chinese independent directors, Hanson (2019) argues that independent directors play a critical role in overseeing the actions of managers in countries where state ownership is high.

Another strand of studies exhibits how other characteristics of directors play a pivotal role in CG development in China. For example, Du et al. (2017) reveal that the probability of earnings management can be reduced by A share listed with more foreign directors. Du et al. (2018) argue that the connections of directors are very crucial for firms to acquire trade credit in China, specifically for those with financial difficulties. He and Luo (2018) also find that Chinese boards with an even number of directors are less effective with higher rate of absence in board meeting and they possess tunneling behaviors. Hu et al. (2020), on the other hand, demonstrate the importance of political connections for Chinese firm boards by employing the negative stock market reaction to the resignation of independent directors with political connections. Meanwhile, Li et al. (2020) show that innovation in Chinese firms can be enhanced by directors with technology knowledge and expertise.

## 2.2 State Ownership

The significant numbers of SOEs in China is a critical difference between China and advanced western countries. The SOEs are unique due to their distinct concentrated ownership by the central government via ministries, the provinces, and cities or local governments. The significant state ownership of listed firms possesses several considerable implications with more focus on control and cash flow rights. Although state ownership has declined substantially after the share split reform in 2006, their shareholdings are still significant, and they remain the largest shareholders in most of the strategically important industry sectors (Ng et al., 2009; Xu & Wang, 1999; Yu, 2013).

The OECD (2015a) published a report revealing that the listed entities of Chinese SOEs represent approximately 50% market capitalization in China and there is an increasing trend of SOEs in other Asian countries being listed in the stock exchanges. Empirical studies have found that there is a negative relationship between state shares and firm performance (Li et al., 2009; Shleifer & Vishny, 1997; Wei et al., 2005). Ng et al. (2009) showed that the association between state ownership and firm performance is convex.

In addition, Li et al. (2008), Menozzi et al. (2011), Wang (2015), and Wu et al. (2018) reveal that political connections play a vital role in the firm performance of SOEs. A substantial state ownership is better than a dispersed ownership structure because of the benefits of political connections and support from the state (Li et al., 2009; Yu, 2013). For example, Shao (2019) argues that the state ownership has a positive effect on firm performance given the ability of Chinese government to wield significant influence on firms.

On the other hand, Jiang and Kim (2015) found that non-SOEs are more superior than SOEs in terms of firm performance. However, Chinese SOEs which rely heavily on bank debt perform better during the financial crisis of 2008 (Liu et al., 2012). This indicates that government via Chinese state-owned banks helps distressed SOEs to alleviate financial constraints by providing them bailout loans during crisis periods.

As a whole, listed SOEs continue to underachieve compared to listed private enterprises (Jiang et al., 2017) in spite of experiencing various reforms such as partial privatization. Jiang and Kim (2015) argue that the underachievement of SOEs is not due to expropriation by state owners because they do not have the strong reason to tunnel.

Occasionally, state owners deviate from the fundamental purpose of maximizing shareholder values which may ruin the interests of minority shareholders because the objective of the state is to uphold social stability. In order to compensate for SOEs, state owners will offer them with preferential bank loans and state subsidies (Gan et al., 2018; Lin & Tan, 1999).

The conflicting relationship between the principal and the agent is the underlying problem that stems in SOEs as asserted by the shareholder theory. Managers could involve in actions for personal gain which are detrimental to the interests of both state and minority shareholders due to the separation of ownership and executive decision-making. Hence, the vertical agency problems faced by Chinese SOEs are by and large identical to those of countries that embrace the Anglo-American CG mechanism such as the United States and United Kingdom (Jiang & Kim, 2020).

OECD (2015b) issued a guideline that includes ensuring an adequate legal and regulatory framework for SOEs, fair treatment of shareholders, relationship with stakeholders, transparency and disclosure, and the responsibilities of boards of SOEs in order to address the challenges faced by the SOEs. Embracing a robust CG framework is necessary to strengthen the CG practices of SOEs (Paul, 2016), especially in the case of China.

## 2.3 Concentrated Ownership

Liang et al. (2020) indicate that most of the Chinese firms have concentrated ownership. Thus, the agency problem in China is considerably contributed by the deeply rooted controlling shareholders. As revealed by Jiang and Kim (2015), a large number of Chinese listed firms have large shareholders who control company operations. Additionally, the shareholding of largest shareholder is far exceeds that of the second largest shareholder in most of these firms (Liu, 2006). Furthermore, Liu and Lu (2007) claimed that the interests of Chinese minority shareholders can be jeopardized by majority shareholders via dishonest acts such as misappropriation or embezzlement as found in Shleifer and Vishny (1997) and Cao et al. (2019).

Unfortunately, minority shareholders, investors, and boards in China are unable to dismiss managers appointed by controlling shareholders for poor performance since the controlling shareholder of a firm is either the government or the single largest owner of the equity stake. Therefore, there are limited approaches to address the agency problem in China. Conversely, large shareholders in Western countries are commonly considered a feasible answer to the vertical agency problem between internal managers and external investors due to the dispersed ownership of listed firms (Edmans, 2014; Edmans & Holderness, 2017).

On the other hand, Xu and Wang (1999) argue that the quality of direct monitoring of the managers can be improved by the existence of a controlling shareholder. In a recent study, it is found that there is a positive relationship between concentrated ownership in Chinese listed firms and firm performance even though not significant (Shao, 2019). Shao (2019) argued that larger shareholders may alleviate agency cost and enhance firm performance.

On top of that, Jiang et al. (2017) reveal that listed firms with controlling shareholders who hold more than 50% equity stake are generally found to perform better. The empirical findings of Cao et al. (2019) and Fang et al. (2018), on the other hand, claim that having a single large shareholder in a Chinese firm is less effective in improving firm performance than having several large shareholders.

## 2.4 *Supervisory Board*

A significant dissimilarity between China and other Western countries such as the United States and the United Kingdom is that China has embraced a two-tier board system which is a unique characteristic of the German model (Dahya et al., 2003). China established two-tier board mechanism comprising of the main board and supervisory board in 1992. The Chinese government plays an important role in the appointment of the supervisors on the supervisory board (Firth et al., 2010; Kato & Long, 2006). The primary responsibility of the supervisory board is to monitor and appraise the performance of senior management. The listed firm must have at least three members in a supervisory board (Firth et al., 2007). In addition, elected labor representatives should constitute at least one-third of the supervisory board membership as required by the CSRC in 2005.

According to Wang (2008), the fundamental difference between Chinese dual board system and German governance mechanism is that the German supervisory board is authorized to appoint and remove the directors on the management board. The German management board consists of only executive directors. Furthermore, there is a hierarchical relationship between the two boards in which the German management board is required to report to the supervisory board. However, such a hierarchical relationship between supervisory and management boards does not exist in China in accordance with the Code of Corporate Governance for Listed Companies released by the CSRC in 2002, and both boards are required to report to the shareholders.

The rationale for a supervisory board is to mitigate agency conflict. Supervisory board members should therefore be elected by shareholders to provide an oversight role in monitoring the performance management board which comprises mostly executive directors. However, the Chinese supervisory boards do not have much authority to execute the oversight role (Clarke, 2006). Therefore, it is difficult for a supervisory board to exert effective supervision of the board of directors in China (Conyon & He, 2011; Wang, 2008). In a nutshell, the chairperson of the Chinese supervisory board should be given comparable authority as the chairperson of the board of directors to strengthen the effectiveness of the supervisory board.

The Chinese market has demonstrated that the supervisory boards are rubber stamp and may not be able to effectively provide monitoring (Dahya et al., 2003; Ran et al., 2015; Xi, 2006) in spite that the 2006 Chinese corporate law reform enhancing the oversight power of supervisory boards, reinforcing their right to administer directors and managers (Ding et al., 2010; Ran et al., 2015). Jiang and Kim (2020) further claim that the monitoring role of supervisory board may be negligible or purposeless since the chairperson of the supervisory board is outranked by the chairperson of the board of directors. Nonetheless, their findings are contradictory to the study of Tusek et al. (2009) in which the effectiveness of supervisory boards was found in European economies.

Furthermore, Jia et al. (2009) show that there is no positive relationship between the size and meeting frequency of Chinese supervisory boards and quality of CG. Therefore, elected labor representatives must be included in the supervisory board to further enhance the monitoring capability of directors which ultimately improves firm performance (Jiang & Kim, 2015). Notably, Farag and Mallin (2017) found that independent nonexecutive directors on the board of director could be a perfect substitute for members of the supervisory board which implies that there might be a conflict or overlap in the oversight roles and duties between the two boards.

On the other hand, Ran et al. (2015) provide evidence supporting the effectiveness of Chinese supervisory boards in which the quality of financial information can be improved by certain attributes of supervisory board members. Shao (2019) also reveals that there is a positive correlation between the Chinese supervisory board and firm performance which is contrary to much of the previous and current literature on the inefficacy of the supervisory board in China.

To recap, the Chinese CG system integrates the German model with the Anglo-Saxon model in which public companies (SOEs and non-SOEs) are required by the Chinese company law to set up a supervisory board. In spite of its two-tier board mechanism, it is much similar to the Anglo-American model given that the role of the supervisory board is evidently proven to be redundant and less effective in general.

## 2.5 Institutional Investors

Both domestic and foreign investor confidence play a critical role in the Chinese capital market considering the increasing importance of the Chinese economy and the growing influential role of Chinese multinational corporations in today's world. La Porta et al. (2000, 2002) reveal that traditional finance theories assert that protection of investor plays a vital role in the development of the capital market. Shleifer and Vishny (1997) acknowledge the potential expropriation of minority shareholders by the deeply rooted majority shareholders. The empirical findings of La Porta et al. (2002) demonstrate that the firm values are higher in economies with better minority shareholder protection. Jiang and Kim (2020) argue that foreign investors are willing to pay premium price for assets with better protection as investors expect more profits to be distributed to them with lesser expropriation by majority shareholders in the long run (Jiang & Kim, 2020).

Historically, China has a weak external CG mechanism due to its undeveloped legal system (Fan & Wong, 2002). The undeveloped Chinese legal system is related to the negligence of Chinese company and securities law in providing civil liability and compensation to a certain extent. Also, processes and particular clauses for enforceable civil actions are not provided. Li et al. (2010) asserted that there is no effective legal framework and governance mechanism to safeguard the interests of minority shareholders and investors in China because institutional investors are not allowed to hold more than 10% shares in a listed firm while every listed firm has a majority shareholder that holds more than 10% of listed firm's shares. Thus, institutional investors in China do not have the incentive and the authority to proactively exercise effective oversight role as compared to the institutional investors in the United States (Firth et al., 2010). A recent study (Su & Alexiou, 2020) also shows that investor protection is considerably poor in Chinese listed firms due to a weak institutional environment.

The US capital market consists of a high percentage of institutional investors that are not associated with the government which is in contrast to the high proportion of state ownership in China. The lack of state control and the high percentage of institutional investors in US firms have direct associations with the standard of US CG. Chung and Zhang (2011) investigate US firms and their empirical findings demonstrating that the proportion of a firm's shares held by institutional investors increases in tandem with the standard of its CG model. Aggarwal et al. (2011), Gillan and Starks (2003), and Pitelis and Clarke (2004) found that growing institutional ownership led to improvements in CG because institutional investors play a key role in maintaining higher standard of firm's CG. Edmans and Holderness (2017) also hold that institutional investor, such as mutual and pension funds, is effective and efficient in monitoring the board performance of listed firm in Western economies.

Empirical studies reveal the importance of institutional investors in the context of Chinese CG. For instance, political pressure is likely to affect local institutional investors and cause them to yield to the state (Huang & Zhu, 2015). Comparably, local mutual funds capitulated on political pressure and supported SOEs during the reform of the spilt-share structure (Firth et al., 2010).

As opposed to prior research, recent studies (Aggarwal et al., 2015; Huang & Zhu, 2015; Liu, 2018) revealed that both mutual funds and foreign institutional investors can help improve the CG standard of a firm in the equity market of China. Meng et al. (2018) show that a firm is more willingly to elect outside directors to the board to enhance the monitoring capability of the board in the presence of foreign investors. In brief, foreign institutional investors have begun to play a more significant role in enhancing the Chinese CG model.

### 2.6 A Comparative Overview of Corporate Governance Systems

The table below summarizes the similarities and differences of each model in relation to CG attributes based on the discussion and analysis in the earlier section of this study (Table 1).

## 3 The Myth of Convergence in CG

There is an increasing debate among practitioners and academicians on the convergence of CG models as a result of firms and capital becoming more global. The convergence of the CG model is the progressive movement of CG policies and practices toward a common point.

Prior studies assert that the universal best CG model does not exist in the world. For instance, the Anglo-American mechanism is being advocated throughout the

**Table 1** Corporate governance: similarities and differences

| Attributes | Anglo-American | German | Chinese |
|---|---|---|---|
| Key stakeholder | Shareholder | Stakeholder | Government: SOEs |
| Board composition | Mostly nonexecutive and independent directors | Employees, founding members, and banks | Less independent directors when ownership concentration is high |
| Board structure | Single | Two-tier | Two-tier |
| Board of directors: insider/outsider | Mainly outsiders | Mainly insiders | Mainly insiders |
| Ownership | Widespread, noncorporate | Concentrated, high corporate, bank, family | Concentrated: high SOEs |
| Legal system | Common law | Civil law | Legal system is relatively weak |

world, but the outcomes of the research that examined against the differences of the country's specific characteristics which include economic conditions, legal systems, ownership structures, and financial markets remain ambiguous. Such equivocal empirical findings are against the total convergence that advocates for full replication of the Western CG model. However, advocates of total divergence are required to pay attention on issues such as global stock exchange, globalization, international private equity funding, global mergers and acquisitions, international trade, and international accounting standards which are in conflict with isolated divergence systems. But the potential negative effects of deglobalization due to trade war between the United States and China as well as the recent COVID-19 pandemic may further hinder the convergence of the CG model.

Generally, there are several loopholes in the US and China's CG systems. The inefficiency of the US CG can be summarised as CEO primacy; non-vigilant directors: the director-manager agency issue; and gaps of fiduciary duties: the directors-shareholder agency issue while key problems of Chinese CG comprise of highly concentrated ownership structure (SOEs and privately owned enterprises), corporate affairs controlled by shareholders and fabrication of financial and non-financial information, inadequate protection of minority shareholders' rights and ineffective monitoring mechanism of independent directors and supervisory boards.

The separation between ownership and control remains the fundamental issue of CG. The central problem of listed companies' governance in the United States is that the current and historical practices of the US corporate laws in relation to the distribution of power between shareholders and the board of directors have granted too much freedom and authority to the management and board which in turn increases embezzlement due to widespread of ownership in the United States.

Roberts et al. (2005) opine that agency theory is influential in the development of studies, reforms, and practices in the discipline of the US CG model. Agency theory highlights the issues arising from the separation of ownership and control which centers on the protection of shareholder rights from self-serving managers (Fama, 1980; Jensen & Meckling, 1976; Shleifer & Vishny, 1997). Sarbanes-Oxley Act 2002 was established in the United States in which governance mechanisms such as board independence and accountability of directors were introduced to resolve the conflict between shareholders and management.

On the other hand, the underlying problem concerning the governance related issues in China is that the board of directors and management are easily influenced and controlled by large shareholders or shareholder groups to attain the only interests of individual shareholder or group of shareholders instead of the interests of all shareholders. Hence, the fundamental CG problem in the United States is the conflicts between shareholders and management, while the oppression of minority shareholders is the core problem that remains unresolved in the current China's CG system.

China has largely adopted the Anglo-American CG model since 1992. However, Young et al. (2008) and Yusuf et al. (2018) argue that the model of agency theory is not applicable to emerging economies in which ownership is largely concentrated.

As of today, SOEs continue to dominate the Chinese economy as a whole (Cumming et al., 2021).

Cuomo et al. (2016) show higher compliance levels with CG codes in Anglo-American countries as opposed to developing countries like China. However, there is a scarcity of data in China which explores why China has lower compliance level with CG codes as most research on CG is mainly based on quantitative methods. In addition, Cuomo et al. (2016) reveal that CG and firm performance relationships have been found to be inconsistent due to conceptual and methodological issues in which the role of a country's institutional environment on the behavior of firm has not been adequately explored.

Ironically, the bankruptcy of Lehman Brothers and the takeover of Fannie Mae and Freddie Mac in the United States were the major causes of the 2008 global financial crisis and indicating that Anglo-American CG practices might not be the model to follow. It is vital for the Chinese government to study and draw on the negative foreign lessons in which the US government employed the administrative intervention and public money to bail out the distressed firms. This course of actions may be taken as "one-off" measure, but it manifests the US CG being ineffective and may be seen as the underlying reason of the crisis.

In Anglo-American countries, insiders and outsiders of firms such as independent directors, auditors, institutional investors, security analysts, and media are considered to be an effective oversight mechanism but remain highly contestable. Dyck et al. (2010) demonstrate that the outcomes generated by some of the above intended monitoring tools in the United States are either slow or unsatisfactory; and in some cases, these monitoring mechanisms fail entirely to uncover corporate deception and scandals. In brief, the efficacy of the Anglo-American CG model remains dubious even in the context of the United States.

On the other hand, an in-depth comprehension of China's unique economic issues, CG mechanisms, institutional framework, business practices, and customs is extremely important for foreign institutional investors and foreign corporate executives. From an academic point of view, Cumming et al. (2021) assert that neither the well-established theory from the West can be directly applied in the Chinese economy nor the empirical findings from emerging economies could explain the distinctive characteristics of the Chinese capital market. Jiang and Kim (2020) further highlight that many researchers misunderstand or are unaware of the dissimilarities between the Western countries and China in the context of economic, regulatory, and cultural frameworks which led them to inaccurately interpret Chinese empirical studies based on the Western theories and perceptions. Consequently, China should establish a CG model conforming to its own reality based on its previous experiences and lessons learned.

# 4 Conclusion and Future Research

The objective of this study is to review the Chinese CG via the thematic lens that is directly relevant to the major considerations of foreign institutional investors and foreign corporate executives which include protection of their firms' interests in China and the extent of transparency and reliability of the second largest economy's governance policy. The selected themes are derived from the major considerations of foreign investors and foreign corporate executives which comprise of board independence, state ownership, concentrated ownership, supervisory board, and institutional investors.

Equally important, the contemporary overview of Chinese capital market and regulatory body, namely, Chinese Securities Regulatory Commission, is also provided in this review which will enhance the target audience's understanding of the distinctive characteristics of financial system and legal framework in China. At the same time, the review has implications for policymakers and regulators as well as contributes to the expansion of CG literature in China.

Agency theory argues that an increasing number of independent directors on a board promotes stronger and better CG. However, this review asserts that Chinese independent directors are short of independence and are unable to carry out their monitoring roles as stipulated in the CG guidelines and company law. The day-to-day operations are continuing dominated by the insider managers. As a result, insider control problem remains rampant in China despite studies showing that the average number of independent directors in listed firms is growing. Accordingly, this study suggests that greater power should be granted to independent directors by the regulations to effectively fulfill their oversight role.

In addition, this review holds that the agency issues concerning SOE managers such as embezzlement are generally much more destructive than the conflict of interest between the state and minority shareholders due to the separation of ownership and management. Thus, it is suggested that Chinese regulators should revisit the monitoring role of independent directors and members of supervisory boards to mitigate the agency problems caused by SOE managers and safeguard the interest of minority shareholders. It is also found that the efficacy of concentrated ownership or controlling shareholders, be it single or multiple, in mitigating agency issues and improving firm performance in China remains inconclusive. Principal-principal conflicts continues to cause high agency costs for Chinese firms. Consequently, it is recommended that Chinese policymakers revise the rules and regulations concerning the protection of minority shareholders including the foreign institutional investors to better protect their interests so as to attract more foreign investment.

With regard to supervisory board, the review shows that supervisory board is ineffective in monitoring the performance of board of directors and managers. Consequently, they are incapable of identifying and resolving expropriation by managers. Even though recent empirical evidence unveils that the supervisory board is able to improve firm performance which is in contrast with most of the prior literature on the ineffectiveness of the Chinese supervisory board, this review suggests that Chinese policymaker and regulator should distinguish and separate the possible overlapping duties and roles of independent nonexecutive directors and members of supervisory board in listed firms.

This review signifies the critical role of institutional investors in the Chinese capital market and CG. This review found that foreign institutional investors have higher impact on state owners and are more active in overseeing SOEs because they are unsusceptible to political pressure. The degree of foreign institutional ownership and its composition are important features of the Chinese CG. Institutional investors are instrumental in causing changes within a firm in the United States, while Chinese institutional investors have limited role to play in administering SOEs. Thus, growth in foreign institutional ownership is important for improving Chinese CG. It is proposed that company and securities laws be enhanced safeguarding both the domestic and foreign institutional investors.

On another note, this study provides a review on whether Chinese CG converge toward the Anglo-American practices or continue to diverge from the Anglo-American model based on the current literature. This review shows that the Chinese CG mechanism possesses some similarities to both Anglo-American and German models while maintaining its unique attributes after two decades of extensive reform through multiple series of regulatory changes and the development of new rules. The existing Chinese regulatory framework, capital market structure, and institutional factors are found to have considerable influence on the implementation of CG procedures, code, and regulation although China has predominantly embraced the Anglo-Saxon CG mechanism. However, the severe damages caused by the 2008 global financial crisis on the world economy has, to an extent, overturned the credence of the market-based Anglo-American CG system being the only effective and efficient universal standard of CG model. At the same time, the German CG model is far from universal norm. Thus, this review concurs with the notion of many researchers that there is no one-size-fits-all CG model given the diverse and dynamic environment of emerging economies. This study, therefore, offers useful empirical guidance concerning CG for Chinese policymakers and regulatory authorities to consider the need to not only enhancing the quality of CG but also reshaping its CG codes and regulations based on their distinctive economic conditions, history, financial system, legal framework, social development, cultural aspects, competition, ownership structure, capital market framework, and industry- and firm-specific characteristics that truly reflect their own realities in order to smoothen their transition from a planned economy to a fully market-oriented economy.

There are limitations in this study. Firstly, this study is based on narrative review method, and this technique has been criticized for its subjective and biased methodology. Another limitation is that the focus of this study is only on China and hence the findings of this study cannot be fully generalized to other emerging economies given that countries have their own distinctive characteristics as revealed in previous empirical studies. However, this review provides general suggestion and empirical guidance for emerging countries with homogeneous institutional elements.

In 2018, the CSRC has issued the revised Guidelines on the Governance of Listed Companies (the first revision since 2002). Major changes to the updated guidelines include the recommendation of board diversity, enhancement of the functions of the audit committee, confining the powers of controlling shareholders, the requirement of environmental, social and governance guidelines and etc. The introduction of the revised CG guidelines, coupled with the expansion of new company law in 2019, is continuously changing the Chinese CG landscape. Hence, this review suggests that further studies are needed to not only investigate the effect of these recent changes on country-level and firm-level outcomes but also to review the themes of this paper (or other relevant themes) using different methodologies to deepen the understanding of foreign investors and foreign corporate practitioners on contemporary CG in China.

## References


Aggarwal, R., Erel, I., Ferreira, M., & Matos, P. (2011). Does governance travel around the world? Evidence from institutional investors. *Journal of Financial Economics, 100*(1), 154–181. https://doi.org/10.1016/j.jfineco.2010.10.018

Aggarwal, R., Hu, M., & Yang, J. (2015). Fraud, market reaction, and the role of institutional investors in Chinese listed firms. *Journal of Portfolio Management, 41*, 92–109. https://doi.org/10.3905/jpm.2015.41.5.092

Al Dah, B. (2018). Monitoring or empowering CEOs? The moderating effect of shareholder rights. *Research in International Business and Finance, 46*, 502–515. https://doi.org/10.1016/j.ribaf.2018.06.006

Allen, J., & Li, R. (2017, December 13). Awakening Governance: ACGA China Corporate Governance Report 2018. *Harvard Law School Forum on Corporate Governance.* https://corpgov.law.harvard.edu/2018/08/25/awakening-governance-acga-china-corporate-governance-report-2018/

Allen, F., Qian, J., & Qian, M. (2005). Law, finance, and economic growth in China. *Journal of Financial Economics, 77*(1), 57–116. https://doi.org/10.1016/j.jfineco.2004.06.010

Bai, C. E., Liu, Q., Lu, J., Song, F. M., & Zhang, J. (2004). Corporate governance and market valuation in China. *Journal of Comparative Economics, 32*, 599–616. https://doi.org/10.1016/j.jce.2004.07.002

Baldenius, T., Melumad, N., & Meng, X. (2014). Board composition and CEO power. *Journal of Financial Economics, 112*(1), 53–68. https://doi.org/10.1016/j.jfineco.2013.10.004

Baumeister, R. F., & Leary, M. R. (1997). Writing narrative literature reviews. *Review of General Psychology, 1*(3), 311–320. https://doi.org/10.1037/1089-2680.1.3.311

Berle, A., & Means, G. (1932). *The modern corporation and private property*. Macmillan.

Cao, F., Peng, S., & Ye, K. (2019). Multiple large shareholders and corporate social responsibility reporting. *Emerging Markets Review, 38*, 287–309. https://doi.org/10.1016/j.ememar.2019.02.004

Chung, K. H., & Zhang, H. (2011). Corporate governance and institutional ownership. *Journal of Financial and Quantitative Analysis, 46*(1), 247–273. https://doi.org/10.1017/S0022109010000682

Clarke, D. C. (2003). Corporate governance in China: An overview. *China Economic Review, 14*(4), 494–507. https://doi.org/10.1016/j.chieco.2003.09.019

Clarke, D. C. (2006). The independent director in Chinese Corporate Governance. *Delaware Journal of Corporate Law, 31*, 125–173.

Conyon, M. J., & He, L. (2011). Executive compensation and corporate governance in China. *Journal of Corporate Finance, 17*(4), 1158–1175. https://doi.org/10.1016/j.jcorpfin.2011.04.006

Costa, S. H. E. (2021). *Two accounting scandals in China in one week burn investors*. Bloomberg. https://www.bloomberg.com/news/articles/2020-04-08/two-accounting-scandals-in-one-week-burn-investors-in-china-inc

CSRC. (2001, January 7). Code of corporate governance for listed companies in China. http://www.csrc.gov.cn/pub/csrc_en/newsfacts/release/200708/t20070810_69223.html

Cumming, D. J., Verdoliva, V., & Zhan, F. (2021). New and future research in corporate finance and governance in China and emerging markets. *Emerging Markets Review, 46*, 100792. https://doi.org/10.1016/j.ememar.2020.100792

Cuomo, F., Mallin, C., & Zattoni, A. (2016). Corporate governance codes: A review and research agenda. *Corporate Governance: An International Review, 24*(3), 222–241. https://doi.org/10.1111/corg.12148

Dahya, J., Karbhari, Y., Xiao, J. Z., & Yang, M. (2003). The usefulness of the supervisory board report in China. *Corporate Governance: An International Review, 11*(4), 308–321. https://doi.org/10.1111/1467-8683.00329

Ding, S., Wu, Z., Li, Y., & Jia, C. (2010). Executive compensation, supervisory board, and China's governance reform: A legal approach perspective. *Review of Quantitative Finance and Accounting, 35*(4), 445–471. https://doi.org/10.1007/s11156-010-0168-1

Du, F., Erkens, D. H., Young, S. M., & Tang, G. (2018). How adopting new performance measures affects subjective performance evaluations: Evidence from EVA adoption by Chinese state-owned enterprises. *Accounting Review, 93*, 161–185. https://doi.org/10.2308/accr-51763

Du, X., Jian, W., & Lai, S. (2017). Do foreign directors mitigate earnings management? Evidence from China. *The International Journal of Accounting, 52*, 142–177. https://doi.org/10.1016/j.intacc.2017.04.002

Dyck, A., Morse, A., & Zingales, L. (2010). Who blows the whistle on corporate fraud? *Journal of Finance, 65*(6), 2213–2253. https://doi.org/10.1111/j.15406261.2010.01614.x

Edmans, A. (2014). Blockholders and corporate governance. *Annual Review of Financial Economics, 6*, 23–50. https://doi.org/10.1146/annurev-financial-110613-034455

Edmans, A., & Holderness, C. G. (2017). Blockholders: A survey of theory and evidence. *The Handbook of the Economics of Corporate Governance, 1*, 541–636. https://doi.org/10.1016/bs.hecg.2017.11.002

Fama, E. F. (1980). Agency problems and the theory of the firm. *Journal of Political Economy, 88*(2), 288–307. https://doi.org/10.1086/260866

Fama, E. F., & Jensen, M. C. (1983). The separation of ownership and control. *Journal of Law and Economics, 26*(2), 301–325. https://doi.org/10.1086/467037

Fan, J. P. H., & Wong, T. J. (2002). Corporate ownership structure and the informativeness of accounting earnings in East Asia. *Journal of Accounting and Economics, 33*(3), 401–425. https://doi.org/10.1016/S0165-4101(02)00047-2

Fang, Y., Hu, M., & Yang, Q. (2018). Do executives benefit from shareholder disputes? Evidence from multiple large shareholders in Chinese listed firms. *Journal of Corporate Finance, 51*, 275–315. https://doi.org/10.1016/j.jcorpfin.2018.06.008

Farag, H., & Mallin, C. (2017). Monitoring corporate boards: Evidence from China. *European Journal of Finance, 25*(6), 524–549. https://doi.org/10.1080/1351847X.2017.1369138

Firth, M., Fung, P. M. Y., & Rui, O. M. (2007). Ownership, two-tier board structure, and the informativeness of earnings – Evidence from China. *Journal of Accounting and Public Policy, 26*(4), 463–496. https://doi.org/10.1016/j.jaccpubpol.2007.05.004

Firth, M., Lin, C., & Zou, H. (2010). Friend or foe? The role of state and mutual fund ownership in the split share structure reform in China. *Journal of Financial and Quantitative Analysis, 45*(3), 685–706. https://doi.org/10.1017/S0022109010000190

Gan, J., Guo, Y., & Xu, C. (2018). Decentralized privatization and change of control rights in China. *Review of Financial Studies, 31*(10), 3854–3894. https://doi.org/10.1093/rfs/hhx100

Gillan, S. L., & Starks, L. T. (2003). Institutional investors, corporate ownership and corporate governance: Global perspectives. In L. Sun (Ed.), *Ownership and governance of enterprises* (pp. 36–68). Palgrave Macmillan. https://doi.org/10.1057/9781403943903_2

Guldiken, O., & Darendeli, I. S. (2016). Too much of a good thing: Board monitoring and R&D investments. *Journal of Business Research, 69*(8), 2931–2938. https://doi.org/10.1016/j.jbusres.2015.12.062

Hanson, A. H. (2019). *Public enterprise and economic development*. Routledge.

He, W., & Luo, J. (2018). Agency problems in firms with an even number of directors: Evidence from China. *Journal of Banking & Finance, 93*, 139–150. https://doi.org/10.1016/j.jbankfin.2018.06.006

Hodgkinson, G. P., & Ford, J. K. (2014). Narrative, meta-analytic, and systematic reviews: What are the differences and why do they matter? *Journal of Organizational Behavior, 35*(1), S1–S5. https://doi.org/10.1002/job.1918

Hu, R., Karim, K., Lin, K. J., & Tan, J. (2020). Do investors want politically connected independent directors? Evidence from their forced resignations in China. *Journal of Corporate Finance, 61*, 101421. https://doi.org/10.1016/j.jcorpfin.2018.11.004

Huang, W., & Zhu, T. (2015). Foreign institutional investors and corporate governance in emerging markets: Evidence of a split-share structure reform in China. *Journal of Corporate Finance, 32*, 312–326. https://doi.org/10.1016/j.jcorpfin.2014.10.013

Jensen, M. C., & Meckling, W. H. (1976). Theory of the firm: Managerial behaviour, agency costs and ownership structure. *Journal of Financial Economics, 3*(4), 305–360. https://doi.org/10.1016/0304-405X(76)90026-X

Jia, C., Ding, S., Li, Y., & Wu, Z. (2009). Fraud, enforcement action, and the role of corporate governance: Evidence from China. *Journal of Business Ethics, 90*(4), 561–576. https://doi.org/10.1007/s10551-009-0061-9

Jiang, F., Jiang, Z., & Kim, K. A. (2017). Capital markets, financial institutions, and corporate finance in China. *Journal of Corporate Finance, 63*, 101309. https://doi.org/10.1016/j.jcorpfin.2017.12.001

Jiang, F., & Kim, K. A. (2015). Corporate governance in China: A modern perspective. *Journal of Corporate Finance, 32*, 190–216. https://doi.org/10.1016/j.jcorpfin.2014.10.010

Jiang, F., & Kim, K. A. (2020). Corporate governance in China: A survey. *Review of Finance, 24*(4), 733–772. https://doi.org/10.1093/rof/rfaa012

Jiang, B. B., Laurenceson, J., & Tang, K. K. (2008). Share reform and the performance of China's listed companies. *China Economic Review, 19*(3), 489–501. https://doi.org/10.1016/j.chieco.2008.02.001

Johnson, S., McMillan, J., & Woodruff, C. (2002). Property rights and finance. *American Economic Review, 92*(5), 1335–1356. https://doi.org/10.1257/000282802762024539

Kato, T., & Long, C. (2006). CEO turnover, firm performance, and enterprise reform in China: Evidence from micro data. *Journal of Comparative Economics, 34*(4), 796–817. https://doi.org/10.1016/j.jce.2006.08.002

Kaye, L. (1995, February 23). Withering state: Fire when ready. *Far Eastern Economic Review, 158*(8), 50–54.

La Porta, R., Lopez-De-Silanes, F., Schleifer, A., & Vishny, R. W. (2000). Investor protection and corporate governance. *Journal of Financial Economics, 58*(1-2), 3–27. https://doi.org/10.1016/S0304-405X(00)00065-9

La Porta, R., Lopez-De-Silanes, F., Schleifer, A., & Vishny, R. W. (2002). Investor protection and corporate valuation. *Journal of Finance, 57*(3), 1147–1170. https://doi.org/10.1111/1540-6261.00457

La Porta, R., Lopez-de-Silanes, F., Shleifer, A., & Vishny, R. W. (1998). Law and finance. *Journal of Political Economy, 106*(6), 1113–1155. https://doi.org/10.1086/250042

Lardy, N. R. (1998). *China's unfinished economic revolution*. Brookings Institution Press.

Latapí Agudelo, M. A., Jóhannsdóttir, L., & Davídsdóttir, B. A. (2019). Literature review of the history and evolution of corporate social responsibility. *International Journal of Corporate Social Responsibility, 4*, 1. https://doi.org/10.1186/s40991-018-0039-y

Li, H., Meng, L., Wang, Q., & Zhou, L. A. (2008). Political connections, financing and firm performance: Evidence from Chinese private firms. *Journal of Development Economics, 87*(2), 283–299. https://doi.org/10.1016/j.jdeveco.2007.03.001

Li, L., & Naughton, T. (2007). Going public with good governance: Evidence from China. *Corporate Governance: An International Review, 15*, 1190–1202. https://doi.org/10.1111/j.1467-8683.2007.00640.x

Li, T., Sun, L., & Zou, L. (2009). State ownership and corporate performance: A quantile regression analysis of Chinese listed companies. *China Economic Review, 20*(4), 703–716. https://doi.org/10.1016/j.chieco.2009.05.006

Li, Y., Wright, M., & Scholes, L. (2010). Chinese management buyouts and board transformation. *Journal of Business Ethics, 95*(2), 361–380. https://doi.org/10.1007/s10551-011-0850-9

Li, Z., Zhou, X., Jung, S., & Li, J. (2020). China's 40-year road to innovation. *Chinese Management Studies, 14*(2), 335–357. https://doi.org/10.1108/CMS-01-2019-0019

Liang, Q., Ling, L., Tang, J., Zeng, H., & Zhuang, M. (2020). Managerial overconfidence, firm transparency, and stock price crash risk: Evidence from an emerging market. *China Finance Review International, 10*(3), 271–296. https://doi.org/10.1108/CFRI-01-2019-0007

Lin, J. Y., & Tan, G. (1999). Policy burdens, accountability, and the soft budget constraint. *American Economic Review, 89*(2), 426–431. https://doi.org/10.1257/aer.89.2.426

Liu, Q. (2006). Corporate governance in China: Current practices, economic effects and institutional determinants. *CESifo Economic Studies, 52*, 415–453. https://doi.org/10.1093/cesifo/ifl001

Liu, C. (2018). Are women greener? Corporate gender diversity and environmental violations. *Journal of Corporate Finance, 52*, 118–142. https://doi.org/10.1016/j.jcorpfin.2018.08.004

Liu, Q., & Lu, Z. (2007). Corporate governance and earnings management in the Chinese listed companies: A tunnelling perspective. *Journal of Corporate Finance, 13*(5), 881–906. https://doi.org/10.1016/j.jcorpfin.2007.07.003

Liu, Y., Miletkov, M. K., Wei, Z., & Yang, T. (2015). Board independence and firm performance in China. *Journal of Corporate Finance, 30*(C), 223–244. https://doi.org/10.1016/j.jcorpfin.2014.12.004

Liu, G., & Sun, P. (2005). Identifying ultimate controlling shareholders in Chinese public corporations: An empirical survey. In D. Brown & A. MacBean (Eds.), *Challenges for China's development*. Routledge.

Liu, C., Uchida, K., & Yang, Y. (2012). Corporate governance and firm value during the global financial crisis: Evidence from China. *International Review of Financial Analysis, 21*(C), 70–80. https://doi.org/10.1016/j.irfa.2011.11.002

Meng, Y., Clements, M. P., & Padgett, C. (2018). Independent directors, information costs and foreign ownership in Chinese companies. *Journal of International Financial Markets, Institutions and Money, 53*, 139–157. https://doi.org/10.1016/j.intfin.2017.09.016

Menozzi, A., Gutierrez Urtiaga, M., & Vannoni, D. (2011). Board composition, political connections, and performance in state-owned enterprises. *Industrial and Corporate Change, 21*(3), 671–698. https://doi.org/10.1093/icc/dtr055

Ng, A., Yuce, A., & Chen, E. (2009). Determinants of state equity ownership, and its effect on value/performance: China's privatized firms. *Pacific-Basin Finance Journal, 17*(4), 413–443. https://doi.org/10.1016/j.pacfin.2008.10.003

OECD. (2015a). *Economic surveys China*. OECD Publishing.

OECD. (2015b). *OECD guidelines on corporate governance of state-owned enterprises*. OECD Publishing. https://doi.org/10.1787/9789264244160-en

Paul, J. (2016). The rise of China: What, when, where, and why? *The International Trade Journal, 30*(3), 207–222.

Pitelis, C., & Clarke, T. (2004). Valuing corporate governance: An introduction. *Corporate Governance: An International Review, 12*, 125–128. https://doi.org/10.1111/j.1467-8683.2004.00350.x

Rajagopalan, N., & Zhang, Y. (2008). Corporate governance reforms in China and India: Challenges and opportunities. *Business Horizons, 51*(1), 55–64. https://doi.org/10.1016/j.bushor.2007.09.005

Ran, G., Fang, Q., Luo, S., & Chan, K. C. (2015). Supervisory board characteristics and accounting information quality: Evidence from China. *International Review of Economics and Finance, 37*(C), 18–32. https://doi.org/10.1016/j.iref.2014.10.011

Roberts, J., McNulty, T., & Stiles, P. (2005). Beyond agency conceptions of the work of non-executive director: Creating accountability in the boardroom. *British Journal of Management, 16*(1), 5–26. https://doi.org/10.1111/j.14678551.2005.00444.x

Rousseau, D. M. (Ed.). (2012). *The Oxford handbook of evidence-based management.* Oxford University Press.

Rousseau, D. M., Manning, J., & Denyer, D. (2008). Evidence in management and organizational science: Assembling the field's full weight of scientific knowledge through reflective reviews. *Annals of the Academy of Management, 2*(1), 475–515. https://doi.org/10.5465/19416520802211651

Shao, L. (2019). Dynamic study of corporate governance structure and firm performance in China: Evidence from 2001-2015. *Chinese Management Studies, 13*(2), 299–317. https://doi.org/10.1108/CMS-08-2017-0217

Shleifer, A., & Vishny, R. W. (1997). A survey of corporate governance. *Journal of Finance, 52*(2), 737–783. https://doi.org/10.1111/j.1540-6261.1997.tb04820.x

Snyder, H. (2019). Literature review as a research methodology: An overview and guidelines. *Journal of Business Research, 104*, 333–339. https://doi.org/10.1016/j.jbusres.2019.07.039

Su, Z., & Alexiou, C. (2020). Equity incentive schemes, investor protection and corporate performance: Evidence from China. *China Finance Review International, 10*(3), 297–322. https://doi.org/10.1108/CFRI-07-2019-0091

The Asian Banker Journal. (2002, December 31). *An intensified will to tackle the problem.* Special Edition, 3–6.

Tusek, B., Filipovic, D., & Pokrovac, I. (2009). Empirical evidence about the relationship between external auditing and the supervisory board in the republic of Croatia. *Ekonomski Pregled Mjesečnik Hrvatskog Društva Ekonomista Zagreb, 60*(1-2), 3–27.

Usman, M., Zhang, J., Farooq, M. U., Makki, M. A. M., & Dong, N. (2018). Female directors and CEO power. *Economics Letters, 165*, 44–47. https://doi.org/10.1016/j.econlet.2018.01.030

Wang, J. (2008). The strange role of independent directors in a two-tier board structure in China's listed companies. In M. Nakamura (Ed.), *Changing corporate governance practices in China and Japan.* Palgrave Macmillan. https://doi.org/10.1057/9780230595156_9

Wang, L. (2015). Protection or expropriation: Politically connected independent directors in China. *Journal of Banking & Finance, 55*(6), 92–106. https://doi.org/10.1016/j.jbankfin.2015.02.015

Wang, X., Xu, L. C., & Zhu, T. (2004). State-owned enterprises going public the case of China. *The Economics of Transition and Institutional Change, 12*(3), 467–487. https://doi.org/10.1111/j.0967-0750.2004.00189.x

Wei, Z., Xie, F., & Zhang, S. (2005). Ownership structure and firm value in China's privatized firms: 1991-2001. *Journal of Financial and Quantitative Analysis, 40*(1), 87–108. https://doi.org/10.1017/S0022109000001757

Wong, G., Greenhalgh, T., Westhorp, G., Buckingham, J., & Pawson, R. (2013). RAMESES publication standards: Meta-narrative reviews. *BMC Medicine, 11*, 20. https://doi.org/10.1186/1741-7015-11-20

World Bank. (2020). Market capitalization of listed domestic companies (current US$) – China. https://data.worldbank.org/indicator/CM.MKT.LCAP.CD?locations=CN

Wu, H., Li, S., Ying, S. X., & Chen, X. (2018). Politically connected CEOs, firm performance, and CEO pay. *Journal of Business Research, 91*, 169–180. https://doi.org/10.1016/j.jbusres.2018.06.003

Xi, C. (2006). In search of an effective monitoring board model: Board reforms and the political economy of corporate law in China. *Connecticut Journal of International Law, 22*(1), 1–46. https://ssrn.com/abstract=977400

Xu, X., & Wang, Y. (1999). Ownership structure and corporate governance in Chinese stock companies. *China Economic Review, 10*(1), 75–98. https://doi.org/10.1016/S1043-951X(99)00006-1

Young, M. N., Peng, M. W., Ahlstrom, D., Bruton, G. D., & Jiang, Y. (2008). Corporate governance in emerging economies: A review of the principal–principal perspective. *Journal of Management Studies, 45*(1), 196–220. https://doi.org/10.1111/j.1467-6486.2007.00752.x

Yu, M. (2013). State ownership and firm performance: Empirical evidence from Chinese listed companies. *China Journal of Accounting Research, 6*(2), 75–87. https://doi.org/10.1016/j.cjar.2013.03.003

Yusuf, F., Yousaf, A., & Saeed, A. (2018). Rethinking agency theory in developing countries: A case study of Pakistan. *Accounting Forum, 42*(4), 281–292. https://doi.org/10.1016/j.accfor.2018.10.002

**Shy Lih Wong** is currently pursuing a doctoral degree in Business Administration at UCSI University, Malaysia. His experience of over 20 years spans many fields, including property development, property investment, construction, fund management, and corporate finance. Prior to pursuing his doctorate, he was a manager in his family-owned firms which are principally involved in housing development and commercial property investment. He holds a Master of Finance and a Bachelor of Business (Economics and Finance) with Distinction from RMIT University in Melbourne, Australia. His current research interests focus on environmental, social, and governance as well as corporate social responsibility.

# The Core Capital Approach to Board: A Lesson from People's Credit Banks in Indonesia

**Yafet Yosafet Wilben Rissy**

## Abbreviations

| | |
|---|---|
| BOCs | Board of commissioners |
| BODs | Board of directors |
| GMS | General meeting of shareholders |
| IFSA | Indonesian Financial Services Authority |
| LLC | Limited liability company |
| MB | Management board |
| PCBs | People's Credit Banks |
| SB | Supervisory board |

## 1 Introduction

Board of a company is one of the most critical corporate governance mechanisms as it tries to optimize and balance shareholders' as well as other stakeholders' welfare and interests (Beiner et al., 2004; Adams & Ferreira, 2007; Fama & Jensen, 1983; Hermalin & Weisbach, 1988; Shleifer & Vishny, 1997; Donaldson & Preston, 1995; Freeman et al., 2004; Freeman & Reed, 1983; Post et al., 2002). But what approach can be deployed when forming a company's board? Up until now, a board and its composition is mainly determined by approaches aimed at compliance with required governance structures either by hard laws or acts or a mandatory driven (the Sarbanes-Oxley Act of 2002; Rose, 2016; Cohen & Qaimmaqami, 2005; Indonesia

Y. Y. W. Rissy (✉)
Faculty of Law, Satya Wacana Christian University, Salatiga, Indonesia
e-mail: yafet.rissy@uksw.edu

 
I. G. Arraiano et al. (eds.), *Corporate Social Responsibility in a Dynamic Global Environment*, CSR, Sustainability, Ethics & Governance,
https://doi.org/10.1007/978-3-031-24647-0_15

Company Law, 2007) or the voluntary establishment of the board governance structure that most suits the needs as indicated by soft laws or codes or a voluntary driven (Arcot et al., 2010; the Cadbury Committee, 1992; the United Kingdom Corporate Governance Code, 2016; the Dutch Corporate Governance Code, 2016; the Germany Corporate Governance Code, 2017). These two approaches generally simply provide general criteria to be met or considered when instituting a board. These two approaches, however, do not offer a clear approach that can be used to determine the board and its composition for big and small enterprises.

In Indonesia, PCBs are categorized as small banks. They are established only to serve small-scale businesses and to provide banking services to rural people (explanatory notes of the Indonesian Financial Services Authority (IFSA) People's Credit Banks' Regulation of 2004). In terms of the establishment of PCBs' board, prior to 2014, there was no valid approach to be used to form PCBs' board structures as well as no rules that forbid family members from becoming members of board of directors (BODs) and board of commissioners (BOCs). As a result, some PCBs have oversized board or smaller board which may lead to inefficiency and ineffectiveness in directing and supervising PCBs' board (Rissy, 2018).

Scholars have been divided on the issues of bigger and smaller size of board. Some experts argue that bigger board might lead to the diffusion of duties. Thus, some members of board might be less productive than others and have difficulty in effectively utilizing their knowledge and skills. There could also be problems of coordination (Hermalin & Weisbach, 2001). Contrary to that, other scholars argue that bigger board might have more people with diverse knowledge, skill, and experience that in turn might enhance the quality of strategic decision-making and enhance the planning process. Bigger board might also have more independent directors who are able to convey more objective judgment to a company's issues and decisions, can provide much sounder advice to the top management, have better access to the resources needed by a company, have strategic external relationships, and more effectively protect shareholders from agency problem (William & Elisabeth, 1989; Daily & Johnson, 1997; Provan, 1980; Golden & Zajac, 2001; Pfeffer, 1972; Byrd & Hickman, 1992; Brickley & James, 1987).

On the other side, some scholars contend that smaller board is much more effective than bigger board as it is less bureaucratic, more functional, and provides better financial reporting monitoring (Xie et al., 2003). Regardless of the disputes among scholars on ideal size of board, experts generally agree that board is tasked to supervise and control managers and to eliminate ineffective management (Fama & Jensen, 1983; Beiner et al., 2004). Board also provides resources for the company (Hillman & Dalziel, 2003) and becomes a potential critical effective internal control (Baysinger & Hoskisson, 1990).

In the case of PCBs, as they are small enterprises, they simply need smaller board. However, the problem in PCBs is that majority of them have bigger size of board. Even it is more complicated if a PCB is a family-dominated company and the owners appoint their own family members or colleagues in the structure (Rissy, 2018). In addition, PCBs' board structures, prior to 2014, were poor and lacked the committees, units, and functions needed to function effectively. At the same time, PCBs were managed under a strong family firm-dominated culture. The majority of BOCs and BODs members were the controlling shareholders, their families or relatives. Thus, they were often seen as family-dominated firms. Also, as there was no separation between control and ownership, some members of PCBs' BODs and BOCs, who were also the controlling shareholders' families or relatives, misused their positions and committed fraud by embezzling depositors' and lenders' money and illegally used the funds for their own personal interests (Okezone, 2016; IFSA, 2015, pp. 108–109; IFSA, 2014; Bank Indonesia, 2011; Joglosemar, 2013; Wartahukum, 2017; Detikcom, 2012; Neracacoid, 2016).

Recognizing that this problem would further exacerbate PCBs' business operations, in 2004, IFSA then explicitly forbade controlling shareholders' family members from holding positions as members of BODs and BOCs (Articles 27–29 of IFSA PCB Regulations of 2014). Later, in 2015, IFSA took a firm step by establishing explicit and detailed requirements for the formation of PCBs' board structures based on PCBs' core capital. PCBs' core capital mainly consists of paid up capital and/plus additional capital reserves (Article 3 of the IFSA Minimum Capital and Core Capital Supply Obligation for PCBs Regulation of 2015). The principle is the greater the core capital is, the bigger (up to the maximum allowed) the composition of BODs and BOCs is, including the need to form committees, units, and functions (IFSA Corporate Governance for PCBs Regulation of 2015; IFSA Risk Management for PCBs Regulation of 2015). The main goal of the core capital approach toward PCBs' board is to provide a framework for PCBs to institute their board that most suits PCBs' need and characteristic as small banks or enterprises based on their core capital. In addition, as PCBs are small banks, PCBs simply need a smaller board to function effectively.

This study, therefore, seeks to explore the significance of the core capital approach as a theoretical framework in establishing PCBs' board, to evaluate the level of PCBs' compliance with the board structures based on their core capital, to identify problems faced by PCBs to comply with their board structures, and finally to find out some solutions to mitigate the problems.

## 2 Methodology

This research had three main research questions: (1) How significant are the contributions of the core capital approach as a theoretical framework toward the establishment of PCBs' board structures? (2) To what extent have PCBs complied with their board structures as required by their core capital? (3) What were the factors that contributed to failures of PCBs to comply with the board structures as required by their core capital?

This research applied a mixed methods approach, combining doctrinal legal research to address research question (1) and a qualitative social research method via in-depth interviews for research questions (2) and (3). Doctrinal legal research is a research approach that offers a systematic explanation of a particular legal category, provides rules-relationship analysis, explains difficult areas of the rules, and possibly predicts future developments (Pearce et al., 1987). In the doctrinal legal research, legal primary documents such as laws/acts and IFSA regulations which dealt with PCBs' boards were the main sources of the research. These various documents were analyzed descriptively to determine the essence and parameters of the laws and regulations in the core capital approach and PCBs' board structures (Hutchinson & Duncan, 2012, pp. 110–111).

To have a better understanding of PCBs' compliance with the required board structures and the barriers to that compliance, the researcher conducted a qualitative empirical research via in-depth interviews (Tracers, 2013, p. 229; Curtis & Curtis, 2011, pp. 29–31) with key participants from IFSA and National PCBs' Association and gathered secondary data from IFSA, Bank Indonesia, and mass media publications.

During the field research, three key participants from IFSA and one key participant from the National PCBs' Association were interviewed. The three key participants from IFSA were Bambang Widjanarko, the Head of the IFSA Region 1, Jakarta and Banten Provinces; Ahmad Fauzie, the Eastern Regional Supervisor Coordinator of IFSA; and Sulistyaningsih, a senior official of the IFSA Region 3, Central Java and Yogyakarta Provinces. The one key participant from the National PCBs Association was the Chairperson of the National PCBs' Association. The key participants from IFSA and the National PCBs' Association had to possess expertise and knowledge on PCBs. In-depth interviews were conducted in Jakarta, the capital city of Indonesia, and in Semarang, the capital city of Central Java Province. The data obtained in these in-depth interviews was coded and analyzed descriptively (Corbin & Straus, 2006, p. 66; Charmas, 2011, pp. 368–369).

## 3 Literature Review

### 3.1 *Approaches to the Board Structures*

Before continuing the discussion on the approaches toward the board and its composition, it is important to define board in an Indonesian context. Indonesia's board structure has traditionally relied upon a statutory regulation which advocates the two-tier board model. Indonesia inherited the two-tier model from the Dutch colonial administration through the enactment of the Indonesian Commercial Code. Sections 44 and 55 of the Indonesian Commercial Code recognize bodies such as the BODs, BOCs, and general meetings of shareholders (GMS) in a limited liability company (LLC). The regulation concerning LLC in the Indonesian Commercial Code was first abolished in 1995. However, the existence of the two-tier board is still recognized in Article 1(2) of the Company Law of 1995 and its successor, Article 1 (2) of the Company Law of 2007. The two-tier structure also applies in countries like the Netherlands, Austria, Finland, Denmark, and Germany (Jungmann, 2006, p. 427). Tricker (2011, p. 157) calls it the "continental European two-tier model."

This model generally has bodies such as the GMS, BOCs, and BODs. Therefore, the term "boards" in this article refers to both BODs and BOCs.

There have been two main approaches to the board structure that are widely accepted: voluntary and mandatory. In the voluntary approach, companies freely set up the board structures, including committees and units that most suit their needs, culture, business scale and scope, environment, and characteristics without weakening the fundamental economic determinants (Guest, 2008, p. 51). This method mainly operates in the United Kingdom (the Cadbury Report of 2012; the 2016 UK Code) and later in the Netherlands (the 2016 Dutch Code) and Germany (the Germany Code of 2017).

In the United Kingdom, in which voluntary principle or comply or explain is used as an approach with respect to the compliance with corporate governance code, including the board's structure and composition, the 2016 UK Code does not provide a valid approach to be used when forming the board. Instead, it only states general principles that "the board should be of sufficient size ... should not be so large, so it can function effectively as to be unwieldy." The board should consist of an appropriate combination of executive and independent nonexecutive directors. This means that companies can voluntarily determine the structure and composition of the boards that most suit their characteristics, culture, and needs (the 2016 UK Code, Supporting Principles B.1). It also specifies three committees needed by a company such as nomination, audit, and remuneration committees (the 2016 UK Code, Provisions B.2.1, C.3.1, and D.1.1).

In the Netherlands and Germany, the governance structure of a company consists of GMS, a supervisory board (SB), and a management board (MB). In these two countries, the codes only take general principles in relation to SB, MB, and committee members and their composition. Taking into account the nature and the activities of the company, the SB should prepare its profile so that it addresses the desired expertise and background, the diversity of its composition, as well as the size and the independence of the SB members. The SB should be composed of members that are able to work "independently and critically vis-à-vis one another, the management board, and any particular interests involved" (the 2016 Dutch Code, Principles 2.1, 2.1.1, 2.1.7). Regarding committees in a company, it is stated that if the SB is composed of more than four members, it should appoint from among its members to be the members of an audit committee, a remuneration committee, and a selection and appointment committee (the 2016 Dutch Code, Principles 2.3.2, 2.3.3, 2.3.5).

In Germany, the SB is comprised of members that have the knowledge, skills, and professional expertise required to appropriately execute their duties. Meanwhile, the MB should be made up of some members and have a chair or spokesperson. Based on the company's specific circumstances and the SB members, the SB shall establish committees where their compositions consist of members of the SB with relevant expertise. There are two committees to be established by the SB: the audit committee and the nomination committee (the 2017 German Code, Principles 4.2.1, 5.3.2, 5.3.3, 5.4.1).

Meanwhile, in the mandatory approach, the companies' governance structures are simply a product of laws (Beiner et al., 2004, p. 4). Generally, the laws have already specifically determined the board structures and composition, including the criteria. In this approach, companies must comply with the board governance structures that have been determined in the laws. Otherwise, they will be punished by the market and the regulator. This approach mainly operates in the United States and Indonesia (Thompson, 2003, pp. 102–104; Aguilera & Jackson, 2010, p. 533; the Sarbanes-Oxley Act of 2002, Cohen & Qaimmaqami, 2005; Article 1 (2) of the Company Law of 2007). In the United States, BODs are composed of executive and nonexecutive directors, and as required by Section 130 of the Sarbanes-Oxley Act of 2002, firms must have three committees: audit, compensation, and governance. All these committees must be entirely composed of independent directors (Chen & Wu, 2016, p. 3).

At this point, it is clear that both in voluntary and mandatory approaches to the board structures, the soft laws (codes) and the hard laws (acts) only specify the types of structures and the criteria to be the members of the board structures, including committees. Unfortunately, they do not explain the basis of what is used as an approach to determine the ideal size of the board structures. Hence, it is expected that the core capital approach to PCBs' board structures, practically, may provide clear guidelines to institute PCBs' board structures and, theoretically, may enrich the existing academic literature in relation to the formation of board structures, especially for small banks like PCBs.

### 3.2 *Previous Studies on Board and PCBs' Board*

Pursuant to Articles 1(3), 1(4), and 5(1) of the Banking Law of 1998, two sorts of banks are recognized: commercial banks and PCBs. Commercial banks and PCBs are conventionally run and based on Sharia principles. The difference, however, is that commercial banks provide payment transaction services, while PCBs do not. While commercial banks can serve all customers, including big, medium, and small companies, PCBs are established only to serve small-scale businesses and to offer banking services to people in rural areas (the explanatory note of IFSA People's Credit Banks Regulation of 2014).

PCBs' main products or services comprise three aspects: loans, savings, and deposits. PCBs are not allowed to get involved in payment systems, joint ventures, insurance businesses, or general interbank recurring order (GIRO) savings. PCBs can only engage in doing exchange rate trades with the permission of IFSA. PCBs are also forbidden from being an acquirer (to use cards as a mode of payment). This means that PCBs are not allowed to deploy or use debit, credit, or ATM cards in their payments and transactions (Articles 13–14, 51 of IFSA People's Credit Bank Regulation of 2014).

Numerous studies across several countries have been conducted in relation to factors that contribute to the formation of board and its composition, the importance of board in directing a company, and the significance of board in maximizing shareholder and stakeholder welfare (Beiner et al., 2004; Pfeffer, 1972; Adams & Ferreira, 2007; Fama & Jensen, 1983; Hermalin & Weisbach, 1988; Shleifer & Vishny, 1997; Donaldson & Preston, 1995; Freeman et al., 2004; Freeman & Reed, 1983; Post et al., 2002).

There have been a number of empirical studies that linked board and composition with firms' valuation and performance both in a stable financial condition and during financial crisis (Huther, 1997; Eisenberg et al., 1998; Postma et al., 2003; Loderer & Peyer, 2002; Mahadeo et al., 2012). There have been also several studies that dealt with board diversity and its independence and the correlation between board diversity and firm performance and corporate social responsibility (Kang et al., 2007; Taljaard et al., 2015; Ibrahim & Hanefah, 2016; Rao & Tilt, 2016; Fraga & Silva, 2012; Terjesen et al., 2016).

In the field of PCBs, it is to be recognized that there have been some research conducted by Duetsche Gesellschaft Technische Zusammenarbeit and Bank Indonesia (2000), Deutsche Gesellschaft für Technische Zusammenarbeit (2002), Lapenu (2000), Meagher et al. (2006), Rissy (2018), and Rissy (2019). Mainly, they focused on PCBs' financial and organizational regulatory frameworks as microfinance institutions and corporate governance standards in PCBs.

It is clear that, however, none of those studies dealt with an approach that can be used to form a board and its composition, especially for small banks like PCBs. Hence, again, this research is critical as it focuses on issues that have not been scrutinized before such as the contribution of the core capital approach as a theoretical framework toward PCBs' board structures, the compliance of PCBs with their governance structures as required by their core capital, and the barriers PCBs face in complying with their board structures based on their core capital and solutions to mitigate the barriers.

## 4 Results and Discussion

### The Significant Contribution of the Core Capital Approach as a Theoretical Framework Toward the Establishment of PCBs' Board Structures

PCBs' core capital mainly consists of paid up capital and additional capital reserves. PCBs are required to have at least 8% minimum core capital as compared to PCBs' risk weight assets. A risk weight asset is the sum of PCBs' balance sheet assets that are weighted according to the level of risk attached to each asset (Articles 3, 4, and 5 of the IFSA Minimum Capital and Core Capital Supply Obligation for PCBs Regulations of 2015).

IFSA requires PCBs to establish their boards based on their core capital. The higher the core capital is, the bigger the size of the boards (up to the maximum allowed) is. This includes the formation of committees, units, and functions. Based on their core capital, PCBs can be categorized as large, medium, and small PCBs. Each of them has different governance structures (Tables 1 and 2).

**Table 1.** The structure and composition of PCBs' BOCs and BODs based on the core capital, in rupiah (1 USD = Rp.13.000)

| Amount of core capital | Number of BOCs | Number of independent commissioners | Number of BODs |
|---|---|---|---|
| Less than 50 billion (US $3,846,154) (small PCBs) | (a) Minimum 2<br>(b) Maximum is the same as the total number of directors | Not required | Minimum 2 |
| At least 50 billion (US $3,846,154) Up to less than 80 billion (US $6,153,846) (medium PCBs) | (a) Minimum 3<br>(b) Maximum is the same as the total number of directors | (a) Minimum 1 | Minimum 3 |
| At least 80 billion (at least US $6,153,846) (large PCBs) | (a) Minimum 3<br>(b) Maximum is the same as the total number of directors | (a) Minimum 50% from the total number of BOCs | Minimum 3 |

Source: Compiled by the author based on the IFSA Corporate Governance for PCBs Regulation of 2015 and the IFSA Risk Management for PCBs Regulation of 2015

**Table 2.** PCBs' committees, units, functions based on PCBs' core capital, in rupiah (1 USD = Rp.13.000)

| Amount of core capital | Committees | Units | Functions |
|---|---|---|---|
| Less than 50 billion (US $3,846,154) (small PCBs) | | | 1. Internal audit function<br>2. Risk management function<br>3. Compliance function |
| At least 50 billion (US $3,846,154) Up to less than 80 billion (US $6,153,846) (medium PCBs) | | 1. Internal audit unit<br>2. Risk management unit<br>3. Compliance unit | |
| At least 80 billion (US $6,153,846) (large PCBs) | 1. Audit committee<br>2. Risk management committee<br>3. Nomination and remuneration committee (optional) | | |

Source: Compiled by the author based on IFSA Corporate Governance for PCBs Regulation 2015 and IFSA Risk Management for PCBs Regulation 2015

Table 1 depicts that based on their core capital, small PCBs should have a minimum number of two members of BODs and two members of BOCs, while medium and large PCBs should have a minimum number of three members of BODs and BOCs. Medium PCBs should have also minimum one number of independent commissioners, and large PCBs should also have minimum 50% independent commissioners from the total number of BOCs. From the composition required, it is obvious that an ideal size of PCBs' board is a smaller board.

Table 2 shows that based on their core capital, large PCBs must establish a risk management committee and an audit committee, and they may opt to form a nomination and remuneration committee. It is BOCs' responsibility to form the committees. Medium PCBs must institute an internal audit unit, a risk management unit, and a compliance unit. Then small PCBs must appoint key executives to implement an internal audit function, a risk management function, and a compliance function. It is BODs' duty to establish the units and to appoint the key executives.

As seen in Tables 1 and 2, IFSA regulations on corporate governance and risk management for PCBs of 2015 clearly require PCBs to form their board structures based on their core capital. This approach was not known prior to 2015. As a result, up until 2015, PCBs' board structures do not represent the real needs. Some have oversized structures, and some are the other way around, small. Another problem was that there was no regulation that required PCBs to have committees, units, and other functions. Thus, practically, the introduction of the core capital approach has provided PCBs concrete guidelines for establishing their board structures.

Furthermore, as explained, to date, the approach to board structures is generally based on the voluntary model (the 2016 UK Code, the 2016 Dutch Code, the 2017 Germany Code) or the obligatory model (the Sarbanes-Oxley Act of 2002, the 2007 Indonesia's Company Law). Both of these approaches generally only provide basic criteria regarding the formation of board structures and committees that are tailored to their needs. Both approaches do not offer a real approach that can be used as guidelines for the formation of board structures. Therefore, theoretically, the presence of the core capital approach can also be acknowledged as a theoretical breakthrough to fill the gaps in the current literature. The theoretical contributions of the capital core approach in the context of PCBs can be seen from two aspects. First, there is the formation of BODs and BOCs, directors, and commissioners, including the introduction of minimum independent commissioners. Second, there is the formation of committees, units, and functions.

In the light of the first aspect, the core capital approach provides a theoretical framework to be deployed when forming PCBs' BODs and BOCs and having a minimum number of directors and commissioners, and for the first time, medium and large PCBs are required to have minimum independent commissioners. It should be acknowledged that the introduction of minimum number of directors and commissioner based on the core capital offers a clear framework for PCBs to form their boards. In relation to the introduction of independent commissioners in PCBs, it should also be recognized that this is a significant milestone in overcoming one fundamental problem in PCBs that have long existed in family-centered businesses (Miller & le Breton-Miller, 2003, p. 127) where there was no separation between ownership and control. This circumstance is known as the unification of ownership and control (the unification thesis). As there was no separation between control and ownership, supervision was quite weak. However, some scholars argue that the unification of ownership and control might lead to better firms' performance. Carney (2005, p. 13), for example, argues that family-controlled firms can result in cost leadership benefits, social capital creation, and opportunistic entrepreneurship processes. Dennis and McConnell (2003, p. 3) also contend that "it is reasonable to presume that greater overlaps between ownership and control should lead to a reduction in conflicts of interest and, therefore, to higher firm value." Unfortunately, this is not the case in PCBs. To some extent, the unification of ownership and control in PCBs leads to value expropriation and potential financial statement fraud. According to Salusra Satria, the Claims and Bank Resolution Executive Director of the Indonesian Deposit Insurance Cooperation (IDIC), the majority of PCB scandals that led to the closure of PCBs were mainly caused by internal fraud, fake credit scandals, and embezzlement of credit installment payments or credit payments by the owners and/or their affiliated management (Kontanonline, 2015).

Here, the shareholders and their affiliated management willfully acted against the law, by taking benefits at the expense of PCBs' lenders and depositors, or even expropriated these non-shareholder stakeholders' interests for their own personal, family, or business goals. In addition, mismanagement by PCB bodies is suspected to be a contributing factor to the closures of some PCBs (Kontanonline, 2015). These facts are in line with the findings by Chen et al. (2007, p. 547) and Yi and Kim (2006, p. 435) that earnings management is most likely found in family-controlled firms.

Arguably, the introduction of independent commissioners in PCBs (independent nonexecutive directors in a one-tier structure) could also lead to better decision-making in dealing with mergers, acquisitions, directors' remuneration schemes, and key executive turnover in PCBs. Hermalin and Weisbach (2001, p. 13) found that even though a higher number of independent directors did not positively contribute to a firm's performance, the existence of independent directors did lead to better decisions in dealing with issues such as acquisitions, executive compensation, and CEO turnover.

With respect to the second aspect, for the first time, IFSA regulations also require medium PCBs to establish at least a risk management unit, a compliance unit, and an internal audit and oblige large PCBs to form a risk management committee and an audit committee. There is also a requirement that small PCBs must appoint key executives to implement the risk management function, compliance function, and internal audit function. Meanwhile, the formation of a remuneration and nomination committees is not a must for large PCBs (Articles 4(2), 42(2), 13(1 a), 59 (1), 13(1, b), 32(1, 2) of the IFSA Corporate Governance for PCBs Regulation of 2015).

The existence of committees, units, and functions in PCBs based on their core capital is a new governance structure and a critical one in PCBs. This policy has changed PCBs' structure from a simple model to complex model (Rissy, 2019). However, this new policy has brought PCBs' board to have the committees, units, and functions as generally recognized by best practice standards on board structures globally (the 2016 UK Code, the 2016 Dutch Code, the 2017 Germany Code, Basel Committee Principles, 2015).

The core capital approach, hence, theoretically, has significantly provided PCBs clear guidelines to institute their boards that meet their needs as small banks. It has eased and helped PCBs to form their boards. The core capital approach has also enriched the existing literature in the way that it provides a framework for PCBs to

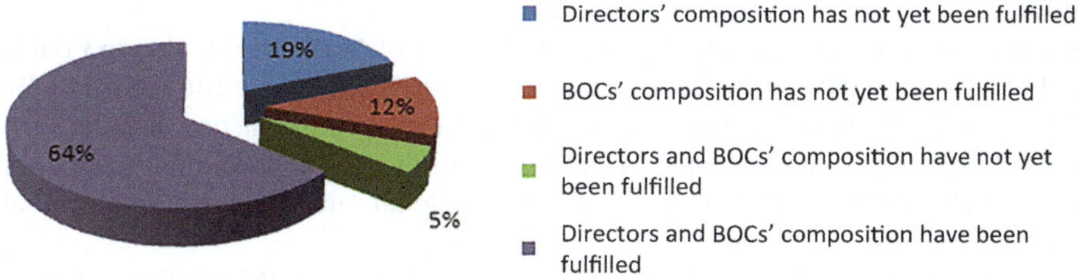


**Graph 1.** The number and percentage of PCBs based on the fulfillment of directors and the board of commissioners in 2015 (IFSA Report 2015)

form their board structures where such an approach has not yet been existed in the current literature. In addition, the core capital approach toward PCBs' board has been also a breakthrough to fill the lack of such an approach toward the formation of boards generally, especially for small banks like PCBs.

### PCBs' Compliance with the Required Board Structures, the Barriers to the Compliance, and Their Solutions

By law, as explained, PCBs must comply with their board structures based on their core capital such as a minimum number of BODs, BOCs, and independent commissioners; the establishment of a risk management committee, an audit committee, a compliance unit, a risk management unit, and an internal audit unit; and the appointment of key executives to implement the compliance, the risk management, and the internal audit functions.

The question is to what extent have PCBs complied with the required board structures? Up to the end of 2015, the total number of PCBs was 1637 (IFSA, 2015). As depicted by Pie Chart 1.1, in terms of the composition of BODs and BOCs, PCBs that have already completed their required composition of BODs and BOCs reached 64% (1048 PCBs). In contrast, 36% of PCBs (589 PCBs) have not met the standards. This means that in terms of the fulfillment of the governance structure for BODs and BOCs, the majority of PCBs have already complied with it. However, the failure rate is still significant, reaching 36% or over for one in three PCBs. Meanwhile, in the same period, no data has been released by IFSA as well as PCBs with respect to PCBs' compliance with other required board structures such as committees and units and the appointment of key executives in small PCBs to implement internal audit function, risk management function, and compliance function. This means that PCBs, based on their core capital, have significantly failed to comply with all the required board structures and composition (Graph 1).

There are three major challenges identified that contribute to the failure of PCBs to comply with the board structures that are based on their core capital. They are small PCBs' lower financial capability, a high number of uncertified PCB directors and commissioners' work competence, and a high failure rate of PCBs' directors, commissioners, and key executives in the fit and proper test conducted by IFSA (the data was processed from in-depth interviews and secondary data from IFSA, Bank Indonesia, and mass media publications).

**Small PCBs' Lower Financial Capability**

Small PCBs dominate the total number of PCBs as of May 2015, with 1463 out of 1643 PCBs or 89.04% being categorized as small PCBs. From the 1463 small PCBs, 749 small PCBs (45.6%) have a core capital under 15 million rupiah, 422 small PCBs (25.7%) have a core capital of at least 3 million up to 6 million rupiah, and 292 small PCBs (28.7%) have a core capital under 3 million rupiah (Finansialbisiniscom, 2015).

Overall, these small PCBs have problems complying with the existing corporate structure composition due to their lower financial capability. To resolve this issue, both Bambang Widjanarko and Ahmad Fausie (in-depth interviews, 2016) suggested that small PCBs should either merge or look for new investors to increase their core capital. According to the Head of IFSA in Malang, Indra Krisna, a merger would be much easier, meaning that PCBs would be under one ownership. A merger, he added, can lead PCBs to fulfill the minimum core capital required, increasing their competence and efficiency (Suaramerdekacom, 2015). At the same time, a merger might increase these small PCBs' financial capability to facilitate the fulfillment of their required board structures and composition.

However, it has to be acknowledged that even though the idea that small PCBs need to merge has been strongly recommended since 1999 (Bank Indonesia PCBs Merger, Consolidation and Acquisition Regulations of 2009), in practice, only a few have decided to do so. For example, in 2009, only 75 PCBs merged into 10 PCBs; in 2010, there were only 31 PCBs that merged into 5 PCBs; in 2011, there were only 55 PCBs that merged into 7 PCBs; in 2012, there were only 13 PCBs that merged into 5 PCBs; and in 2015, there were only 8 PCBs that merged into 2 PCBs (Bank Indonesia Annual Banking Report, 2009, 2010, 2011, 2012, 2015. The data was processed by the author). Hence, the effort to increase PCBs' financial capability to complete their board structures and composition through mergers, especially for small PCBs, would be a huge challenge.

**The High Number of Uncertified PCB Directors and Commissioners' Work Competence**

IFSA requires all PCB directors and commissioners to have a work competency certification conducted by a professional certification institution registered at IFSA (Article 2 (a, b) of the IFSA PCB Directors and Commissioners' Work Competencies Certification Regulation of 2015). In practice, the certification is conducted by an institution called the Micro Finance Financial Institution Certification that operates in cooperation with the PCBs Association. The certification of PCB directors was initiated in 2002, starting for the first time in 2004 and reaffirmed 10 years later in 2014 (Article 26 of the IFSA Directors and Commissioners in Public Companies Regulation of 2014). According to the Chairperson of the PCBs Association, Joko Suyanto, the certification of directors has experienced substantial progress and has no significant problems. Up until January 2016, 8158 existing PCB directors had already been certified (Bisniscom, 2016).

However, the certification of the existing PCB commissioners, which started for the first time in 2014 (Article 28 (5, 6) of the IFSA PCB Regulation of 2014), is far from ideal. There have been only 1329 existing PCB commissioners certified (Bisniscom, 2016). By 2015, there were 1637 PCBs across Indonesia. If, on average, each PCB has a minimum of two or three commissioners, the existing PCB commissioners are 3274. This means that the majority of existing PCB commissioners are not yet certified.

In some provinces, the gap between certified and uncertified commissioners is quite high. In South Celebes Province, for example, by March 2015, according to the Chairperson of the South Celebes PCBs Association, Aris Patau, only 10 out of the 58 existing PCB commissioners or only 17.2% were certified (Sindonewscom, 2015). Meanwhile, the percentage of uncertified existing PCB commissioners was even higher, reaching 69.1% in the same period. According to the Chairperson of the Central Java PCBs Association, Dadi Sumarno, only 50 out of 360 existing PCB commissioners have been certified (Suaramerdekacom, 2015). At this stage, it is obvious that the higher percentage of uncertified existing PCB commissioners has been a crucial problem in fully completing the structure and composition, especially with respect to PCB commissioners.

To mitigate this issue, it is recommended that the Certification Institution and the PCBs Association accelerate the certification program across Indonesia to at least three times a year in each province. The more often the certification program is held, the quicker the PCB commissioners will be certified, making them legally eligible to hold this position. Meanwhile, for PCBs, it is suggested that they allocate more funds to finance BOCs' and directors' work competence certification.

**The High Failure Rate of PCB Directors, Commissioners, and Key Executives in the Fit and Proper Test Conducted by IFSA**

The regulations also require that IFSA conduct a fit and proper test for director and commissioner candidates, existing directors and commissioners, and key executives. A director or commissioner candidate must meet three criteria: have proven integrity, be competent, and have a good financial reputation (Articles 2 and 27 of the Bank Indonesia Fit and Proper Test Regulation of 2012). A fit and proper test can also be given to existing directors, commissioners, and key executives if there is an indication that these parties have problems with their integrity, competence, and/or financial reputation while holding these positions. They must receive approval from IFSA, and they are not allowed to commence employment until an approval from IFSA is granted (Articles 26 (1, 2) and 39 (a)–(h) of the Bank Indonesia Fit and Proper Test Regulation of 2012).

In practice, not all directors and commissioners who take this test succeed. From 2014 to 2015, approximately one-third of the total participants (director and commissioner candidates) failed this test (IFSA, 2015). In 2014, from the 249 director candidates who took the test, 162 (67.1%) of them passed the test, and 82 (32.9%) of them failed. In the same year, there were 202 commissioner candidates who took the test. The result was that 133 (65.8%) of them passed and 69 (24.2%) of them failed (IFSA, 2015). The data was processed by the author). The 24.2–32.9% failure rate of the existing directors and commissioners is still too high. This means that the existing directors and commissioners are not properly trained. Hence, to lower this failure level, PCBs should provide suitable fit and proper test training for directors and commissioners. In addition, the overall number of existing directors and commissioners who take the test is not enough to meet the needs of PCBs.

In 2015, from 134 director candidates who took the test, 102 (76.1%) passed and 32 (23.9%) failed. Still in 2015, there were 81 commissioner candidates who took the test: 64 (79%) passed and 17 (21%) failed (IFSA, 2015. The data was processed by the author). The total number of existing directors who took the test in 2015 was 7, but only 1 (14.3%) passed and 6 (85.7%) failed. A similar result was found with the existing key executives who participated in the test in 2015. Out of six existing key executives who took test, only one (16.7%) passed and five (83.3%) failed (IFSA, 2015). The 21–23.9% failure rate of commissioner and director candidates who took the fit and proper test is also quite high. Even the failure rate for key executives, who took the same test, is very high. This data also shows that one-third of PCB directors and commissioners and three-quarters of key executive candidates are not yet ready to participate in the fit and proper test. Therefore, the fit and proper test has also significantly contributed to the failure of PCBs to complete their required director and commissioner board structures.

To overcome this problem, it is suggested that PCBs should be careful in meeting all the administrative criteria and do the self-assessment professionally. Only those who meet the criteria are proposed to participate in the fit and proper test as director, commissioner, and key executive candidates. By doing so, PCBs can accelerate the supply of qualified director, commissioner, and key executive candidates.

## 5 Conclusions and Recommendations

The existing IFSA regulations on corporate governance and risk management for PCBs of 2015 require PCBs to establish and complete their board structures based on their core capital. PCBs, based on their core capital, must institute BODs and BOCs and have minimum and maximum members of BODs and BOCs including minimum number of independent commissioners. Large PCBs must institute a risk management committee and an audit committee and may opt to form remuneration and nomination committee. Medium PCBs are obliged to have an internal audit unit, a risk management unit, and a compliance unit. These units are instituted by and under BOD supervision. Small PCBs are required to appoint key executives to implement a compliance function, a risk management function, and an internal audit function.

Theoretically, the core capital approach toward PCBs' board structure is a genuine effort to fill the absence of the current literature to deal with the basis for determining PCBs' board. It significantly provides a clear theoretical framework for PCBs in determining their board, including committees, units, and functions. It is seen as a strategic scheme to have an ideal size of PCBs' board as small banks. It is also an original model since it has not been addressed by both voluntary and mandatory approaches toward board as depicted in several international codes of corporate governance such as the 2016 UK Code, the 2016 Dutch Code, the 2017 Germany Code, acts like the 2007 Indonesian Company Law, and the 2002 Sarbanes-Oxley Act as well as several scholars like Adams and Ferreira (2007), Hermalin and Weisbach (1988, 2001), Byrd and Hickman (1992), Brickley and James (1987), and Xie et al. (2003).

However, practically, PCBs do struggle to comply with the required board structures. PCBs are recommended to accelerate and increase the number of existing directors and commissioners or candidate directors and commissioners to take part in the work competence certification. PCBs should professionally train and prepare their directors, commissioners, and key executive candidates to take fit and proper test conducted by IFSA, to conduct a proper self-assessment before proposing their director, commissioner, and key executive candidates to participate in the fit and proper test. The PCBs Association and the Certification Institution should certify more PCB directors, commissioners, and key executives, and small PCBs should also consider merging to increase their core capital and financial capability.

**Practical Implication**

This study has provided a basis for developing a model called the core capital approach to board, especially in PCB case in Indonesia. It is obvious that, theoretically, the core capital approach has bridged the gap in the theoretical field in dealing with a valid approach that can be deployed when forming a board in PCBs as this approach has not been explained yet by both mandatory approach and voluntary approach to board. From a practical perspective, the implication might be that the core capital approach may be further developed or modified by other small banks or small enterprises to assist them forming their board that most suits their needs and characteristic. In a wider context, the approach could also be considered by nonbank small and medium enterprises or other small and medium microfinance institutions.

## References


Adams, R. B., & Ferreira, D. (2007). A theory of friendly boards. *Journal of Finance, 62*(1), 217–250.

Aguilera, R. V., & Jackson, G. (2010). Comparative and international corporate governance. *The Academy of Management Annals, 4*(11), 485–556.

Arcot, S., Bruno, V., & Faure-Grimaud, A. (2010). Corporate governance in the UK: Is the comply or explain approach working? *International Review of Law & Economics, 30*(2), 193–201.

Bank Indonesia. (2009). *Annual Banking Report*. Bank Indonesia.

Bank Indonesia. (2010). *Annual Banking Report*. Bank Indonesia.

Bank Indonesia. (2011). *Annual Banking Report*. Bank Indonesia.

Bank Indonesia. (2012). *Annual Banking Report*. Bank Indonesia.

Bank Indonesia and Deutsche Gesellschaft für Technische Zusammenarbeit. (2000). Legislation, regulation and supervision of microfinance institutions in Indonesia. Retreieved May 1, 2018, from https://www.findevgateway.org/sites/default/files/mfg-en-paper-legislation-regulation-and-supervision-of-microfinance-institutions-in-indonesia-2000.pdf

Bank Indonesia Director Decree Number 32/52/KEP/DIR, dated 14 May 2009 Concerning merger, Consolidation and Acquisition of PCBs.

Bank Indonesia Regulation Number 14/09/PBI/2012 Concerning Fit and Proper Tests in People's Credit Banks.

Basel Committee on Banking Supervision. (2015). Corporate governance principles for banks. Retrieved September 12, 2017, from https://www.bis.org/publ/bcbs294.pdf

Baysinger, B., & Hoskisson, R. (1990). The composition of boards of directors and strategic control: Effects on corporate strategy. *The Academy of Management Review, 15*(1), 72–87.

Beiner, S., Drobetz, W., Schmid, M., & Zimmermann, H. (2004). *An integrated framework of corporate governance and firm valuation - Evidence from Switzerland* (ECGI - Finance Working Paper No. 34/2004, pp. 1–54). Retrieved August 9, 2017, from https://ssrn.com/abstract=489322 or https://doi.org/10.2139/ssrn.489322

Bisniscom. (2016). Retrieved July 17, 2016, from http://finansial.bisnis.com/read/20160114/90/509550/perbarindo-keluarkan-12.207-serifikat

Brickley, J. A., & James, C. M. (1987). The takeover market, corporate board composition, and ownership structure: The case of banking. *The Journal of Law and Economics, 30*(1), 161–180.

Byrd, J. W., & Hickman, K. A. (1992). Do outside directors monitor manager?: Evidence from tender offer bids. *Journal of Financial Economics, 32*(2), 195–221.

Cadbury Committee. (1992). *Report on the committee on the financial aspect of corporate governance*. Retrieved September 8, 2017, from http://www.ecgi.org/codes/documents/cadbury.pdf

Carney, M. (2005). Corporate governance and competitive advantage in family-controlled firms. *Entrepreneurship Theory and Practice, 29*(3), 249–265.

Charmas, K. (2011). Grounded theory methods in social justice research. In N. K. Denzin & Y. S. Lincoln (Eds.), *Qualitative research*. Sage Publication.

Chen, K. D., & Wu, A. (2016). *The structure of board committees* (Harvard Business School, Working Paper 17-032, pp. 1–37).

Chen, K. Y., Elder, R. J., & Hsieh, Y. M. (2007). Corporate governance and earnings management: The implication of corporate governance best-practice principles for Taiwanese listed companies. *Journal of Contemporary Accounting & Economics, 3*(2), 73–102.

Cohen, A., & Qaimmaqami, D. (2005). The US Sarbanes-Oxley Act of 2002: Summary and update for non-US issuers. *International Journal of Disclosure and Governance, 2*(1), 81–106.

Corbin, J., & Straus, A. (2006). *Basic of social research*. Sage Publication Inc..

Curtis, B., & Curtis, C. (2011). *Social research: A practical introduction*. Sage Publication Inc..

Daily, C. C., & Johnson, J. L. (1997). Sources of CEO power and firm financial performance: A longitudinal assessment. *Journal of Management, 23*(2), 97–117.

Dennis, D. K., & McConnell, J. J. (2003). International corporate governance. *Journal of Financial and Quantitative Analysis, 38*(1), 1–36.

Deutsche Gesellschaft für Technische Zusammenarbeit. (2002). *Regulatory requirements for microfinance: A comparison of legal framework in 11 countries worldwide*. Retrieved November 5, 2017, from http://www.bu.edu/bucflp/files/2012/08/Regulatory-Requirements-for-Microfinance.pdf

Donaldson, T., & Preston, L. E. (1995). The stakeholder theory of the corporation: Concepts, evidence, and implications. *Academy of Management Review, 20*(1), 65–91.

Dutch Corporate Governance Code Monitoring Committee. (2016). *The Dutch Corporate Governance Code*. Retrieved April 3, 2017, from http://www.mccg.nl/download/?id=3367

Eisenberg, T., Sundgren, S., & Wells, M. (1998). Larger board size and decreasing firm value in small firms. *Journal of Financial Economics, 48*(1), 35–54.

Fama, E. F., & Jensen, M. C. (1983). Separation of ownership and control. *The Journal of Law and Economics, 26*(2), 301–325.

Financial Reporting Council. (2016). *The UK Corporate Governance Code*. Retrieved May 6, 2017, from https://www.frc.org.uk/Our-Work/Publications/Corporate-Governance/UK-Corporate-Governance-Code-April-2016.pdf

Finansialbisiniscom. (2015). Retrieved August 12, 2017, from http://finansial.bisnis.com/read/20150821/90/464595/perkuat-permodalan-ojk-dorong-bpr-merger-atau-akuisisi

Fraga, J. B., & Silva, V. A. B. (2012). Board diversity and firm performance: An empirical investigation in the Brazilian market. *Brazilian Business Review, 1*, 55–77.

Freeman, R. E., Andrew, C. W., & Parmar, B. (2004). Stakeholder theory and "the corporate objective revisited". *Organization Science, 15*(3), 364–369.

Freeman, R. E., & Reed, D. L. (1983). Stockholders and stakeholders: A new perspective on corporate governance. *California Management Review, 25*(3), 88–106.

Golden, B. R., & Zajac, E. J. (2001). When will boards influence strategy? Inclination × power = strategic change. *Strategic Management Journal, 22*(12), 1087–1111.

Guest, P. M. (2008). The determinants of board size and composition: Evidence from the UK. *Journal of Corporate Finance, 14*(1), 51–72.

Hermalin, B. E., & Weisbach, M. S. (1988). Endogenously chosen boards of directors and their monitoring of the CEO. *The American Economic Review, 88*(1), 96–118.

Hermalin, B. E., & Weisbach, M. S. (2001). *Boards of directors as an endogenously determined institution: A survey of the economic literature* [NBER Working Paper Series 8161, pp. 1–41].

Hillman, A. J., & Dalziel, T. (2003). Boards of directors and firm performance: Integrating agency and resource dependence perspectives. *The Academy of Management Review, 28*(3), 383–396.

Hutchinson, T., & Duncan, N. (2012). Defining and describing what we do: Doctrinal legal research. *Deakin Law Review, 17*(1), 83–119.

Huther, J. (1997). An empirical test of the effect of board size on firm efficiency. *Economics Letters, 54*(3), 259–264.

Ibrahim, A. H., & Hanefah, M. H. (2016). Board diversity and corporate social responsibility in Jordan. *Journal of Financial Reporting and Accounting, 14*(2), 279–298.

Indonesian Financial Services Authority. (2015). *Annual Report*. Indonesian Financial Services Authority.

Indonesian Financial Services Authority Regulation Number 20/POJK.01/2014 Concerning People's Credit Banks.

Indonesian Financial Services Authority Regulation Number 33/POJK.04/2014 Concerning Directors and Commissioners in Public Companies.

Indonesian Financial Services Authority Regulation Number 44/POJK.03/2015 Concerning Work Competencies Certification for Members of Directors and Board of Commissioners of People's Credit Banks and Sariah Financing People's Banks.

Indonesian Financial Services Authority Regulation Number 5/POJK.03/2015 Concerning Minimum Capital Supply Obligation and Minimum Core Capital Obligation for People's Credit Banks.

Jungmann, C. (2006). The effectiveness of corporate governance in one-tier and two tier board systems – Evidence from the UK and Germany. *European Company and Financial Law Review, 3*(4), 426–474

Kang, H., Cheng, M., & Gray, S. J. (2007). Corporate governance and board composition: Diversity and independence of Australian boards. *Corporate Governance: An International Review, 15*(2), 194–207.

Kontanonline. (2015). Retrieved August 9, 2016, from http://keuangan.kontan.co.id/news/lps-minta-bpr-merger-untuk-tekan-fraud

Lapenu, C. (2000). *The role of the state in promoting microfinance institutions* (FCND Discussion Paper No. 89, pp. 1–42). Retrieved April 12, 2018, from https://pdfs.semanticscholar.org/b31d/278e7c05697f5747724594d3a36e4d8425b3.pdf?_ga=2.225482073.213309550.1574827441-804000972.1566466141

Loderer, C., & Peyer, U. (2002). Board overlap, seat accumulation and share prices. *European Financial Management, 8*(2), 165–192.

Mahadeo, J., Soobaroyen, T., & Hanuman, V. (2012). Board composition and financial performance: Uncovering the effects of diversity in an emerging economy. *Journal of Business Ethics, 105*(3), 375–388.

Meagher, P., Campos, P., Christen, R. P., Druschel, K., Gallardo, J., & Martowijoyo, S. (2006). *Microfinance regulation in seven countries: A comparative study*. The IISH Center, University of Maryland. Retrieved August 2, 2017, from http://www.microfinancegateway.org/sites/default/files/mfg-en-paper-microfinance-regulation-in-seven-countries-a-comparative-study-2006_0.pdf

Miller, D., & Le Breton-Miller, I. (2003). Challenge versus advantage in family business. *Strategic Organization, 1*(1), 127–134.

Pearce, D., Campbell, E., & Harding, D. (1987). *Australian law schools: A discipline assessment for the Commonwealth Tertiary Education Commission*. Australian Government Publishing Service.

Pfeffer, J. (1972). Size and composition of corporate board of directors: The organization and its environment. *Administrative Science Quarterly, 17*(2), 218–228.

Post, J. E., Preston, L. E., & Sachs, S. (2002). Managing the extended enterprise: The new stakeholder view. *California Management Review, 45*(1), 6–28.

Postma, T. J. B. M., van Ees, H., & Sterken, E. (2003). Board composition and firm performance in the Netherlands. *Eastern Economic Journal, 29*, 41–58.

Provan, K. G. (1980). Board power and organizational effectiveness among human service agencies. *Academy of Management Journal, 23*(2), 221–236.

Rao, K., & Tilt, C. (2016). Board diversity and CSR reporting: An Australian study. *Meditari Accountancy Research, 24*(2), 182–210.

Rissy, Y. Y. W. (2018). *Corporate governance in people's credit banks in Indonesia: A study of the standards, model and compliance* [PhD Thesis, School of Law, Faculty of Law, Queensland University of Technology, Australia]. Retrieved October 5, 2018, from https://eprints.qut.edu.au/120982/

Rissy, Y. Y. W. (2019). Corporate governance in people's credit banks in Indonesia: A challenge for a better future. In R. Schmidpeter, N. Capaldi, S. O. Idowu, & A. S. Herrera (Eds.), *International dimensions of sustainable management: Latest perspectives from corporate governance, responsible finance and CSR* (pp. 163–174). Springer.

Rose, C. (2016). Firm performance and comply or explain disclosure in corporate governance. *European Management Journal, 34*(3), 202–222.

Shleifer, A., & Vishny, R. W. (1997). A survey of corporate governance. *The Journal of Finance, 52*(2), 737–783.

Sindonewscom. (2015). Retrieved February 5, 2015, from https://ekbis.sindonews.com/read/962182/34/perbarindo-tingkatkan-kompetensi-sdm-perbankan-1423488504>

Suaramerdekacom. (2015). Retrieved March 7, 2015, from http://berita.suaramerdeka.com/direksi-dan-komisaris-bpr-wajib-bersertifikasi/

Taljaard, C., Ward, M. J. D., & Muller, C. J. (2015). Board diversity and financial performance: A graphical time-series approach. *South African Journal of Economic and Management Sciences, 18*(3), 425–448.

Terjesen, S., Couto, E., & Francisco, P. (2016). Does the presence of independent and female directors impact firm performance? A multi-country study of board diversity. *Journal of Management & Governance, 20*(3), 447–483.

The Regierungskommission Deutscher Corporate Governance Kodex. (2017). German Corporate Governance Code. Retrieved February 21, 2018, from http://www.dcgk.de//files/dcgk/usercontent/en/download/code/170214_Code.pdf

The Sarbanes-Oxley Act of 2002.

Thompson, R. B. (2003). Corporate governance after Enron. *Houston Law Review, 40*, 99–118.

Tracers, M. (2013). Qualitative interviewing methods. In M. Walter (Ed.), *Social research methods* (3rd ed.). Oxford University Press.

Tricker, R. I. (2011). *Corporate governance: Principles, policies, and practices* (2nd ed.). Oxford University Press.

William, L. J., & Elisabeth, M. (1989). *Pawns or potentates: The reality of America's corporate boards*. Harvard Business School Press.

Xie, B., Wallace, N. D., & DaDalt, P. J. (2003). Earnings management and corporate governance: The role of the board and the audit committee. *Journal of Corporate Finance, 9*(3), 295–316.

Yi, C. H., & Kim, J. B. (2006). Ownership structure, business group affiliation, listing status, and earnings management. *Contemporary Accounting Research, 23*(2), 427–464.

**Yafet Yosafet Wilben Rissy**, Ph.D. (AFHEA), is an Associate Professor at Faculty of Law, Satya Wacana Christian University, Indonesia. He holds his Bachelor of Laws from the University of Nusa Cendana; Master of Development Studies from Satya Wacana Christian University, Indonesia; Master of Laws from Faculty of Law, University of Canberra, Australia; and Ph.D. from School of Law, Queensland University of Technology, Brisbane, Australia. He teaches some units in corporate governance, corporations law, intellectual property rights law, and insolvency law. His research and publications mainly focus on topics related to the units he teaches.

# Lessons in Financial Governance from Bowen's Family Systems Model

**Paul R. Sachs and Shame Mugova**

## 1 Introduction

Financial operations are central to a business's stability and sustainability. Nowadays, there is greater recognition that a business has responsibility beyond delivering financial results to its owners or shareholders. Nonetheless, financial data and reporting of that data continue to be the primary means of measuring business performance.

Business financial data and reporting have become more complex due to the amount of information available through technology and the speed at which such information can be accessed. So, paradoxically, at the same time that a business's responsibility is seen as broader than just its financial outcomes, the expectations for financial data and analysis are even greater. Business leaders and financial specialists are challenged to manage this paradox.

Family systems theory and practice provides a unique perspective from which to meet this challenge. Family systems theory is concerned with many aspects of interpersonal behavior, and these are felt to be pertinent to understanding financial governance within the context of business operations. Social communication, social influence, culture, and decision-making are some of these aspects.

In this paper, Bowen's family systems model is examined. Elements of the model suggest ways to manage the challenges of financial governance and to stimulate further research.

P. R. Sachs (✉)
Bala-Cynwyd, PA, USA

S. Mugova
Birmingham City University, Birmingham, England
e-mail: 21556779@dut4life.ac.za

 
I. G. Arraiano et al. (eds.), *Corporate Social Responsibility in a Dynamic Global Environment*, CSR, Sustainability, Ethics & Governance,
https://doi.org/10.1007/978-3-031-24647-0_16

## 2 Systems Approach

If one were going to understand how an automobile engine works, one could carefully examine each of the components of the engine, the engine block, the valves, the cylinders, the spark plug, the carburetor, and so forth. These parts could be described and measured in detail. Yet, one would not come close to understanding how an automobile engine works. One would need to observe these parts in action – watch, listen to, and measure them while the engine is running – in order to truly understand how the engine works.

Likewise, when one is trying to understand how a group of people behave, interviewing individual members of the group provides little information about how the group actually operates. Much more useful is to observe the group in action.

There are characteristics of systems that apply whether speaking about mechanical or social systems. One has just been mentioned – the whole is greater than the sum of the parts. The system is a thing unto itself. It is connected to and influenced by yet distinct from the individual parts of the system.

A second characteristic of systems is the drive for homeostasis. Actions cause reactions. The purpose of any action or reaction is to maintain the stability of the system. In mechanical systems, these actions and reactions can be readily understood once one understands the different components of the system. The automobile's internal combustion engine uses controlled explosions to generate power. Through the steps of intake, compression, combustion, and exhaust, the engine delivers energy to the mechanical parts which cause the vehicle to move.

In a social system, these actions and reactions are not so readily ascertained. Nonetheless, they exist. The buildup of pressure in a social system, for example, from an interpersonal conflict, will be discharged in some way so that the social system can rebalance in order to continue its operations.

Tatar et al. (2019) have usefully observed that conflicts have affective and cognitive components. Moreover they note that conflict is not necessarily destructive. Conflict can gestate new ideas. The discharge of pressure from resolving a conflict productively can move the group forward, yielding not just new ideas but also new or different relationships among group members.

### *2.1 Social Systems: Families and Businesses*

The family is the first social system to which a person belongs. Regardless of the characteristics of that family, it remains the first place in which a person learns about interactions with other people. Learning in the family often occurs by modeling – children watching and imitating the actions of their parents. The modeling can occur as intentional instruction or unintentional observation. Social learning is powerful because it is interactional and multisensory and engages cognitive and emotional skills. It is exactly for these reasons that so much value is placed on having a stable family environment with well-guided parents to facilitate the development of the child.

The previously mentioned characteristics of systems apply to families (Kerr, 1991). To wit, the family is an entity unto itself, apart from the individuals who comprise the family. Families have tasks or responsibilities. In response to the environment in which a family functions, families will face changes and challenges. Each family will have characteristics; one might say personalities, based on how the family members interact with each other and manage day-to-day challenges.

Through the course of time, the family inevitably faces changes. The stable family responds to these changes in order to maintain equilibrium or order. Child-rearing, for example, involves adapting to the needs and capacities of the growing child (and the fatigued and aging parent). The family also interacts with larger society to mediate relationships with the parents' place of employment, schools, economic communities, and faith communities. Some of these interactions will require the family to adjust its behaviors. The departure of a child to school, a parent's new job or unemployment, an illness or death of a family member require changes in the family's behavior. Family functioning is always dynamic, never static.

The business enterprise is also a dynamic social system, more than the sum of its component parts, which seeks to manage the challenges to its operations in order to survive and thrive. It is noted that many businesses began as family run enterprises which then developed beyond the family. Although this paper focuses on family theory, such family-owned and family-run businesses present special considerations that will not be the focus of this paper.

Extending the metaphor from mechanical systems a bit further, just as pressure in the engine drives a vehicle forward, so does pressure in the business. Innovation incubators, strategic planning, budgeting, and corporate communications all involve interactions between individuals and groups of people; many times the discussions in these groups can get heated. Ideally, however, they move the business forward with new ideas and effective operational decisions.

## *2.2 The Financial System and Financial Operations*

Financial data and reporting continue to be the primary means of measuring business performance. The responsibilities of a business's financial department include forecasting and planning, investment and financing decision, coordination and control, interface with financial markets, and risk management (Brigham et al., 1999, p. 8).

Financial operations in a business enterprise have concrete and symbolic aspects. Concrete financial operations refer to tasks related to the responsibilities noted above. Although these are described by specific numbers, they necessarily involve assumptions in which one may have varying degrees of confidence. Creating a budget in an established business, for example, may be based on historical information, but it will still require forecasting assumptions about the budget year ahead and assessments of risks.

Financial operations have symbolic value. The importance of financial data and the specialized knowledge required to prepare and interpret such data can confer a higher degree of authority to financial personnel than personnel in other departments.

Financial outcomes carry with it other symbolic meanings dependent on the culture of the organization. The meaning that the business attributes to money, financial dominance, or financial stability will affect the role that fiscal operations have within the business and the governance of financial operations.

## 2.3 *Financial Governance Concepts*

Financial governance refers to a set of rules and regulations regarding how financial data is collected and reported and the actions that are taken with respect to this data, such as budgeting and investment.

Finance, perhaps more than any other aspects of business, is subject to regulation from the community or political entity in which it operates. Any business owes its existence to being able to fill a societal need. As a result, there is a necessary interplay between the business and the surrounding community. The local community, the state, or the country each set expectations, sometimes in the form of laws and regulations, for the business. The business in turn influences the community by the goods and services produced and by the employment of community members. Regulations, ideally, structure the relationship between business and community.

A structure to financial governance gives the business assurance that data are accurate and meaningful in order to guide business operations. Several criteria are needed for effective financial governance of which the most prominent are independence, transparency, and a code of ethics.

The financial operations must be able to operate independently in order to collect and report data. At the same time, the financial operations must be transparent so that the recipient understands the way in which data were obtained. Underlying both the independence and transparency is a code of ethics that sets a standard for honesty and thoroughness in reporting data.

The board of a business has ultimate responsibility for the integrity of the business and the financial department's adherence to these aforementioned principles of independence, transparency, and ethics.

Several financial issues are worthy of special attention.

1. Budgeting

   Budget formulation is essential for effective business operations. Budgeting requires detailed collaborations between work groups and decision-makers. The allocation of limited resources, often due to capital inadequacy, cash flow problems, and liquidity needs, is factor which influences the budgeting decisions and the overall governance of financial resources.

Budgeting in corporate businesses is a team effort involving people from all levels of the organization. Board of directors or top management is ultimately responsible for approving or denying budget recommendations. Organizational members are highly interdependent. However, through budgeting, divisions and groups compete over scarce resources and fight to determine organizational direction. Work on budgets can test and disrupt this interdependence (Etherington & Tjosvol, 1998). The accounting team's recommendations may carry more weight because they deal with payroll, accounts payable, and accounts receivable. Managers are confronted with many conflicts when they use budgets. Their discussion of opposing views may have a substantial impact on whether the conflicts contribute to the budget process productively or destructively (Etherington & Tjosvol, 1998).

An effective business leader enables the financial department to develop autonomy, allowing that department to provide meaningful, objective information for financing and investment decisions. The finance department must provide valuable information for capital expenditure budgets and operational budgets to assess viability and profitability of new projects. The auditing function also requires independence and autonomy to carry out its functions effectively. Financial governance requires internal control systems which require independence for effective risk management. Financial reporting is another function which requires independence; undue influence by the owners in small businesses usually results in manipulated financial statements.

2. Corruption

   Corruption is a form of dishonest and fraudulent act through violation of business ethics, policies, and procedures. Corruption includes a wide range of unethical behavior, but the most glaring results of corruption usually involve financial activities. Corruption may result from inadequate or ineffective financial governance (Ivanyna et al., 2018)

   Corruption involves not following procedures and policies. It is common in both private and public sectors. In 2016 public authorities in Zimbabwe corruptly disregarded the outcome of a tender process for a diesel power plant and awarded the contract for the Dema Diesel Power Project to Sakunda Holdings which did not even take part in the bidding process (Magaisa, 2018). In fact, the tender was won by an American company called APR Energy Holdings. Sakunda did not have experience in power generation, but APR was later sidelined in favor of Sakunda after intervention by the President's Office. The Sakunda Holdings owner Kudakwashe Tagwirei is a ruling party associate. Sakunda subcontracted a British multinational called Aggreko plc, which has vast experience in the business. Aggreko had participated in the tender process and had therefore lost the bid to its American rival, APR.

3. Informal economy and regulation

   Organizational financial policy helps to monitor financial activities, provides internal control, and ensures the business operations comply with legal requirements. Informal economy is a phenomenon where businesses do operations which cannot be governed by the state and thus cannot easily be assessed to pay taxes. The size and importance of the informal economy varies across the globe (Ouédrago, 2017). Although the informal economy is an important area of attention for financial governance, it will not be the focus of this paper.

With this overview of systems, financial operations, and governance, we now consider the Bowen family systems model.

## 3 Bowen's Family Systems Model

Murray Bowen was an American psychiatrist who developed his systemic family theory over several decades in observation and with families of individuals with severe and persistent mental illness. Bowen drew on biological sciences in trying to understand family functioning. His work contrasted with an intrapsychic approach to mental illness where the problem is thought to reside within the disturbed individual. Rather, Bowen looked at the system in which the behavior was expressed, particularly the family system, in order to understand its development and ultimately to seek the amelioration of the presenting problem.

Because of this systemic or milieu approach, it is felt that Bowen's theory offers useful insights into understanding the business enterprise. There are eight concepts that are associated with Bowen's theory. They are as follows: triangles, family emotional process, family projection process, differentiation, multigenerational transmission, emotional cutoff, sibling position, and societal-emotional process (The Bowen Center, n.d.).

These concepts will be reviewed. Those that are felt to be most relevant to a discussion about financial governance will be discussed in more detail.

1. *Triangles*: A dyad, a two person interaction, is the essential social group. But, Bowen describes triangles as "the molecule of emotional systems" (2004, p. 424) because they best show the interplay of forces between people.

   In business, as in any social system, there are many triangles. One example of importance to governance is the relationship between the Board Chair, CEO, and COO. In terms of financial operations, one might consider the Finance Committee Chair or the Financial Officer and the various triangles within the financial department. At a macro level, the relationship between prime ministers and finance ministers has been examined (Herzog & Jankin Mikhaylov, 2019) as they affect financial governance.

   As a further example, the bankruptcy of the Enron Corporation in 2001 was the largest in Wall Street history at the time. Much of the review of the case has focused on problematic financial governance and accounting. These included Enron's use of mark to market (MTM) accounting and off balance sheet special purpose vehicles (SPVs). The handling of these legitimate practices was given the patina of approval by Arthur Andersen audits. Further, there was blatant misrepresentation in financial reports and alleged embezzlement by Enron leadership.

Taking a systems approach, one cannot view Enron's accounting practices in isolation. Certainly they occurred during a period of lower regulation and at the end of the speculative dot-com era. In addition, however, Enron leadership allowed, facilitated, and participated in the practices that led to the downfall of the company and losses to the shareholders. The triangle of Ken Lay, Jeffrey Skilling, and Andersen Consulting is worthy of further examination in analyzing this case.

Frequently, business problems are recognized at a high level (e.g., reduced revenue or employee turnover). To understand and ameliorate the problem, one must examine the problem in more detail. Interpersonal triangles provide a structure for this examination.

2. *Family emotional process*: Emotional aspects of relationships are particularly intense in family systems given the biological and social connections between family members. The same intensity is seen in business relationships. Ideally, a degree of safety or order exists in the social system which allows each person to express emotions and to have their emotional status understood by others.

When such an environment does not exist, emotions can be expressed in less adaptive ways. Bowen identifies four possible outcomes:

(a) Emotional distance between individuals
(b) Dysfunction in one member of the social group
(c) Outright conflict
(d) Projection onto another family member

Item d, projection onto another family member, is important enough that Bowen accords it discussion as a separate concept.

With respect to the first three outcomes (a, b, and c above), individuals may resolve their conflict by avoiding each other or by having outright conflict (items a and c).

For example, in business operations, one can observe avoidance or conflict between staff members or between department heads. A department manager and a financial analyst may avoid contact with each other. Or the two may express overt conflict. In monthly financial reviews, a manager's department may not have met its financial goals. The manager may seek to blame the finance department by saying that the financial targets were set too high for that time period. The financial analyst disagrees and notes that the operations manager ignored important data that the analyst provided.

In this example, the manager's blaming of the financial analyst is based on a specious argument. The budget creation should involve collaboration with the financial analyst which is then approved at a higher level. When budget processes go awry, the cause may be due to avoidance or conflictual interactions between the financial and operational personnel. Maybe they were avoiding each other to avoid an unrelated interpersonal problem, or maybe they did not have the skills or authorization to solve the problem. The effective leader brings the parties together in order to address an issue.

What may be less evident is the source of the interpersonal problem between financial and operational personnel. This scenario speaks to Bowen's Item "b" above. The dysfunction of one individual is an attempt to resolve an emotional situation.

In a nonbusiness example, two spouses may disagree about how money is spent. One spouse is more careless with spending than the other resulting in a conflict between the two. They are unable or unwilling to work out the conflict. Suddenly, one spouse develops physical symptoms – such as severe headaches or back pain – or may take a severe fall resulting in a broken bone. The physical symptoms have the effect of reducing conflict between the spouses. They now focus on the ill spouse's health and recovery rather than the spending conflict they have.

In a business organization, problems with production or product quality can result in low sales and returns. The finance department which is in charge of cash flow will experience liquidity constraints. In the course of this conflict, other problems may emerge which distract from the core conflict of poor quality of production. The business then focuses on the problematic cash flow rather than addressing the problems with the production system.

3. *Family projection process*: Another means of resolving problems is to project the problem onto a third party (Item "d" noted in the previous section). A prototypical example is the child who manifests a behavior problem when the "real" problem is that the parents do not get along. Intervention that is focused on ameliorating the child's problem behavior will be unsuccessful because the problem child is merely expressing the projected conflict of the parents.

   In business operations, the finance department may be a common target for projected conflict. For example, conflicts between marketing and production departments lead to an unhealthy financial position. This conflict is evident in statements of cash flow and income causing the focus to be on the financial department when the "real" problem is the conflict between the marketing and production departments.

   In corporate governance issues of fair compensation and remuneration, including pay disputes, will end up manifesting through the statement of cash flow and income statement. The firm can only solve the liquidity constraints through resolving the initial dispute.

   The most common problem of financial operations is the allocation of scarce resources. While waste and losses might have been incurred by other cost centers, the finance department is often blamed for the problems – the focus becoming on cash flow rather than the core wasteful production processes in other departments. Interventions focused on the "problem child" finance department will be misguided and ineffective in correcting the core problem.

4. *Differentiation*: Differentiation describes the process by which a person separates emotional and intellectual aspects of thinking. The well-differentiated person is able to distinguish the two aspects of thinking in the course of their interactions

with others. The undifferentiated person does not distinguish the two areas as well.

Thus, a person who is well differentiated is able to recognize when their decisions and actions are affected by emotions and intellect. The undifferentiated person is less aware of the distinction. They may presume that feelings, for example, are facts.

For example, Manager A makes the statement, "I know Manager B doesn't like me." The differentiated person can provide examples of behaviors they experienced which showed the dislike. The undifferentiated person will simply say that they feel this way, without any specific evidence. To the undifferentiated person, their feeling is as good as fact.

Differentiation across people is understood to be on a continuum. People do not just fall into a differentiated or undifferentiated category. Moreover, there are times, particularly when a person is angry or distressed, when even the most differentiated person is overwhelmed by emotions and unwilling or unable to consider the facts of the situation.

In a business environment, the level of differentiation of a leader can have an impact on how decisions are made and how communication occurs. It is argued here that a well-differentiated leader will more often make decisions based on data than a less well-differentiated leader. The data may be financial or some other outcome measure.

A not well-differentiated leader is apt to have more conflicts with the financial department or the financial committee of the board. That leader does not focus on objective facts and finds the attention to such details restrictive. Such a leader is apt to be swayed as much by feelings as facts.

This point is not to minimize the importance of experience and intuition in a leader's decisions. Certainly there are times when an effective leader knows that something "feels" right or wrong and yet cannot articulate the reasons for that feeling. But, a leader who is not able to differentiate between thought and feeling is likely to provide ambiguous or unpredictable direction to others and create conflict rather than resolve it.

The concept of differentiation should also be considered in the creation of competent boards that are capable of objective and independent judgment.

An area for further research could examine the budgeting process in light of the characteristics of the operational and financial leaders. One would hypothesize that a differentiated leader, led by facts, will lead a team to produce a more accurate budget than a leader who is less differentiated. Moreover, because a budget is a projection and cannot account for unexpected events, and because unexpected business events can lead to emotional reactions, one would also hypothesize that the differentiated leader would manage the unexpected events more effectively – not radically changing budgets midstream without an analysis of data.

5. *Multigenerational transmission*: Bowen noted that emotional patterns are passed on across generations, just as are behavioral patterns. In particular, Bowen focuses on the multigenerational transmission of differentiation patterns. People

that are lower in differentiated category may choose partners who are similar to them in this characteristic. What results is a cascade of lower differentiated children of this couple who are even more vulnerable to emotional disturbance.

In the business environment, this concept is useful in considering the history of the organization and changes in leadership. When board members choose new leaders, are they perpetuating or enhancing patterns that previously existed? For example, the Boeing Company ousted its CEO in December 2019 due to problems with the 737 plane. Yet, they appointed an insider as the new CEO. Will this person be able to correct the problems which led to the 737 design and construction errors? Or will the new CEO perpetuate an existing culture?

6. *Emotional cutoff*: As previously mentioned in connection with Item # 2 above, family emotional process, one way of handling a problem in a system is to cut off contact with the other party. Bowen felt that this process was important enough to warrant consideration as a concept in its own right.

   Emotional cutoff may be manifest in physical separation between parties or manifest in superficial relationships between the parties. This problem is evidenced in a particular challenge among large, geographically far-flung business organizations where physical separation prevents the resolution of other operational conflicts.

   The COVID-19 crisis provides a further challenge and case example. Not only are multinational businesses affected by physical separation. Even local businesses have mandated employees to work remotely. Financial review meetings or top-level management review about business development decisions must now occur via technology. What impact this will have on how the key players in the decisions are perceived and how decisions are made is something future research will need to examine.

7. *Sibling position*: In the families on which he based his work, Bowen observed that there were consistent differences among siblings based on their birth-order position. To be sure, Bowen's observations were based on a limited sample of families in an American society. Any parent can attest to the reality of sibling rivalry, despite efforts to avoid it.

   With respect to the business environment, different departments may be seen as siblings in a way, each vying for the attention and resources of the business leader or the business as a whole. For purposes of the current paper, however, this concept is not seen as especially useful for understanding financial operations and governance.

8. *Societal-emotional process*: Bowen noted that nonfamily systems are affected by the same dynamics as family systems. Indeed that is the key premise of this paper. In working with delinquent youth, for example, Bowen observed that the judicial system behaved in ways that were parallel to the parents of such youth – such as, giving mixed messages about responsibility and setting limits inconsistently.

   This point is fundamental to the present paper. To wit, the business operations evince the same system processes one observes in families.

   Again, we may reference the COVID-19 pandemic which shows this connection keenly. The pandemic has forced families and businesses to manage their operations differently. Businesses have had to adapt more than ever to the family demands of their workforce, given that childcare setting, schools, and universities were closed. Youth were forced to be at home with the incumbent childcare duties now falling on parents in addition to parents' usual work responsibilities. Some job positions lent themselves to remote work more than others, and some business operations were able to maintain their activities more reliably than others.

### 3.1 *Interventions in Bowen's Model*

Bowen's model is utilized in several interventions with families which can be adapted to the business environment (Kott, 2014). Observing interactions among family members is essential to developing interventions that help them resolve a given presenting problem. Thus, Bowen's model includes all family members in some aspect in the treatment.

Emotions and how they are expressed and understood are important foci of interventions. An ongoing goal is to help family members to increase their sense of differentiation. To this end, family members are guided to make "I" statements in the therapy sessions. This guidance encourages each family member to identify and own their thoughts and feelings and to distinguish between what is a thought and what is a feeling. Moreover, this action facilitates a process by which family members can differentiate themselves as individuals from the family as a whole.

As multigenerational transmission of patterns of emotional expression is part of the model, the Bowen method looks for past influences on the current situation. At times, these antecedents can lead family members to take a larger, less personalized view of a problematic situation and consider new ways of handling it. A child, for example, may take a parent's behavior less personally when the child learns that the parent had similar challenges with their own parent (the child's grandparent).

Applying these approaches to a business, an organizational consultant enlisted to help address a problem involving financial operations or governance should look at the business as a whole and include many individuals and departments in the problem resolution. One should not presume that just because the presenting problem is stated in financial terms that it is purely a financial problem.

Dealing with the issue of differentiation in a business environment presents special challenges. The business as a whole operates with a mission and vision which applies to all departments and employees. Yet, within the overarching mission, each department must have a degree of autonomy and independent voice.

Business leadership which does not allow for autonomy could stifle differentiation and also be blind to underlying conflicts that are disrupting business operations. Therefore, the consultant might ask each department or employee what is their mission and vision. Can they own the organizational mission or are there unique aspects which define their own work not captured in the overall business mission?

Organizational history and culture are important considerations when understanding a presenting organizational problem and seeking to ameliorate it. The role that the financial department plays historically in the organization and, as noted, the significance (concretely and symbolically) that is attributed to financial gains are important areas for assessment before determining an intervention.

Part of the history involves understanding specific players and their relationships with each other. Is a given department head or financial leader part of a core leadership group that has been with the company many years? Or is that person in the role through an acquisition or merger? These aspects may have an impact on the nature of the presenting problem and may suggest means of resolving the problem.

Financial governance issues vary widely. Nonetheless, the following assessment questions are offered as a starting point for systemic interventions to remedy a financial governance problem or improve operations.

1. Statement of the problem or area for improvement: The statement should be summarized in a short paragraph. The statement should be obtained independently from key stakeholders (upper management, financial management, operations, labor). Besides differences in types of problems noted, one would also note if one stakeholder sees a problem and another does not.
2. Statement of goal: Where would the organization like to be with respect to the problem or improvement area? This answer may be included in # 1, but if not, it should be asked.
3. Minimal outcome: What is the one thing that the organization would like to see that, if one does see it, will indicate that the organization is moving in the right direction? This question focuses on two things: breaking down a large goal into smaller components and establishing the standards on which trust in the intervention process can be built.
4. Financial personnel and history: Who are the key persons involved in financial operations and governance and what is their history with the company? This information will provide background on any past relationships that may have a bearing on the presenting issue.
5. What is the role of the financial department? The question may appear self-evident. But, the department may play different roles in different organizations in addition to the obvious one of gathering and analyzing financial data. Another way to ask the question is: what would be different if there were no financial department or if the financial department were half as big or twice as big as it is now? Again, this answer should be provided by multiple stakeholders.
6. Financial stability of the company: Information about the current financial status of the company and its goals and threats. These factors may figure into the context of the presenting problem.
7. How are decisions regarding expanding, maintaining, or reducing operations made in the company and what is the role of the financial department and financial governance in such decisions? These decisions ought always to involve financial data, but the weight that such data carries in the decision will differ across situations and across businesses.
8. How often do operations and financial personnel meet and in what context? There are likely to be many different meetings of varying size and degree of importance.

Understanding the different reactions to the various meetings can also provide context for understanding the presenting problem.

## 4 Conclusion and Suggestions for Future Research

The present paper posits that Bowen's family systems model can be usefully applied to understanding financial governance. In presenting this point, the paper outlined elements of Bowen's model and suggested examples within business operations where these elements are evident.

The model suggests that management strategies and organizational development interventions use a systemic approach as opposed to a top-down one. The systemic approach involves incorporating all levels of the organization, assessing the level of differentiation and autonomy among leaders and department heads, and being mindful of both concrete and symbolic aspects of financial governance.

The perspective offered in this paper complements other work in behavioral economics which examines the impact of cognition and social influence on decision-making strategies. One area for further research, therefore, would involve examining the interface between social system factors described here and the foci of behavioral economists.

Other areas that follow from the present paper include the following:

- Characteristics of business leaders (differentiated or non-differentiated) and their financial governance styles. It is proposed that leaders who show a higher degree of differentiation (as defined by Bowen) will be more data-driven and more flexible in managing financial downturns.
- The connection between organizational culture and characteristics and how governance problems are resolved. Mitchell (2017), for example, has discussed the differing solutions to organizational conflict (avoidance, competition, accommodation, collaboration, and compromise). Each of these solutions may be more or less common based on the characteristics of financial governance in an organization.

As noted earlier, this paper did not focus on other important factors in financial governance including corruption and the informal economy. These areas need to be examined with the same systemic rigor proposed here.

Lastly this paper was prepared and accepted to the 2020 conference prior to the onset of the global COVID-19 pandemic. Although a few comments have been added here with regard to the impact of the pandemic on business operations, much more work is needed.

This future research will be important for understanding how business operations responded to the stressors of the pandemic and the strengths and weaknesses that the pandemic revealed. Moreover, it will be important to enable planning for future catastrophic events. In this regard, Bowen's model focused as it is on a system in continual response to change can be a strong foundation for future research.

## References

Bowen, M. (2004). *Family therapy in clinical practice*. Rowman & Littlefield.

Brigham, E. F., Gapenski, L. C., & Ehrhardt, M. C. (1999). *Financial management: Theory and practice*. Dryden.

Etherington, L., & Tjosvol, D. (1998). Managing budget conflicts: Contribution of goal interdependence and interaction. *Canadian Journal of Administrative Sciences/Revue Canadienne des Sciences de l'Administration, 15*(2), 142–151.

Herzog, A., & Jankin Mikhaylov, S. (2019). Intra-cabinet politics and fiscal governance in times of austerity. *Political Science Research and Methods, 8*(3), 409–424. https://doi.org/10.1017/psrm.2019.40

Ivanyna, M., Moumouras, A., & Rangazas, P. (Eds.). (2018). *The macroeconomics of corruption: Governance and growth*. Springer.

Kerr, M. (1991). Family systems theory and therapy. In A. S. Gurman & D. P. Kniskern (Eds.), *Handbook of family therapy* (Vol. 1, pp. 226–264). Brunner-Mazel.

Kott, K. (2014). Applying Bowen theory to work systems. *Organizational Development Practitioner, 46*(3), 76–82.

Magaisa, A. (2018). *A Murky tale: The Dema Power Project Revisited*. Retrieved February 18, 2020, from https://www.bigsr.co.uk/single-post/2018/10/22/BSR-A-murky-tale-The-Dema-Power-Project-Revisited

Mitchell, D. (2017). Responses to organizational conflicts. In A. Farazmand (Ed.), *Global encyclopedia of public administration, public policy, and governance*. Springer. https://doi.org/10.1007/978-3-319-31819-8_3078.1

Ouédrago, I. M. (2017). Governance, corruption, and the informal economy. *Modern Economy, 8*(2), 256–271.

Tatar, A., Oznacar, B., Öĝün, M. N., & Debeş, G. (2019). Interorganizational conflicts and solution strategies. *Revista San Gregorio, 36*, 375–401.

The Bowen Center (n.d.). *Eight concepts*. Retrieved November 28, 2019, from https://thebowencenter.org/theory/eight-concepts/

**Paul R. Sachs** is an independent consultant, trainer and coach, and a licensed psychologist. He holds a Ph.D. in clinical psychology from Vanderbilt University and an MBA from Saint Joseph's University. He has worked over 20 years as a nonprofit executive. He is the author of one book, coeditor of one book, and author of several chapters and articles in the fields of rehabilitation psychology and corporate social responsibility. He can be reached at paul@reidsachs.com.

**Shame Mugova** is a financial analyst, a lecturer in finance at the Birmingham City University (England), and a research fellow at the Durban University of Technology, from which he holds a Ph.D. in finance. He is editor of two books and author of numerous chapters and professional articles. His research interests are corporate social responsibility, corporate governance, and trade credit. He can be reached at mugovas@gmail.com.

# Part IV
# CSR and Performance: Reporting and Evaluation

# Mandatory Nonfinancial Disclosures: The State of Art

**Eva Cerioni, Alessia D'Andrea, Marco Giuliani, and Stefano Marasca**

## 1 Introduction

During the past decades, the disclosure of nonfinancial information has attracted the attention of academics and practitioners (Jackson et al., 2020; Mion & Loza Adaui, 2019; Du Plessis, 2016). Corporate responsibilities to society have increased significantly, and previous mandatory reporting models have been revised in line with a more pluralistic approach, which takes stakeholders, sustainability, business ethics, and transparency into account (Mion & Loza Adaui, 2019; Frías-Aceituno et al., 2013).

In fact, in recent decades, due to the growing complexity of the business world, new reporting requirements have been added, through a patchwork of laws, regulations, and standards (Carini et al., 2021; Venturelli et al., 2020; Girella et al., 2019; Mion & Loza Adaui, 2019; Baldini et al., 2018; De Villiers & Maroun, 2018; Fasan & Mio, 2017; Monciardini et al., 2017). The influence of mandatory requirements on the company's disclosure practices has been extensively discussed in the literature (Mion & Loza Adaui, 2019; Hoffmann et al., 2018).

Some scholars have highlighted that mandatory nonfinancial disclosure (NFD) "increases the comparability of reports among firms as well as over time" (Cordazzo et al., 2020, p. 3473) and "promotes better comparability but may also help prevent some companies from 'greenwashing' or deliberately manipulating stakeholders' perceptions through discretionary reporting" (Siew et al., 2013, p. 20). Concerning this, Jackson et al. (2020) examined "whether mandatory NFD might help to increase the stringency of CSR practices (as with government regulation) while

E. Cerioni (✉) · A. D'Andrea · M. Giuliani · S. Marasca
Department of Management, Università Politecnica delle Marche, Ancona, Italy
e-mail: e.cerioni@univpm.it; a.dandrea@univpm.it; m.giuliani@univpm.it; s.marasca@univpm.it



I. G. Arraiano et al. (eds.), *Corporate Social Responsibility in a Dynamic Global Environment*, CSR, Sustainability, Ethics & Governance,
https://doi.org/10.1007/978-3-031-24647-0_17

maintaining the high degree of flexibility associated with business self-regulation." Adopting an international horizon, Monciardini et al. (2017) highlight how different non-similar countries' regulatory demands could reduce the competitiveness of businesses and therefore call for harmonization in this respect. Therefore, although there is high potential of mandatory reporting to allow comparability, there is a silent request for harmonization at the regulatory level to put all companies in front of the same requests.

Our research aims to investigate the convergence in the regulatory provisions of the different countries regarding NFD. It also investigates whether there is a move toward harmonization of the efforts conducted by the regulatory bodies, whether they are governmental or market regulation on the NFD. This objective is linked to the desire to understand whether the de facto harmonization that exists on the application of the GRI and the IIRC by companies (Cerioni et al., 2021; Doni et al., 2020; Assonime, 2017; Tschopp & Huefner, 2015; Tschopp & Nastanski, 2014) is accepted also at the regulatory level.

To achieve this aim, a manual text analysis (Loughran & McDonald, 2016; Steenkamp & Northcott, 2007) of legislation of different countries has been performed. The text analysis does focus on different dimensions to have an overall picture of the regulation. This is not based on quantitative analyses (Venturelli et al., 2020) but on in-depth textual analysis to identify processes of convergence at an analytical content (Brännström & Giuliani, 2009).

The structure of the paper is the following: First, a literature review on the main contributions on mandatory NFD, then, the design of the study, and the results of the research are provided. In the final sections, the authors attempt to analyze the case findings, present the theoretical arguments of the study, and lastly, propose future research opportunities.

## 2 Background

Most of the NFD literature has initially focused on voluntary disclosure, but during the last years, the scholars have explored the NFD under a mandatory regime (Korca et al., 2021; Mion & Loza Adaui, 2019; Stubbs & Higgins, 2018; Chelli et al., 2018; Schneider et al., 2018; Du Plessis, 2016).

Some scholars (Doni et al., 2020) criticize the mandatory regime because of the delicate issue of "responsibility," which should be a voluntary and not obligatory choice for companies; this at the aim not to invade their "ethical space" and not risk becoming a burden, therefore forcing companies to produce NFDs required by the legislature but of poor quality.

On the other hand, Du Plessis (2016) said that "hard law" may be the only alternative if corporations do not disclose nonfinancial information that will enable investors and the public to determine whether corporations are "good corporate citizens" (p. 74). Part of the literature focuses on the degree of regulation detail: Poorly detailed legislation leads to poor comparability between company reports

(Jackson et al., 2020; Bini et al., 2017; Hüfner, 2007), while too detailed legislation limits the discretion of the management (Baldwin et al., 2011). Jackson et al. (2020) said that "'soft' forms of government regulation may trigger other types of isomorphic processes, as suggested by neo-institutional theory." By this, mimetic isomorphism acts like a coercive force in the voluntary adoption of NFR, and also in the context of mandatory reporting, the coercive force of the law prevails (La Torre et al., 2018).

Other studies focus on the impact of mandatory regimes on the company's disclosure practices. Ioannou and Serafeim (2017) investigated, in China, Denmark, Malaysia, and South Africa, the impact of regulatory changes, and it sought to understand if mandatory disclosure exerts transparency on sustainability disclosure and if regulation affects firm valuations and organizational practices through assurance. Their finding explains that after regulatory changes occur, disclosure increases, and companies are more likely to seek assurance on their environmental, social, and governance disclosure. Stubbs and Higgins (2018) investigated stakeholders' perspectives on the role of regulatory reform in integrated reporting. The findings have shown that, while there is a little appetite for regulatory reform, half of the investors support mandatory IR because, in their experience, voluntary sustainability reporting has not led to more substantive disclosures or increased the quality of reporting. Other scholars have investigated the impact of NFD regulation on practice in terms of quantity and quality of information disclosed (Korca et al., 2021). Chelli et al. (2018) examined how French and Canadian firms changed their reporting practices in reaction to the promulgation of laws and regulations in their respective countries. Their findings reveal a very low level of substantive disclosure in both countries. Schneider et al. (2018) have described which type of mandatory reporting on environmental and social matters currently exists in Canada, the United States, and the EU, and they have encouraged research into the mandatory reporting of environmental and social matters. Jeffery (2017) focused on the implementation of the European Directive by the United Kingdom, Germany, France, and Italy: Differences occurred regarding report contents and penalties linked to noncompliance with the law setting. Hoffmann et al. (2018) study the impacts of mandatoriness in the NFD and state that the introduction of a reporting mandate may at first lead to a decrease in the overall quality of the reporting due to the number of new journalists but will probably increase over time. A recent study by Venturelli et al. (2019) notes the differences in financial disclosure between British and Italian listed companies, highlighting the importance of national and subnational policies to encourage greater compliance with recent requests for nonfinancial disclosures in the directive 2014/95/EU (Venturelli et al., 2019).

Other studies believe that the mandatory publication of NFDs could lead to an improvement in information systems and the possibility of disseminating shared models to be adopted. Jackson et al. (2020) said that "NFD policy leads to more homogeneous CSR activities within a country and reduces the gap between 'best practice' firms and 'typical' firms" (p. 334). But from their findings, "it is unclear whether or not mandatory NFD will increase CSR stringency but increase the rigidity of CSR practices" (Jackson et al., 2020).

Monciardini et al. (2017) highlight that "in the context of (almost) borderless markets, business is increasingly reluctant to further impositions of social and environmental regulations that – this is the argument – increase 'red tape' and diminish 'competitiveness' against companies based in other, more business-friendly, regulatory regimes" (p. 13).

This leads to the request for harmonization at the regulatory level to put all companies in front of the same requests.

No research has broadened the perspective to the several countries in the world in a comparability manner. Practitioners, scholars, and consulting firms are clamoring for the ability to make spatial and temporal comparisons of performance on the sustainability of businesses (Cerioni et al., 2021; Badia et al., 2020; KPMG, 2020; WBCSD, 2018, Eurosif, 2018; Boiral & Henri, 2017; Cohen et al., 2015).

Several government initiatives aim to improve the comparability of NFD. These kinds of initiatives have taken place in Europe (European Commission, 2020a, 2020b), the United States (SASB, 2016; Council of Institutional Investors, 2019), Australia (Parliament of Australia, 2020, www.aph.gov.au/), and Asia (HKEx Environmental, Social and Governance Reporting Guide, 2015). An international consultation on NFD by the European Commission (EU) revealed that the majority of nonfinancial information reported by companies is lacking in terms of comparability (European Commission, 2020a). The EU has launched a project. The project consists of the creation of "future standards on non-financial reporting" and the desire to "require some or all companies within the scope of the legislation to use a non-financial reporting standard" (European Commission, 2020b). The Standards Board (SASB), regarding the US NFD regulation, highlighted investors' demand for greater comparability of information on sustainability (SASB, 2016). The US Council of Institutional Investors, more recently, defines "clearer and comparable information on key ESG factors" and believes that "The adoption by investors and issuers of common ESG reporting standards would represent a highly significant market improvement" (CIII, 2019). In Australia, the minister recommends considering an "agreed reporting framework (...) for allowing for greater comparability"(Parliament of Australia, 2020, www.aph.gov.au/). Besides, in the Asian continent, there is a strong propensity to report nonfinancial information using uniform guidelines to allow for international comparability (HKEx Environmental, Social and Governance Reporting Guide, 2015).

After giving a brief overview of the demands in terms of harmonization of nonfinancial disclosures at the regulatory level, it is interesting to understand what the state of the art is today in the world.

## 3 Methodology

The object of the analysis is represented by the current mandatory legislation on nonfinancial information published by different countries.

To achieve this aim, a manual text analysis (Loughran & McDonald, 2016; Steenkamp & Northcott, 2007) of legislation of different countries has been performed. The sampled countries included are found in the KPMG report (2020). In particular, the analysis is focused on the countries and jurisdictions with reporting rates of sustainability higher than 90%. The countries are Japan, Mexico, Malaysia, India, the United States, Europe (Sweden, Spain, France, the United Kingdom, Germany) South Africa, Taiwan, Australia, and Canada. Within these countries, the regulations relating to all listed companies were analyzed. We choose this category of companies because especially for large publicly listed companies, communicating the environmental, social, and governance (ESG) policies is a strong request by investors and the community (Du Plessis, 2016). To investigate the regulatory system, the Carrots & Sticks database is taken as a reference.

These regulations have been identified as follows: The list of countries' regulations was downloaded from the database available on the Carrots & Sticks website, updated to March 20, 2021, and includes the mandatory regulations on NFD for each country. The selection carried out 40 mandatory regulations (Table 1).

Of the 40 regulations in the initial sample, 1 regulation was discarded because it was forthcoming, and another 4 were excluded from the analysis sample because they were replaced by subsequent regulations.

The text analysis does focus on different dimensions: recipients of the standard, users, the purpose of the standard, principles to be followed for the preparation of the report, the position of the report, type of information, and the indication of the expected report structure with indicators and mandatory framework. Moreover, this is not based on quantitative analyses (Venturelli et al., 2020) but on in-depth textual analysis to identify processes of convergence at an analytical content (Brännström & Giuliani, 2009).

To analyze regulations on NFD, a manual text analysis of the countries' regulations has been performed as it allows understanding each element that composed the regulations (Loughran & McDonald, 2016; Steenkamp & Northcott, 2007).

The choice not to use software (e.g., which are typical for content analysis, as Wordstat 7, Nvivo, TLab) derives from the fact that it is important to read all the regulations and find the key elements during the reading.

From the text analysis, information relating to the dimension considered by the legislation distinguished between single dimension and triple dimension was inserted. This parameter was used to divide the sample analyzed into two subgroups: regulations of three dimensions and regulations of a single nonfinancial dimension.

The sample analyzed is therefore composed of 13 standards with a triple focus on nonfinancial information (ESG), while 22 provide for the disclosure of a single dimension. The single dimensions considered are governance (which includes remuneration and anti-corruption) and environmental.

**Table 1** The sample of analysis

| Country | Type of istrument | Link to regulations | Date of publication |
|---|---|---|---|
| Australia | Codes, guidance, and questionnaires | Regulatory Guide 247 on operating and financial review (OFR) | 2013 |
| Australia | Public law and regulation | Energy Efficiency Opportunities Act, 2006 (amended in 2007) | 2007 |
| Australia | Public law and regulation | Listing Rule 4.10.3, Australian Stock Exchange (ASX) | 2003 |
| Australia | Codes, guidance, and questionnaires | The Australian Securities and Investments Commission (ASIC) Section 1013DA Disclosure Guidelines | 2003 |
| Canada | Public law and regulation | National Instrument 58-101 Disclosure of Corporate governance practices | 2014 |
| Canada | Self-regulation | The TSX Timely Disclosure Policy | 2004 |
| EU | Public law and regulation | EU Directive Non-financial and Diversity Information Disclosure 2014/95/EU | 2014 |
| EU | Public law and regulation | EU Modernisation Directive 2003/51/EC | 2003 |
| EU | Public law and regulation | EU Pollutant Release and Transfer Register (PRTR) (REGULATION 166/2006/EC) | 2006 |
| EU | Public law and regulation | Shareholders Rights Directive (EU) 2018/828 | 2017 |
| Hong Kong | Self-regulation | HKEX Listing Rules Disclosure of Financial Information | 2015 |
| India | Codes, guidance, and questionnaires | SEBI/HO/CFD/CMD/CIR/P/2017/10 Integrated Reporting by Listed Entities | 2017 |
| India | Public law and regulation | National Voluntary Guidelines on Social, Environmental and Economic Responsibilities of Business Revision | 2016 |
| India | Self-regulation | Circular No. CIR/CFD/CMD/10/2015 Format for Business Responsibility Report (BRB) | 2015 |
| India | Self-regulation | Circular No. CIR/CFD/POLICY CELL/2/2014 Equity Listing Agreement | 2014 |
| India | Self-regulation | SEBI Committee on Corporate Governance, Quarterly compliance report | 2003 |
| Japan | Public law and regulation | Act on Promotion of Female Employment | 2015 |
| Japan | Public law and regulation | Law concerning the Rational Use of Energy, Act on Promotion of Global Warming Countermeasures, 1979. | 1979 |
| Japan | Public law and regulation | Mandatory GHG Accounting System | 2005 |
| Malesia | Codes, guidance, and questionnaires | Sustainability Reporting Guide | 2015 |
| Malesia | Self-regulation | Amendments to Main Market Listing Requirements relating to Sustainability Statement in Annual Reports and Issuance of the Sustainability Reporting Guide and Toolkits | 2015 |
| Malesia | Codes, guidance, and questionnaires | Malaysian Code on Corporate Governance | 2012 |
| Malesia | Public law and regulation | Main Markets Listing Requirements CSR description | 2007 |
| Messico | Public law and regulation | Climate Change law, 2012. | 2012 |
| Messico | Public law and regulation | Federal Law on Anti-corruption Practices in Public Contracts | 2012 |
| Messico | Other e.g. action plan, strategy, programme, voluntary initiative | National Emissions Register (RENE) | 2014 |
| South Africa | Self-regulation | Johannesburg Stock Exchange Listing Requirement 2010. | 2010 |
| South Africa | Self-regulation | JSE Amendments Listing Requirement | 2017 |
| Taiwan | Self-regulation | Corporate Governance Roadmap 2018-2020 | 2018 |
| Taiwan | Type of instrumentSelf-regulation | Taiwan Stock Exchange Corporation Rules Governing the Preparation and Filing of Corporate Social Responsibility Reports by TWSE Listed Companies | 2019 |
| Taiwan | Self-regulation | Corporate Governance Best Practice Principles for TWSE/TPEx Listed Companies | 2018 |
| Taiwan | Self-regulation | Taiwan CSR best practice principals for SITEs/SICEs, futures commissions merchants, Security Firms | 2004-2015 |
| United States of America | Public law and regulation | Amendment to S-K Regulation: Pay Ratio Disclosure Rule | 2015 |
| United States of America | Public law and regulation | Dodd-Frank Wall Street Reform and Consumer Protection Act (includes 2012 conflict minerals rule) | 2010 |
| United States of America | Guidelines and standards for non-financial reporting | Guidance on Pay Ratio Disclosure | 2017 |
| United States of America | Self-regulation | NYSE Section 3 Corporate Responsibility | 2014 |
| United States of America | Self-regulation | Regulation S-K | 2010 |
| United States of America | Public law and regulation | Clean Water Act (CWA) | 1972 |
| United States of America | Type of instrumentOther e.g. action plan, strategy, programme, voluntary initiative | Mandatory Greenhouse Gas Reporting Rule | 2014 |

The regulations excluded from the sample are highlighted in gray
The regulations highlighted in green are related to the triple dimension ESG

# 4 Finding

The analysis of the sample analyzed provides as a first step the distinction between the regulations that mainly deal with a single topic and the regulations that govern multiple dimensions: environmental, social, economic, and governance. All the analyzed dimensions of the regulations are analyzed separately to be able to compare the approach to the theme of each of them. The considered topics are the purpose of the legislation, principles, type of information requested, structure of the report, indicators, framework, users, capitals, and position of the report.

## *4.1 Analysis of the Multidimensional Sample*

**Purpose of the Legislation**
The first topic analyzed is the purpose for which the regulation is born. Five out of 13 regulations explicitly indicated the purpose for which the norm was born. These are the legislations of Australia, Malaysia, and South Africa. Australia aims to ensure that all listed companies disclose the information they need to investors. Malaysia, on the other hand, describes that "the obligations under the Listing Requirements relating to sustainability are aimed at achieving the following objectives: (a) improving the quality of sustainability-related practices and reporting of our listed issuers, (b) aiding listed issuers to meet sustainability expectations of their stakeholders, (c) attracting funds with a sustainability focus into the Malaysian capital market, and (d) facilitating more listed issuers to qualify for FTSE4Good Bursa Malaysia Index and other international sustainability indices."

Meanwhile, South Africa focuses on the desire to highlight to stakeholders the creation of sustainable value by companies.

**Users**
After the purposes, the second topic analyzed is the clarification of the main user of the envisaged information. Regarding this topic, 9 out of 13 regulations explicitly indicate who the users of the mandatory information are. Of these eight, they generally refer to the category of corporate stakeholders, while only one says that they mainly address investors (Australia).

**Framework**
Concerning the more or less mandatory standard to be followed for the construction of the information, 10 out of 13 regulations deal with the issue. But most of these indicate that you use any internationally recognized framework or standards without specifying one in particular. Malaysia suggests the national guide Sustainability Reporting Guide while leaving companies free to adopt other international frameworks such as the GRI or United Nations Guiding Principles Reporting Framework. In the guide, we read: "There is no standard 'one size fits all' approach to sustainability governance. An organization's culture, needs, sector, size, sustainability-

related risks and opportunities, and maturity in responding to sustainability matters will influence how sustainability governance is considered. Given this, you are encouraged to adopt an approach that is fit for purpose."

India is responsible for the use of the IIRC, while South Africa requires the use of the GRI and the IIRC.

**Principles**

The fourth topic analyzed is whether or not there are principles for drafting the report by the legislation. Only 3 out of 13 regulations provide for the definition of NFD drafting principles. Among these is Australia, which provides within the norm the guiding principle (The Australian Securities and Investments Commission (ASIC) Section 1013 Disclosure Guidelines, 2003). The principles are to disclose what is "known," disclose what a person would "reasonably require," assess "reasonableness" from the client's perspective (more marketing means more disclosure), explain all terms (including industry jargon) used, explain all policies and approaches used, describe any specific measures used, consider referring clients to secondary sources, and identify any external providers or rating mechanisms.

Hong Kong, on the other hand, indicates the following principles: materiality and consistency. India, in its 2017 legislation (SEBI/HO/CFD/CMD/CIR/P/2017/10 Integrated Reporting by Listed Entities) which refers to the principles of the IRC Framework, lists the drafting principles in the standard: "The content of the report and how information is to be presented: Strategic focus and future orientation, connectivity of information, stakeholder relationships, materiality, conciseness, reliability and completeness, Consistency and comparability."

**Type of Information**

The fifth topic concerns the detail of the information that the standard requires disclosure. As many as 11 out of 13 regulations provide more or fewer details of the information to be contained within the NFD. In Australia, the Australian Stock Exchange (ASX) (2003) indicates two macro topics: environmental risk and social risks. These two macro issues to be dealt with do not provide in detail what types of information to disclose but leave flexibility to the company. Also, in Australia, the Product Disclosure Statement (PDS) is expected to be published, which requires the provision of environmental, ethical, and social information relating to the investment in question, not specifying which ones. In Canada, the Timely Disclosure Policy (2004) indicates that material information should also include information relating to environmental or social impacts on the company's activity. Much more detailed is the European Directive of 2014 which lists the minimum content of the NFD: "As a minimum, environmental, social and employee matters, respect for human rights, anti-corruption and bribery matters, including: (a) a brief description of the undertaking's business model; (b) a description of the policies pursued by the undertaking in relation to those matters, including due diligence processes implemented; (c) the outcome of those policies; (d) the principal risks related to those matters linked to the undertaking's operations including, where relevant and proportionate, its business relationships, products or services which are likely to cause adverse (e) impacts in those areas, and how the undertaking manages those risks; non-financial key performance indicators relevant to the particular business."

Hong Kong plans to include information on governance such as "'(i) a disclosure of the board's oversight of ESG issues; (ii) the board' SESG management approach and strategy, including the process used to evaluate, prioritize, and manage material ESG-related issues (including risks to the issuer's businesses); and (iii) how the board reviews progress made against ESG-related goals and targets with an explanation of how they relate to the issuer's businesses."

India envisages a high level of detail for the type of information to be included in the report such as the use of polluting resources, anti-corruption, and gender information for the person. In Malaysia, the norm refers only to ESG information but without further details. In South Africa, the updated version of the King Code indicates the macro topics to be contained in the NFD report: leadership, ethics and corporate citizenship, strategy, performance and reporting, governing structures and delegation, governance functional areas, and stakeholder relationships.

Finally, Taiwan indicates in great detail for all macro-information to be included the information that must be disclosed but only for companies in specific sectors while referring to the GRI standard for other companies not covered by the disciplined sectors.

**Structure of the Report**

Another topic analyzed is the indication in the legislation of a mandatory structure for the drafting of the nonfinancial report. Regarding this topic, 6 out of 13 regulations provide a structure of the report to be published. In Australia, the drafting of the PDS is given the sections of which the document should be composed. In India, for the sections of the document, please refer to the IIRC Framework or the Business Responsibility Reporting Format included in the suggested guidelines to respond to the requirements of the legislation (scheduled sections: general disclosures, management and process disclosures, principle wise performance disclosure). For each section, its content and indicators shall be described. In Malaysia, the structure includes "(a) the governance structure in place to manage the economic, environmental, and social risks and opportunities; (b) the scope of the Sustainability Statement and basis for the scope; (c) material sustainability matters; (d) policies to manage these sustainability matters; (e) measures or actions taken to deal with these sustainability matters; and (f) indicators relevant to these sustainability matters which demonstrate how the listed issuer has performed in managing these sustainability matters." Finally, for South Africa, please refer to the IIRC Framework for sections to follow to structure the report.

**Indicators**

Entering the details of the indication by the regulation of the punctual indicators to be reported 7 out of 13 regulations provide within the regulations the more or less punctual indication of the nonfinancial indicators to be reported. In detail, Europe does not indicate the list of indicators to be included but specifies "With a view to Facilitating the disclosure of non-financial information by undertakings, the Commission should prepare non-binding guidelines, including general and Sectoral non-financial key performance indicators." Hong Kong divides the NFD into two macro arguments: social and environmental. Each argument indicates the general disclosures and KPIs that companies should insert even if they do not indicate the obligation to report them all. India indicates all the indicators that the nonfinancial report should contain. Malaysia expects that the regulation will refer to the national guide "Sustainability Reporting Guide" which provides specific indicators for each reporting area. Finally, South Africa envisages in the regulation the indication of the indicators concerning "good governance" while referring to the GRI for other environmental and social indicators.

**Position of the Report**

As regards where to include this type of nonfinancial information, 10 out of 13 regulations in the text give indications about this. The number of regulations that provide for the choice to include the information in the annual report or to insert it in a separate document is equally divided, while the other five regulations stipulate that it should be included in the annual report.

## *4.2 Analysis of the Single-Dimensional Sample*

**Governance (Remuneration – Women)**

About governance, 12 out of 22 regulations focus on this topic, 2 of which focus particularly on the remuneration of employees and boards (Amendment to S-K Regulation: Pay Ratio Disclosure Rule and Guidance on Pay Ratio Disclosure in the United States of America) and 2 on gender issues (National Instrument 58-101 Disclosure of Corporate Governance Practices in Canada and Act on Promotion of Female Employment in Japan). The **purpose** for which the regulation is born is indicated by half of the standards of the sample and can be summarized with two keywords: to ensure good and efficient governance. On the size of **users** only in five cases, there is a specific reference to such subjects who are in all case shareholders and investors. Concerning the **framework** to be used to communicate governance information, only three regulations explicitly indicate them. Australia refers to the OFR Regulatory Guide, while in Malaysia, the standard is to include such information in the integrated report using an international shared framework without specifying one in particular. While Japan envisages the use of guidelines for the drafting of the employment plan for women, a particular focus of the legislation is under consideration. The governance regulations do not provide for any particular **principles** for the drafting of reports. On the other hand, 8 out of 12 regulations define its content on the **type of information** to be communicated. Australia lists as its main content the following: operations, financial position, business strategies, and prospects for future financial years. The European legislation (Shareholders Rights Directive (EU) 2018/828) provides a description of policies as well as a detailed report on the remuneration of board members. India has a "compliance status on eight subclauses, namely, Board of Directors, Audit Committee, Shareholders/

Investors Grievance Committee, remuneration of directors, board procedures, management, shareholders, and report on corporate governance." Taiwan does not envisage information to be disclosed but lists the practices and behaviors that listed companies would be appropriate for governance. For example we read, "the Board of Directors of a TWSE/TPEx listed company, in consideration of the company's scale and type of operations and the number of its board members, may set up functional committees for auditing, remuneration, nomination, risk management, or any other functions and, based on concepts of corporate social responsibility and sustainable operation, may set up environmental protection, corporate social responsibility, or other committees and expressly provide for them in the articles of incorporation." In the United States, NYSE collects a list of practices that increase the quality of corporate governance: ". . . high-quality board and succession planning to ensure a board works effectively as a team. It goes on to explore a range of topics that a board must address if it is to enable the company to achieve its full potential including strategy, risk management, communicating with shareholders, and overseeing an effective ethics and compliance program." Always in the United States, the Pay Ratio Disclosure Rule and Guidance on Pay Ratio Disclosure indicate the information/indicators to be reported (e.g., *non-US employees*, *median employee*, and *independent contractors*). Only Malaysia legislation provides sections into to articulate the structure of the Integrated Report; no other legislation provides a standard report in the structure to be used. Also concerning the **indicators** to be reported, the regulations of Australia provide a section with a list of the indicators that should be properly reported. So does the Pay Ratio Disclosure that lists the indicators (*median of the annual total compensation of all employees of a registrant excluding the chief executive officer, the annual total compensation of that registrant's chief executive officer, and the ratio of the median of the annual total compensation of all employees to the annual total compensation of the chief executive officer*). Finally, on the governance report, 6 out of 12 regulations predict its **position**. Three of these – the United States, Taiwan, and Malaysia – plan to include them in the annual report. A separate document and publication on the company's website are planned in Europe and Australia.

**Environment**

As the only reference dimension, 10 out of 22 regulations deal only with the environmental issue. Some regulations focus only on emissions, whereas others on the use of certain resources such as water, energy, or minerals. Almost all of the regulations, seven out of ten, explain the **purpose** of the norm. Among the purpose of the norm are highlighted: efficient use of energy for Australia's Legislations, reduce pollution for Europe Union (EU), Mexico and America 2006 legislation. All regulations are not aimed primarily at individual **users** except in the case of Mexico, where it is indicated that the reports resulting from the application of the legislation are addressed to the federal prosecutor for environmental protection. Concerning **frameworks** or standards, only four regulations from two countries (Mexico and the United States) refer to specific standards for the drafting of these specific reports: Official Mexican Standards and Best Available Monitoring Methods (BAMM) for the United States. No legislation refers to a specific standard known as Carbon Disclosure. About the monitored size of drafting **principles**, only Mexico in the Climate Change Law of 2012 lists the principles that generally guide national climate change policy and which should therefore be inspired by companies. Among these, we read: "I. Sustainability in the exploitation or use of ecosystems and their natural elements; II. Co-responsibility between state and society in general, in the implementation of mitigation and adaptation actions to the negative effects of climate change; III. Attention, (. . .) to climate change; IV. Prevention (. . .) V. Adoption of production and consumption models by the public, social and private sectors to move towards a low carbon economy (. . .) VII. Participation of citizens in the formulation, implementation, monitoring and evaluation of the National Strategy, plans and programmes for mitigation and adaptation to the effects of climate change; VIII. Environmental liability (. . .); IX. The use of economic tools to mitigate, adapt and reduce vulnerability to climate change (. . .) X. Transparency, access to information and justice (. . .) XI. Conservation of ecosystems and their biodiversity (. . .) XII. Commitment to the economy and national economic development (. . .)." All the regulations list the **type of information**, some in a very detailed manner (such as Japan and Mexico), while others only indicate the macro-categories as the emissions of different pollutants then leaving the company much flexibility in the type of indicators to provide. The strictest legislation is the Mandatory Greenhouse Gas Reporting Rule of the United States, which lists all the indicators to be reported. Australia, Mexico, and America provide a **structure** for reporting such information, and among these are obvious common sections such as that of the general information of the company but specific sections such as "a description of the way in which the corporation has carried out, during the period, the proposal in its approved assessment plan for assessing the opportunities for improving the energy efficiency of its group" requested by Australia. Finally, six out of ten regulations indicate **where** to provide such information, and the whole of it provides for the preparation of separate reports; on these, there is alignment between the regulations.

## 5 Discussion and Conclusion

Our research aimed to investigate the convergence in the regulatory provisions of the different countries regarding NFD. Because, given the growing demand for convergence by scholars and professionals to obtain comparable nonfinancial information (Cerioni et al., 2021; Badia et al., 2020; KPMG, 2020; WBCSD, 2018; Eurosif, 2018; Boiral & Henri, 2017; Cohen et al., 2015), it is important to understand what is expected at a regulatory level from different countries (Cerioni et al., 2021; Cosma et al., 2020; La Torre et al., 2018; Sierra-Garcia et al., 2018). It would also investigate whether there is a move toward harmonization of the efforts conducted by the regulatory bodies, whether they are governmental or market regulation on the NFD. The main results are the following: The main results highlight the absence of similarities between the different legislations. Most of the regulations have a flexible approach leaving a wide margin of choice to the companies. The results showed a heterogeneous situation: While it is true that the NFD today finds a regulatory base in most countries of the world, each legislation differs from the others. There are situations of convergence regarding the position of the report and the wide flexibility on the choice of framework. On the position, a result to be highlighted is how ESG information is aimed both to include it in the annual report and to separate reporting. For the one-dimensional information, the request to report it separately prevails. We can say that for ESG information although in a few cases the standard explicitly refers to the GRI and the IIRC Framework and this is in line with de facto harmonisation by companies in the adoption of these frameworks (Cerioni et al., 2021; Doni et al., 2020; Assonime, 2017; Tschopp & Huefner, 2015; Tschopp & Nastanski, 2014). In relation to the aims specified in the rules, in some cases emerges the prevalence of accountability policy, and that is to have to report something more than financial-economic information, but at the same time in other situations emerges the vision "forward-looking" and therefore the need to undertake extra-financial efficient management for the creation of a sustainable value of the company that lasts over time safeguarding the environment in which it operates and therefore the attention in the present and future to pollution.

The need for harmonization can therefore also be refuted at a regulatory level, except for specific sectors or small groups of countries.

The need of harmonization appears to be an invitation to policymakers to transpose virtuous behaviour and to implement the desired harmonization of iure.

Many studies discuss the best approach (principle or rules-based) and the degree of freedom that should be left to the preparations to make the NFD useful. But our results show that where there is flexibility, there is no comparability. To date, regulations are mainly flexible in every element envisaged, and this leads to a lack of comparability. That is why the path should be that of homogeneous regulation widespread if the objective is to standardize documents and comparability of this as is the case with the financial statement. But if there is no harmonization at a regulatory level, how do we think to find a common approach by companies? In any event, the drive for harmonization and comparability will have to deal with the considerations that an organization's culture, needs, sector, size, sustainability-related risks and opportunities, and maturity in responding to sustainability matters will influence how sustainability governance is considered differently in any organizations.

So, we think that it will always be necessary to be aware of the culture of sustainability within organizations so that companies will not just adapt their current reporting practices to comply with the regulations and maintain a "business as usual" approach (Dumay & Hossain, 2019).

The greatest limit of the presence of research is the sample of analysis. We decided to limit the analysis to listed companies only and ten countries with high sustainability rates (KPMG, 2020).

Research perspectives are represented by extending the analysis to other countries of the world, and it could be included in the sample of all regulations applied to all types of companies.

## References

Assonime. (2017). Circolare 13/2017 Gli obblighi di comunicazione delle informazioni non finanziarie (Mandatory reporting of non-financial information). Available at: www.assonime.it/attivita-editoriale/circolari/Pagine/Circolare-13-2017.aspx

Badia, F., Bracci, E., & Tallaki, M. (2020). Quality and diffusion of social and sustainability reporting in Italian public utility companies. *Sustainability, 12*(11), 4525.

Baldini, M., Dal Maso, L., Liberatore, G., Mazzi, F., & Terzani, S. (2018). Role of country-and firm-level determinants in environmental, social, and governance disclosure. *Journal of Business Ethics, 150*(1), 79–98.

Baldwin, R., Cave, M., & Lodge, M. (2011). *Understanding regulation: Theory, strategy, and practice*. Oxford University Press.

Bini, L., Dainelli, F., & Giunta, F. (2017). Is a loosely specified regulatory intervention effective in discipline management commentary. The case of performance disclosure indicator. *Journal of Management and Governance, 21*(1), 63–91.

Boiral, O., & Henri, J.-F. (2017). Is sustainability performance comparable? A study of GRI reports of mining organisations. *Business & Society, 56*(2), 283–317.

Brännström, D., & Giuliani, M. (2009). Accounting for intellectual capital: A comparative analysis. *Vine, 39*, 68–79.

Carini, C., Rocca, L., Veneziani, M., & Teodori, C. (2021). Sustainability regulation and global corporate citizenship: A lesson (already) learned? *Corporate Social Responsibility and Environmental Management, 28*(1), 116–126.

Cerioni, E., D'Andrea, A., Giuliani, M., & Marasca, S. (2021). Non-financial disclosure and inter-company comparability: Customisation, harmonisation or standardisation? *Sustainability, 13*, 1177.

Chelli, M., Durocher, S., & Fortin, A. (2018). Normativity in environmental reporting: A comparison of three regimes. *Journal of Business Ethics, 149*(2), 285–311.

Cohen, J. R., Holder-Webb, L., & Zamora, V. L. (2015). Nonfinancial information preferences of professional investors. *Behavioral Research in Accounting, 27*(2), 127–153.

Cordazzo, M., Bini, L., & Marzo, G. (2020). Does the EU Directive on non-financial information influence the value relevance of ESG disclosure? Italian evidence. *Business Strategy and the Environment, 29*(8), 3470–3483.

Cosma, S., Venturelli, A., Schwizer, P., & Boscia, V. (2020). Sustainable development and European banks: A non-financial disclosure analysis. *Sustainability, 12*, 6146.

Council of Institutional Investors (CIII), Subcommittee hearing letter (final), July 9 2019., www.cii.org

De Villiers, C., & Maroun, W. (2018). The future of sustainability accounting and integrated reporting. In *Sustainability accounting and integrated reporting* (pp. 163–170). Routledge.

Doni, F., Bianchi Martini, S., Corvino, A., & Mazzoni, M. (2020). Voluntary versus mandatory non-financial disclosure: EU Directive 95/2014 and sustainability reporting practices based on empirical evidence from Italy. *Meditari Accountancy Research, 28*(5), 781–802.

Du Plessis, J. (2016). Corporate governance, corporate responsibility and law: Disclosure of non-financial information: A powerful corporate governance tool. *Corporate Responsibility and Law: Disclosure of Non-financial Information: A Powerful Corporate Governance Tool, 34*, 16–14.

Dumay, J., & Hossain, M. A. (2019). Sustainability risk disclosure practices of listed companies in Australia. *Australian Accounting Review, 29*(2), 343–359.

European Commission. (2020a). Summary report of the public consultation on the review of the non-financial reporting directive.

European Commission. (2020b). Revision of the non-financial reporting directive – Inception impact Assesment.

Eurosif; A.C.C.A. (2018). *What do investors expect from non-financial reporting 2013*. https://www.accaglobal.com/uk/en/technical-activities/technical-resources-search/2013/august/investors-and-non-financial-reporting.html

Fasan, M., & Mio, C. (2017). Fostering stakeholder engagement: The role of materiality disclosure in integrated reporting. *Business Strategy and the Environment, 26*(3), 288–305.

Frías-Aceituno, J. V., Rodríguez-Ariza, L., & García-Sánchez, I. M. (2013). Is integrated reporting determined by a country's legal system? An exploratory study. *Journal of Cleaner Production, 44*, 45–55.

Girella, L., Zambon, S., & Rossi, P. (2019). Reporting on sustainable development: A comparison of three Italian small and medium-sized enterprises. *Corporate Social Responsibility and Environmental Management, 26*(4), 981–996.

HKEx Environmental. (2015). Social and governance reporting guide. https://www.hkex.com.hk

Hoffmann, E., Dietsche, C., & Hobelsberger, C. (2018). Between mandatory and voluntary: Non-financial reporting by German companies. *Nachhaltigkeits Management Forum|Sustainability Management Forum, 26*(1-4), 47–63.

Hüfner, B. (2007). The sec's MD&A: Does it meet the informational demands of investors? *Schmalenbach Business Review, 59*, 58–84.

Ioannou, I., & Serafeim, G. (2017, May 1). *The Consequences of Mandatory Corporate Sustainability Reporting*. Harvard Business School Research Working Paper No. 11-100, Available at SSRN: https://ssrn.com/abstract=1799589

Jackson, G., Bartosch, J., Avetisyan, E., Kinderman, D., & Knudsen, J. S. (2020). Mandatory non-financial disclosure and its influence on CSR: An international comparison. *Journal of Business Ethics, 162*(2), 323–342.

Jeffery, C. (2017). *Comparing the implementation of the EU non-financial reporting directive in the UK, Germany, France and Italy*. Retrieved from https://ssrn.com/abstract=3083368

Korca, B., Costa, E., & Farneti, F. (2021). From voluntary to mandatory non-financial disclosure following Directive 2014/95/EU: An Italian case study. *Accounting in Europe, 18*(3), 353–377.

KPMG. (2020). Towards consistent and comparable ESG reporting. https://home.kpmg/au/en/home/insights/2020/03/towards-consistent-and-comparable-esg-reporting.html

La Torre, M., Sabelfeld, S., Blomkvist, M., Tarquinio, L., & Dumay, J. (2018). Harmonising non-financial reporting regulation in Europe. *Meditari Accountancy Research, 26*(4), 598–621.

Loughran, T., & McDonald, B. (2016). Textual analysis in accounting and finance: It's a survey. *Journal of Accounting Research, 54*(4), 1187–1230.

Mion, G., & Loza Adaui, C. R. (2019). Mandatory nonfinancial disclosure and its consequences on the sustainability reporting quality of Italian and German companies. *Sustainability, 11*(17), 4612.

Monciardini, D., Dumay, J., & Biondi, L. (2017). Integrated reporting and EU law. Competing, converging or complementary regulatory frameworks?. *Competing, converging or complementary regulatory frameworks.*

Parliament of Australia. (2020). *Chapter Seven - Sustainability reporting: Current legislative and market requirements.* https://www.aph.gov.au/

SASB. (2016). *Business and financial disclosure required by regulation S-K – The SEC's concept release and its implications*. https://www.sasb.org/wp-content/uploads/2016/09/Reg-SK-Comment-Bulletin-091416.pdf

Schneider, T., Michelon, G., & Paananen, M. (2018). Environmental and social matters in mandatory corporate reporting: An academic note. *Accounting Perspectives, 17*(2), 275–305.

Sierra-Garcia, L., Garcia-Benau, M. A., & Bollas-Araya, H. M. (2018). Empirical analysis of non-financial reporting by Spanish companies. *Administrative Sciences, 8*(3), 29.
Siew, R. Y., Balatbat, M. C., & Carmichael, D. G. (2013). The relationship between sustainability practices and financial performance of construction companies. *Smart and Sustainable Built Environment, 2*(1), 6–27.
Steenkamp, N., & Northcott, D. (2007). Content analysis in accounting research: The practical challenges. *Australian Accounting Review, 17*(43), 12–25.
Stubbs, W., & Higgins, C. (2018). Stakeholders' perspectives on the role of regulatory reform in integrated reporting. *Journal of Business Ethics, 147*(3), 489–508.
Tschopp, D., & Huefner, R. J. (2015). Comparing the evolution of CSR reporting to that of financial reporting. *Journal of Business Ethics, 127*(3), 565–577.
Tschopp, D., & Nastanski, M. (2014). The harmonisation and convergence of corporate social responsibility reporting standards. *Journal of Business Ethics, 125*(1), 147–162.
Venturelli, A., Caputo, F., Leopizzi, R., & Pizzi, S. (2019). The state of art of corporate social disclosure before the introduction of non-financial reporting directive: A cross country analysis. *Social Responsibility Journal, 15*(4), 409–423.
Venturelli, A., Pizzi, S., Caputo, F., & Principale, S. (2020). The revision of nonfinancial reporting directive: A critical lens on the comparability principle. *Business Strategy and the Environment, 29*(8), 3584–3597.
WBCSD; PwC. (2018). *Enhancing the credibility of non-financial information: The investor perspective*. https://www.wbcsd.org/Programs/Redefining-Value/External-Disclosure/Assurance-Internal-Controls/Resources/Enhancing-the-credibility-of-non-financial-information-the-investor-perspective

**Eva Cerioni** is a research fellow in accounting since June 2018. She earned a Ph.D. in "business economics" on March 21, 2018. In 2014, she obtained a degree in management and law from the Università Politecnica delle Marche (Italy), Department of Management. Her research field regards management accounting and integrated reporting.

**Alessia D'Andrea** is a researcher in accounting since January 2018. She earned a Ph.D. in "business economics" on February 27, 2009. She teaches business economics since 2013 at the Università Politecnica delle Marche (Italy), Department of Management. Her major research areas and interdisciplinary projects are on cost accounting tools and on the development of economic and business models for the social and healthcare sector and on the implementation and development of socio-environmental reporting documents for public and private companies.

**Marco Giuliani** is a full professor of accounting at the Università Politecnica delle Marche (Ancona, Italy), Department of Management. His main research interests are in financial accounting, auditing, intellectual capital reporting, and nonfinancial reporting/disclosure. He is member of national and international research groups on intellectual capital and company valuation. He is a member of the editorial boards and of the advisory boards of several national and international journals.

**Stefano Marasca** is professor of business economics in the Department of Management of Università Politecnica delle Marche (Ancona, Italy) since 2000. In the last years, he taught in the following courses: strategic control, international financial accounting, management accounting, and business economics of public and nonprofit companies. He is an author of numerous publications, both national and international, and a coordinator of interuniversity and local research projects on topics of annual report and integrated reporting, of company evaluation and intangible assets, and of strategic control and management control. The case studies have regarded both private companies (profit and nonprofit) and public companies and, in particular, health organizations.

# Toward a Novel Approach to Companies-Stakeholder Relationship: Modeling IBEX35 Long Run Value Creation

**Manuel Moreno, Elena Mañas, Oscar Montes-Pineda, and Beatriz Fernández-Olit**

## 1 Introduction

The good performing scores offered by the sustainability ratings and indexes such as the Dow Jones Sustainability Index or FTSE4Good regarding CSR management highly differ with the opinions of some authors emphasizing that companies' nonfinancial reports can be seen as window dressing or cosmetics (Durand et al., 2019; Friede, 2019; Ikram et al., 2019; Porter & Kramer, 2006). They argue that companies' interest in responding to a social demand, which calls for the introduction of CSR in their management, doesn't correspond to the notion held by governance bodies about the ability of CSR to create a strategic advantage. Thus, they would only devote the necessary resources to avoid damaging their reputation rather than boosting their relationship with the stakeholders in a more decisive way. Short-termism may be behind this difficulty (Barton, 2017; Fink, 2018; Friede, 2019), and some authors suggest that a refocusing of companies toward long-termism could facilitate a way out of this situation (Barton, 2017; Porter & Kramer, 2011; Schoenmaker & Schramade, 2019; Sewchurran et al., 2019).

Investors seem to need a financial benefit to engage with ESG issues (Amel-Zadeh & Serafeim, 2018), which is also credible enough to overcome the widespread perception that CSR may be harming the profitability of their investments (Jáuregui, 2019; Karpoff, 2021). This would require overcoming a defensive view of CSR, i.e., one in which CSR is valued insofar as it protects the company (and its investments) against potential reputational risk, stemming from poor ESG management (Hsu

M. Moreno (✉) · E. Mañas · O. Montes-Pineda · B. Fernández-Olit
University of Alcalá, Alcalá de Henares, Spain
e-mail: manuel.moreno@actuarios.org; elenamanas@uah.es; oscar.montes@uah.es; beatriz.olit@uah.es

 
I. G. Arraiano et al. (eds.), *Corporate Social Responsibility in a Dynamic Global Environment*, CSR, Sustainability, Ethics & Governance,
https://doi.org/10.1007/978-3-031-24647-0_18

et al., 2019). In the face of defensive CSR, investors would not engage in its development beyond proper legislative compliance, international standards, or rating agency requirements and would consider it an expense. Thus, shareholders would behave indifferently, ignoring whether or not a security belongs to a sustainability index such as the Dow Jones Sustainability Index (DJSI) (Durand et al., 2019).

To overcome this defensive view of CSR, transparency in corporate reporting is necessary to demonstrate that the use of resources dedicated to ESG management is efficient from a financial point of view (although its benefits may only appear in the profit and loss account). This transparency should be able to overcome investors' natural risk aversion, related to the short-termism of financial markets, and facilitate their engagement with sustainable investment (Sulkowski et al., 2018). Furthermore, this transparency should allow for the use of long-term analysis techniques (where historical data may not be valid because of new situations), such as fundamental analysis (Lo, 2017), business case analysis (Carroll & Shabana, 2010), or scenario analysis (Sewchurran et al., 2019).

On the other hand, and despite the criticisms of several authors on ESG reporting, in addition to the knowledge of how prepared companies are to improve their relationship with stakeholders and the more executive and financial control issues, it is necessary to know where companies are in relation to the development of the relationship with the aforementioned stakeholders under ESG aspects. This should include not only a defensive view of CSR but also a value-creating approach based on active dialog with stakeholders (Porter & Kramer, 2006; Sulkowski et al., 2018).

Finally, it seems necessary to look more closely at the question of the involvement of stakeholders in an active dialog with the company with which they interact. As there is no economic relationship with many of them (apart from shareholders, workers, customers, and suppliers), one might wonder why they would actively participate in a dialog with the company. It should be borne in mind that the aim is to move beyond a defensive vision of repairing possible damage caused by the company and to focus more on processes of value creation and innovation. The answer could be found in issues such as the company's culture and more specifically its purpose. Materializing the company's culture in the purpose for the development of sustainability is mentioned by authors such as Clark et al. (2015), Porter and Kramer (2006), Fink (2018, 2019), Business Roundtable (2019), or Freudenreich et al. (2020).

In short, the issues to be addressed by a possible model for assessing the relationship between companies and stakeholders would be as follows:

- Where companies are in relation to the development of the relationship with the stakeholders
- The readiness of companies to improve their relationship with stakeholders
- Executive issues, budgets, controls, etc.
- Company culture and specifically the purpose of the company

The aim of this paper is to understand the status of the relationship with stakeholders in companies and to test its potential to create long-term financial value from ESG-based development. To this end, the following hypothesis is put forward:

"In practice, companies approach CSR and their relationship with stakeholders more as a defensive tool for the management of their reputation than as a clear commitment to the creation of shared value, so that the management model for the relationship with stakeholders is at an incipient stage of development."

To test the hypothesis, an analysis model has been designed based on Freeman and Velamuri (2008) and the AA1000SES standard, which includes both the principles and the executive process for managing the relationship with stakeholders. Once the model has been defined, it is applied to the IBEX35 companies, based on the information available in their annual reporting. Finally, the general conclusions are addressed.

## 2 Background

This section focuses on locating contributions on the management of the relationship with stakeholders under an idea of value creation, through the analysis of academic literature, reference standards, and the models used by rating agencies.

In Europe, the nonfinancial information of listed companies was scarce and heterogeneous before 2018, since reporting was not required. This situation changed the year when the Directive 2014/95/EU on the disclosure of nonfinancial information and information on diversity came into force. Since then, business reports on nonfinancial information have become available in a much more complete and homogeneous way, which allows working with more robust and standardized information, directly from companies.

Prior to 2018, the academic literature offers interesting references that deal with the relationship between companies and their stakeholders. Among them, the contribution of Freeman and Velamuri (2008) points out a scale assessing the relationship with the stakeholders and the elements needed to improve the position of the companies on the scale. Also interesting are the contributions of Mattingly and Berman (2006) and Choi and Wang (2009), which are used by Mena et al. (2019), more oriented to analyze marketing strategies based on the relationship with the stakeholders, and Kumar et al. (2021) oriented to the development of a scale of factors with the objective of analyzing an approach toward the stakeholders in companies. On the other hand, Pfitzer et al. (2013) and Font et al. (2016) carry out a materiality analysis that is later used by Torelli et al. (2020) to study the relationship processes with stakeholders with the aim of completing Freeman's (1984) stakeholder theory.

Furthermore, Berman et al. (1999), Choi and Wang (2009), or Mattingly and Berman (2006) advanced this line of research more empirically using information of the KLD database. However, this group of studies had shortcomings offering both a limited information on the stakeholders (where among its factors for analysis only employees and customers appear) and working with indirect information.

Since 2018, after the proliferation of nonfinancial business information, a more systematic search of academic literature was carried out, considering the period

**Table 1** Main contributions from the academic literature on managing the relationship with stakeholders under a value-creation approach. 2018–2021 WoS database vs. dataset

| Authors | Publications | Proposals |
|---|---|---|
| Torelli et al. (2020) | The Materiality Assessment and Stakeholder Engagement: A Content Analysis of Sustainability Reports | *Three categories for stakeholder engagement* in the shape of a scale |
| Freudenreich et al. (2020) | A Stakeholder Theory Perspective on Business Models: Value Creation for Sustainability | *Framework* that sets out joint value-creation processes with *five critical stakeholders* |
| Stocker et al. (2020) | Stakeholder Engagement in Sustainability Reporting: A Classification Model | Model to *classify stakeholder relations* according to the strategy adopted and the degree of commitment to the relationship |
| Kumar et al. (2021) | Difference in Stakeholder Engagement Approach of Small and Medium Enterprises and Large Companies and Its Performance Implications | *Own indicators scale*, addressing all aspects required in this work (but limited)<br>*Do not use indirect databases* |

Source: Own elaboration

2018–2021 and based on the WoS database and using the following keywords: stakeholders' assessment and model. Results are presented in Table 1, in which the main theoretical contributions are summarized.

Torelli et al. (2020) highlight the usefulness of defining a stakeholder engagement process and its implementation in the executive processes of companies, which is in line with the objective of this paper. Kumar et al. (2021) propose the following set of elements, which would meet several of the requirements outlined in the introduction, although they are considered limited to comprehensively address the relationship with stakeholders as visualized in (Table 2).

Freeman and Velamuri (2008) present a real model for the relationship with the stakeholder for which they even set out a scale of four levels through which the company must advance if it wants to reach the optimum. In addition, they present ten principles that companies should promote if they want to reach the maximum on the scale mentioned above. No practical application of this model has been presented. Table 3 presents this model supplemented by a fifth level that reflects the view of Sulkowski et al. (2018).

In addition to the outstanding academic contributions, it is also interesting to consider the aspects provided by the AA1000SES and GRI standards.

The AA1000SES standard (Accountability, 2015) focuses mainly on priority stakeholders and does so with a primarily defensive orientation (aimed at mitigating negative impacts). In other words, limitations are observed in the AA1000SES standard if it is analyzed from a value-creation approach that considers the innovation provided from the dialog with all the stakeholders.

The GRI, for its part, in its document GRI 101: Fundamentals 2016 (GRI, 2018) acknowledges taking into account for its materiality the ESG impacts that substantially influence the valuations and decisions of the stakeholders, but, nevertheless, this standard does not take into account the relationship with the stakeholder that

**Table 2** Elements of stakeholder's relationship management used by Kumar et al.

| | | |
|---|---|---|
| 1 | My company is actively involved in constructive dialog with stakeholders (local government, regulators, local community/ organizations) to identify issues that matter to them | The stage where companies are in relation to the development of their relationship with stakeholders |
| 2 | The stakeholder management efforts of my companies are generally aimed at managing good citizenship image of the company | |
| 3 | My company tries to find ways to reinforce company's business by advancing social conditions | Preparing companies to improve their relationship with the stakeholders |
| 4 | Corporate philanthropy in my company has clear measurable goals and results are tracked over time | Executive issues, budgets, controls, etc. |
| 5 | My company takes pride in its positive involvement in the community | Company culture and its purpose |
| 6 | My company has invested in social aspects in ways that improve its competitiveness | Executive issues, budgets, controls, etc. |
| 7 | My company attempts to incorporate social dimensions that will create social impact to its business decision | Company culture and its purpose |
| 8 | My company's business activities are aimed at integrating business with social needs | |

Source: Kumar et al. (2021) and own elaboration

**Table 3** Levels of commitment to the stakeholder approach

| |
|---|
| Level 1 – Basic value proposition |
| Level 2 – Sustained stakeholder cooperation |
| Level 3 – An understanding of broader societal issues |
| Level 4 – Ethical leadership |
| Level 5 – An open, innovative, and active relationship with all stakeholders |

Source: Adapted from Freeman and Velamuri (2008) and Sulkowski et al. (2018)

goes beyond the focus on negative impacts and includes the capacity for value creation from a constructive dialog.

# 3 A Model for Understanding How Companies Are Dealing with Stakeholder Relations

## 3.1 Justification

Of the contributions discussed in the previous section, Freeman and Velamuri (2008) seem to be the most comprehensive. However, no ideas are provided on executive issues nor on entrepreneurial purpose. As their practical application has not been detected, their potential for use with data extracted directly from companies is unknown. For the incorporation of the more executive issues in the model, we start from the basis provided by the AA1000SES standard, as this is a standard directly focused on the executive issue.

The two aforementioned references are oriented and completed to give them a value-creation orientation based on the active relationship with the stakeholders. Under this approach, rather than obtaining a ranking, the model is intended to provide an understanding of what is really happening in terms of management and the relationship with stakeholders in companies.

With respect to the rest of the literature analyzed, this approach (Freeman and AA1000SES) is more complete than that of Torelli et al. (2020), although they share the idea of the degrees of relationship with the stakeholders and the use of direct sources of information. Freudenreich et al. (2020) share the idea of joint value creation but limit its application to a specific and limited number of stakeholders. They also share with these authors the need for the existence of a joint purpose that facilitates the relationship with stakeholders. Kumar et al. (2021) share the idea of developing their own scale of items that draw on direct sources of information, although in reality their aim is not to develop a model but to analyze the differences in the interpretation of CSR between companies, depending on their size, and link them to the type of performance obtained. Other authors such as Berman et al. (1999), Choi and Wang (2009), or Mattingly and Berman (2006) are further away from what is intended here, working with indirect sources and using only partial information (that are provided by these indirect sources) on the relationship with the stakeholders.

For the analysis required in this article, the model proposed by Stocker et al. (2020) would be the most applicable, together with that of Freeman and Velamuri (2008). This model allows the categorization of the relationship with stakeholders according to the strategy followed and the degree of commitment to the relationship, resulting in a classification with nine different categories. In this way, it is easy to pigeonhole the status of the relationship with the stakeholders with the information obtained in the corporate reports of each company into one of the nine categories. However, the Freeman and Velamuri (2008) option has been retained as it lacks the simplification offered by Stocker et al. (2020), making it possible to work with more detailed information and retain all possible nuances. The aim is not to classify but to understand what is going on.

## 3.2 Model Settings

On the basis of the above, an analysis based on blocks has been proposed: The first attempts to assess the situation of the relationship with the stakeholders on a scale of four degrees of commitment to the stakeholder theory, obtained from Freeman and Velamuri (2008); the second analyzes the predisposition of an organization to improve on the scale by following ten principles, also proposed by Freeman and Velamuri; finally, and as a third block, an assessment is incorporated on the executive level exposed by the company in the management of the relationship with the stakeholders, based on the AA1000SES standard. A fourth block is also taken into account, focusing on the purpose, but trying to use the information that may already be contained in the previous three blocks.

## 3.3 *The Objective*

The objective of the four information blocks is complex. It is not a question of measuring the state of the relationship on a scale but rather of understanding the internal dynamics that may be hindering it, as well as the aspects that can enhance it. All this in a framework in which the requirements of rating agencies could be making companies more focused on responding to their demands – under a biased concept of sustainability (Escrig-Olmedo et al., 2019) – than on dialog with stakeholders. Andreu et al. (2018) also indicate that not all leading DJSI companies give relevance to stakeholders. This leads to the need to be very delicate with the collection of information and its treatment in the model.

The model reflects Porter and Kramer (2006) on the need to move beyond CSR that operates outside the business to CSR that is integrated into business strategies and objectives on its own merits. The aim is to justify the integration of economic, environmental, ethical, and social considerations into business strategy to create value. Thus, the conceptually described methodology goes beyond the mitigation of negative ESG impacts and encompasses a whole strategic process of value creation.

## 3.4 *Selected Items and Their Association with Fixed Categories*

The aim is to select elements or items that are sufficiently concrete to be assessed through the information available in corporate annual reports. These items are distributed among the four categories identified (Fig. 1):

1. Degrees for an efficient relationship with stakeholders (Category 1)
2. Enabling elements of the relationship (Category 2)
3. Executive management of the relationship with the stakeholders (Category 3)
4. A purpose for the relationship with stakeholders (Category 4)

These factors and items in each category are presented and defined below.

**Category 1: Scale of Engagement**
Following Freeman and Velamuri (2008), value creation from stakeholder theory is achieved by the company working together with its stakeholders.

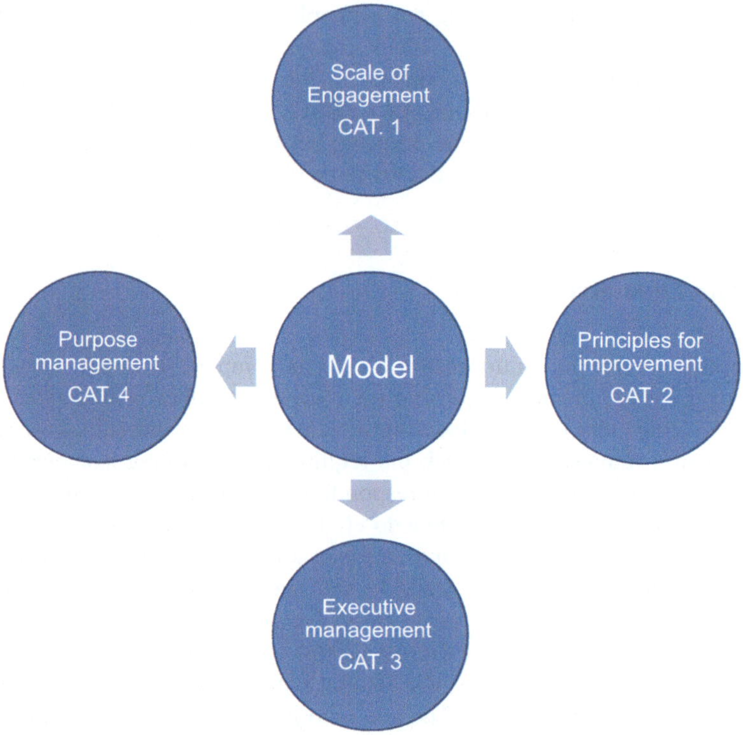


**Fig. 1** Categories determined by the model to encompass the factors and items. Source: Own elaboration

To achieve this, it is necessary to share, to a greater or lesser degree, a basic set of values that facilitates points of convergence and finally takes the form of different degrees of commitment. For these authors, the degrees of commitment would be the following (in order from least to most): the core value proposition, sustained consideration of stakeholders, understanding of broader societal issues, and ethical leadership. Finally, Sulkowski et al.'s (2018) consideration of interest in an open and innovative relationship with all (not just the most obvious) stakeholders has been included.

**Category 2: Principles for Improvement**

In addition to the degree of commitment achieved with the stakeholder theory, it is worth assessing, following Freeman and Velamuri (2008), the willingness of companies to improve on this scale of commitment and reach the highest category, in other words the good business practices that would allow them to improve their relationship with stakeholders and, consequently, their potential to create value from sustainability.

This section may be more reliable than the previous one. The information available in company reports to assess the degrees of commitment to the stakeholder theory could present doubts in its reliability. This is advocated by Schoenmaker and Schramade (2019), for whom only the readiness of companies to transition to a new model could be analyzed.

In order to assess this section, the ten principles proposed by Freeman and Velamuri (2008) are used as a starting point, with further refinements and complements by the review of other authors that have been specifically identified and associated with the corresponding factor in Table 5 (Andreu, 2017; Argandoña, 2010; Carroll & Shabana, 2010; Porter & Kramer, 2011 and Sulkowski et al., 2018).

**Category 3: Executive Management**
The AA1000SES standard introduces many practical and management issues in the relationship with the stakeholders but focuses mainly on priority stakeholders and under a fundamentally defensive orientation (aimed at mitigating negative impacts). In other words, limitations are observed if analyzed from a value-creation approach from the dialog with all the stakeholders. In fact, if hypothetically a company following the AA1000SES approaches were to be placed in the degrees of relationship with the stakeholders shown in Table 3, this company would not exceed level 2 of the five levels presented (it would be focused on maintaining continuity in the relationship with the stakeholders, but without reaching the next level: ethical leadership). Figure 2 shows a sequence of executive processes in the form of a circular diagram.

The AA1000SES standard obliges companies adhering to this standard to report publicly on their stakeholder engagement process. To this end, organizations should integrate public reporting on stakeholder engagement with the organization's other forms of public reporting (e.g., in annual reports). This information should also be independently assured. Table 7 contains the selected items for the assessment of the executive management of stakeholders.

**Category 4: Purpose Management** Following the analysis of the three previous blocks, it has been possible to observe how the question of purpose recurrently appears as a key element for the good functioning of the relationship with the stakeholders. Likewise, the demand by some actors in the economic world to incorporate a social purpose in the management of companies with a view to the long term (Business Roundtable, 2019; Fink, 2018, 2019) suggests that this shared purpose with the stakeholders could be one of the keys to the sustainable behavior of companies (Lleo et al., 2021), as well as improving the possibilities of creating financial value from CSR. From a practical point of view, it could be facilitating the involvement of stakeholders in a joint value-creation process. It should not be forgotten that the majority of stakeholders are not linked to the organization for economic reasons, so finding another type of motivation that invites them to collaborate with the company beyond a defensive strategy (response to negative impacts) may be key. It is a question of moving toward a bottom-up management model, for which in many cases there is no remuneration relationship (stakeholder).

It is not intended to open a new category of items, since there are already many of them related to the purpose of the company in the three previous categories. The aim

**Fig. 2** Executive process for stakeholder engagement. Source: Own elaboration

is to have a criterion and a systematic way of analyzing it, as it seems to be relevant in the processes of creating value from sustainability.

Some recent references describe some of the characteristics that this purpose should have. For the Canvas (2019) analysis, the purpose should be able to make the stakeholders perceive not only that the company is working to mitigate its negative impacts but also that they would miss it if it ceased to exist. For GlobeScan and SC Johnson (2019), purpose should be social and would form part of the elements capable of achieving greater engagement with stakeholders along with ethics and transparency. And the World Economic Forum (cited in Canvas, 2019) qualifies the social purpose by recalling that, in order to generate a positive impact, it should contribute to the "sustainability of the community rather than simply the sustainability of the business."

From a practical point of view, Chevreux et al. (2017) present two pillars for the implementation of purpose in companies: knowledge of purpose and contribution (conversion of purpose into action). Based on this classification, an attempt has been made to reclassify the selected items in the above categories between the two pillars mentioned (Table 6).

Finally, Lleo et al. (2021) present a methodology to assess the status of the application of purpose in organizations and its adherence by both internal and external stakeholders, which incorporates a third pillar: the degree of internalization, which determines how purpose influences the motivation of stakeholders. According to Dewettinck and Defever (2020), purpose must be created and transferred from the individual (in this case the stakeholders) to the organization. In order to do so, their (the stakeholders) beliefs and motivations must be taken into account.

Lleo et al.'s (2021) idea is that the implementation of purpose in a company would be effective to the extent that it clarifies the direction (knowledge of purpose), excites and motivates people (internalization of purpose), and is operationalized in the daily life of the organization (contribution of purpose). These three factors should work together in harmony to be effective. This idea is also advocated by the Purpose Strength Model (Chinchilla et al., 2019) which is defined by the variables knowledge of purpose (head), identification with purpose (heart), and contribution to purpose (hands) (Table 7).

## 3.5 *Feeding the Model*

In the elaboration of this model, a list of 47 basic elements of information or items to be assessed has been designed, which should make it possible to know the relationship with the stakeholders, based on the information contained in the companies' nonfinancial reports. In turn, these items are aggregated into 35 factors and 4 categories (those already mentioned). It should be noted that this information is in a process of improvement and progress in the degree of maturity. For example, the European Directive requiring disclosure of nonfinancial information (Directive 2014/95/EU) is already being updated. Thus, it is considered that the best information will be provided by the most up-to-date reports.

The 47 items have always been approached in such a way as to avoid duplication, so that the assessment of each item always provides something different from the rest, although often, given the limitations of the information contained in the annual reports, the differences between one item and another can be found in the assessment elements chosen to obtain the maximum score.

The guarantees offered by this way of proceeding are to be found in the theoretical foundations worked on, in the establishment of a common valuation system for all companies, and in the large number of items that allow the maximum amount of information available to be collected in the annual reports. However, for a final analysis, it seems more appropriate to look at aggregate results than at specific items in individual companies. [1]

[1] This point is finally dispensed with for a possible analysis of the purpose.

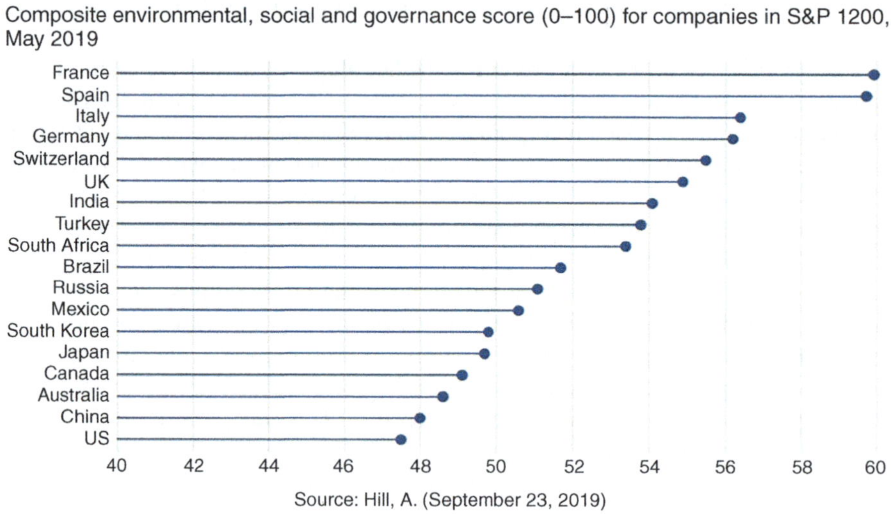


**Fig. 3** Scores obtained by companies in the ASG SandP1200 index. Breakdown by country. Source: Hill (2019)

# 4 Application of the Model to IBEX35 Companies

## *4.1 The Sample*

IBEX35 companies appear to be doing well in CSR management, and many are ranked among the best in the world, according to the Dow Jones Sustainability Index (2020)[2] and the Hill (2019) (Fig. 3). Therefore, the results obtained from the application of the model to this sample can be a benchmark for other countries.

The list of companies that made up the sample for this study is that of the companies that formed part of the IBEX35 in March 2020, as listed in Table 8. As can be seen in this table, the sample is divided by sector.

With respect to the temporal scope, only 1 year has been selected for several reasons:

1. The breadth of the information to be obtained and its constant evolution with reports that are not entirely homogeneous among themselves make it reasonable to extract information from the most recent year possible, declining the possibility of carrying out historical analysis.
2. The relationship with the stakeholders does not seem to have much past, and therefore the most evolved data are the most recent ones. This problem has already been addressed in other works (Kumar et al., 2021; Mir, 2004) in which a set of items have been designed which, being information on aspects

[2] https://www.spglobal.com/esg/csa/yearbook/ranking/ (retrieved 1/9/2020)

related to management, business culture, and organizational aspects, cannot be assessed by means of quantitative variables.

## 4.2 Sources and Measurement Criteria

We have worked with the corporate reports for 2019, which are available on their corporate websites. However, it should be borne in mind that, although nonfinancial information must be reported in the annual reports, its content lacks the same level of rigor as the financial aspects. For example, its assurance is not necessarily carried out by an auditor, or the format of the report, apart from the basic contents required by law, does not necessarily follow a common format in all companies. The GRI standard is widely used for these reports. The following documents were used: mission/purpose, policies (sustainability/CSR, relationship with stakeholders, director remuneration, shareholder communication, shareholder involvement, compliance, corporate culture/CSR, innovation), code of ethics, reports (integrated annual report, sustainability, nonfinancial information, corporate governance, remuneration, relationship with stakeholders, community investment, CA/committees, climate finance), sustainability plans, regulations (sustainability committee, board of directors), shareholder meeting agreements, and votes cast. In addition, an attempt has been made to locate on its corporate website the relevant information that has not been found in its official documents, although always as a secondary and subsidiary source.

In this context, and for each of the three categories, the items to be assessed have been selected using the information contained in the annual reports of the IBEX35 companies. For each item, the criteria for its evaluation were specified, and possible duplications with other items were analyzed. This process has been very laborious, and its difficulty, in addition to the selection of the items themselves, has been in the search for specific elements to evaluate each of the items. An assessment has been made in the context of all the information available in the corporate reports. Where differences between items are slight, this has been explicitly mentioned.

The rating scale used for each of the items ranges from 0 to 2: 0 for noncompliance, 1 for some indication of compliance or partial compliance, and 2 for compliance. For example, the item "(2.8) Negotiate with primary and secondary stakeholders," which is defined as "managers meet, interact, and negotiate with less legitimate stakeholders, but who have the power to affect the company," is evaluated as follows: 0 if there are no relationships, 1 if there are relationships but not desirable, and 2 if relationships are established and are optimal in terms of their ability to create value for long term.

In addition, two other criteria have been followed:

1. If the senior management of the companies understood that the origin of sustainable value creation is in the relationship with the stakeholders, they would expose it in their reports with transparency as an asset of the company (it could be considered part of the Social and Relational Capital considered by the IIRC,

2013). This premise assumes that if there is no information on any of the items, it will be understood that either the company does not practice it or it is unaware of its advantages, so for the purposes of this research, it would mean the same thing. In this case, this item would have the lowest possible score. This reasoning would be supported by the recommendation of the standard (AA1000SES) proposing that the implementation of the plans for the relationship with the stakeholders should be recorded in a transparent way. Thus, they should be visible in the nonfinancial reports of IBEX35 companies and also the GR101 document when it says that the processes for dealing with stakeholders should be adequately documented.

2. The items have been assessed optimistically in a situation of doubt or reasonable doubt in the companies. The aim is to eliminate possible author biases tending to fuel controversy in the companies in order to justify the hypothesis. Moreover, the scant information that can be observed on the relationship with the stakeholders in IBEX35 companies means that, on occasions, valuations have been based on reasonable indications or interpretations after reading all the available information. These reasonable indications or interpretations will be assessed positively.

As a result, a data matrix of 35 rows (35 companies studied) by 47 columns (47 items assessed for each company) has been obtained.

The process of obtaining the information necessary to complete the matrix was very complex and laborious. Furthermore, once an initial general assessment had been made by company, a second review was carried out, by item, in an attempt to homogenize the evaluations in the face of possible changes in the assessments as more knowledge was acquired and as a measure to try to combat this bias.

### *4.3 Model Match*

It is not possible to carry out historical correlation studies with any valid benchmark as no recognized benchmark with historical data has been identified. The data obtained by the model does not help either as it only refers to the year 2019. Nor is it possible to compare the static results referring to a specific year with any benchmark on the relationship with the stakeholders, since, as already mentioned, it does not exist or is not known.

With regard to the categories referring to the potential to improve the relationship with the stakeholders and the one referring to the purpose, there are also no benchmarks against which to compare the results obtained.

With respect to the category assessing the degree of relationship with stakeholders, there may be a benchmark that provides a reasonable comparison and allows some conclusion to be drawn about the validity of the model. The benchmark could be the result of the ratings given by SAM to IBEX35 companies. On average, the companies that are rated obtain an average score of 69.71 (ratings obtained on 1/92020).

The items to compare with this result are those most directly related to a defensive vision, shared by the GRI, of trying to mitigate the negative impacts produced to the stakeholders. The average score of the ratings obtained by the model (in items 1.1–1.3) for IBEX35 companies that are assessed by the SAM (28 out of 35 are rated) is 70.83.

This comparison makes sense as the SAM assessments score ESG aspects in terms of their ability to create value. This reinforces the idea of the validity of the constructed model.

## 5 Results

Below are some initial results obtained using descriptive statistical techniques (average and spider graphs). The aim of these techniques is to determine the degree of compliance of the various IBEX35 companies, in general and by sector, in each of the different categories of the model.

For each category, a series of derived variables will be constructed by averaging the scores obtained in certain factors, which will subsequently be presented in the graphs in the form of the average achieved by each of the 7 sectors and by the 35 companies as a whole. The factors that are in turn subdivided into two items have previously been averaged in order to have the same weight as in the other items. In addition, the figures are expressed in base 100.

**Category 1: Scale of Engagement**

From the factors that form part of this category, presented in Table 4, the following three derived variables are constructed, the results of which, averaged by sector, are presented in Graph 1:

- *Overall average score*: the result of averaging the scores obtained in the five factors (1.1, 1.2, 1.3, 1.4, and 1.5)
- *Average defensive ratio*: result of averaging the scores of factors 1.1, 1.2, and 1.3, which are those factors within the *Category 1: Scale of Engagement* that have a purely defensive capacity in the face of reputational risk
- *Average creative ratio*: result of averaging the scores of factors 1.4 and 1.5, which are those reflecting the ability to create value

The results obtained show that the *overall average* variable exceeds 50% for the IBEX35 companies as a whole (58.57%), although the best result is obtained for the *average defensive ratio* variable, with a result of 70.95%, which is close to that obtained by the SAM rating agency (although in this case SAM only evaluates 28 companies out of the total of 35 in the IBEX index). The *average creative ratio* variable barely reaches 40%. In terms of sector analysis, the oil and energy sector scores the best.

**Table 4** Factors and items integrated in Category 1: Scale of Engagement

| | Factors | Items to be assessed |
|---|---|---|
| 1.1 | Core value proposition | Companies at this level make value propositions to the stakeholders. These are relationships that can be considered defensive for good neighborliness |
| 1.2 | Sustained stakeholder cooperation | The purpose of the relationship is based on the previous point, incorporating a sense of temporal stability to the value propositions. It goes beyond the idea of concrete proposals by sharing the efforts between the stakeholders in the long term |
| 1.3 | Understanding of broader social issues | It also includes the consideration of social concerns that go beyond the scope of the company. To this end, the company must also consider those stakeholders that are more distant and with a more questionable legitimacy |
| 1.4 | Ethical leadership | It incorporates the company's exercise of ethical leadership with the proactive exercise of positive moral ideas. It incorporates into the dialog a sense of active ethics on the part of the company with the capacity to lead the relationship with the stakeholders |
| 1.5 | Open and innovative active relationship with all stakeholders | It incorporates the establishment of active, open, and innovative relationships with all stakeholders with the capacity to add value |

Source: Own elaboration

These results are in line with the findings of the literature review. The need for the energy sector to advance in its CSR, given its delicate situation in the face of the need to address an energy transition, was already advanced by Stocker et al. (2020), and a CSR approach oriented toward the mitigation of negative impacts on stakeholders (associated with the variable defensive stakeholder relationship) rather than toward a value-creation objective.

**Category 2: Principles for Improvement**

From the factors that form part of this category, presented in Table 5, the following derived variable is constructed, whose average results by sector are illustrated in Graph 2:

- *Potentiality*: the result of averaging the scores obtained in the 17 factors (2.1–2.17) relating to the relationship enhancer elements with the stakeholders

For some authors (Schoenmaker & Schramade, 2019), the potentiality represented here (readiness for transition to a new model) is the only reliable element for the analysis. In contrast to the previous concept, its average does not exceed 50% (it stands at 43.15%). By sector, it continues to show that the oil and energy sector is the most developed.

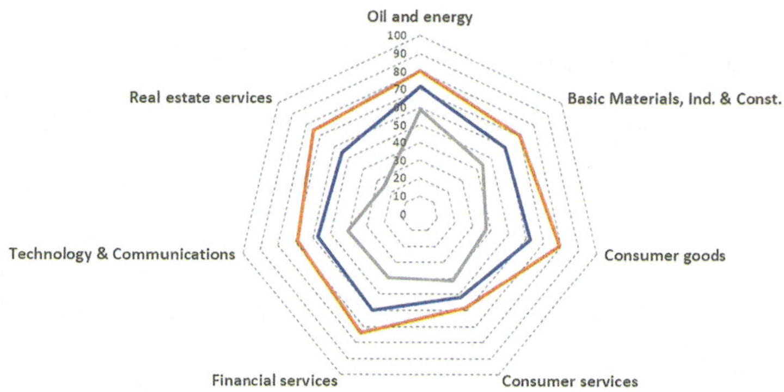


Averages are presented in base 100 for ease of comparison

| | Global | Oil and energy | Basic Materials, Ind. & Const. | Consumer goods | Consumer services | Financial services | Technology & Communications | Real estate services |
|---|---|---|---|---|---|---|---|---|
| **Overall average score** | **58,57** | 71,67 | 60,00 | 62,50 | 51,67 | 60,00 | 57,50 | 55,00 |
| **Average defensive ratio** | **70,95** | 80,56 | 70,59 | 79,17 | 58,33 | 73,81 | 68,75 | 75,00 |
| **Average creative ratio** | **40,00** | 58,33 | 44,12 | 37,50 | 41,67 | 39,29 | 40,63 | 25,00 |

**Graph 1** Category 1: Scale of Engagement. Source: Own elaboration

**Category 3: Executive Management**

Based on the factors that form part of this category, presented in Table 6, the following derived variable is constructed, whose results by sector can be seen in Graph 3.

- *Executive management*: the result of averaging the scores obtained in the 13 factors (3.0–3.12) relating to the elements associated with executive development for the relationship with the stakeholders

This category is the one that obtains the lowest measurements in line with the scarcity of concrete information provided in annual corporate reports on specific aspects of management. This is despite the recommendations to publish this information made by the AA1000SES standard. Globally, the average is 23.90%, and it is once again the oil and energy sector that stands out the most in terms of the rating obtained (37.5%).

**Category 4: Purpose Management**

From the factors that form part of this category, presented in Table 7, the following three derived variables are constructed, whose results averaged by sector are presented in Graph 4:

- *Average knowledge ratio*: result of averaging the scores obtained in 2.4.b, 2.7.a, 2.7.b, 2.17, and 3.6.a regarding the knowledge of the company's purpose
- *Average contribution ratio*: the average of the scores in 1.4, 2.7.c, 2.10, and 3.6.b, relating to the conversion of purpose into action in the company

**Table 5** Factors and items integrated in Category 2: Principles for Improvement

| | Factors | Items to be assessed |
|---|---|---|
| 2.1 | The interests of the stakeholders are jointly managed over time | Companies should maintain a balance between different stakeholders over time |
| 2.2 | Take into account the complexity of stakeholders | Consideration of stakeholders as complex, multi-stakeholder elements and their cultural differences. The assessment takes into account the different claims of each stakeholder in dialog with them **(2.2 a).** Also assessed, on average with (2.2 a), is the consideration of the cultural differences of the stakeholders **(2.2 b)** |
| 2.3 | Seek solutions that satisfy different stakeholders simultaneously | The aim is to economize efforts |
| 2.4 | Engage in intensive dialog with stakeholders and not only with those who are friendly or close to them | To measure this principle, companies are scored by averaging several aspects: the intensity of the dialog **(2.4 a)**, the honesty and openness of the dialog **(2.4 b)**, the inclusion of the least legitimized stakeholders **(2.4 c),** and the inclusion of the stakeholders most critical of the company's activity **(2.4 d)** |
| 2.5 | Willingness to resolve disputes with stakeholders before they are resolved in public bodies | Negotiation is preferred to legislative imposition. To measure this principle, companies are scored on the basis of an average on two aspects: **(2.5 a)** attention to stakeholders' complaints (ethical channel) **and (2.5 b)** maintenance of a public-private agenda |
| 2.6 | Using marketing techniques for understanding key stakeholders | Leverage existing marketing techniques to better address and understand the relationship with stakeholders. To measure this principle, the three aspects are averaged: **(2.6 a)** ranking and prioritization of stakeholders, **(2.6 b)** use of marketing techniques in the relationship with stakeholders, and **(2.6 c)** preferential attention to key stakeholders |
| 2.7 | The value-creation process of the company puts the stakeholders in a better position | The company works for its stakeholders and never trades off the interests of one against the other on a continuous basis over time. To measure this principle, the three aspects are averaged: **(2.7 a)** consideration in the purpose/mission of sustainability, **(2.7 b)** alignment of the relationship with the stakeholders with the purpose of the company, and **(2.7 c)** CEOs' attendance to this purpose as well |
| 2.8 | Negotiate with primary and secondary stakeholders | The company also engages with less legitimized stakeholders |

(continued)

**Table 5** (continued)

| | Factors | Items to be assessed |
|---|---|---|
| 2.9 | Constantly monitor and redesign processes with stakeholders | This is a correction to changing business circumstances |
| 2.10 | Acting under a purpose that satisfies the commitment to the stakeholders | The idea is that a purpose that incorporates aspects that are not purely economic can help in the relationship |
| 2.11 | Consider long term in CSR plans and/or objectives (Porter & Kramer, 2011) | Directive 2017/828 proposes as long term to periods between 7 and 10 years |
| 2.12 | Use innovative solutions in approaching the relationship with stakeholders (Sulkowski et al., 2018) | Innovation affects not only the outcomes of the relationship with stakeholders but also how the relationship itself is addressed |
| 2.13 | Conduct cost-benefit analysis based on the relationship with the stakeholders (Carroll & Shabana, 2010) | Use a cost-benefit analysis as in any other functional area of the company |
| 2.14 | Promote a "provocative" relationship with stakeholders (Sulkowski et al., 2018) | It is about the company promoting the relationship with its stakeholders as it is a source of value creation |
| 2.15 | Benefits are produced, for the company and the stakeholders, both internal and external to the company (Porter & Kramer, 2011) | Involves consideration of benefits external to the enterprise |
| 2.16 | From dialog with the stakeholders come solutions for the company to mitigate its negative ESG impacts (Argandoña, 2010) | The company should spend on correcting its negative impacts even if they do not generate benefits in the short term |
| 2.17 | Existence of a company culture that facilitates the relationship with stakeholders (Andreu, 2017) | It means that relations with stakeholders are seen as something positive in the company, which is derived from specific aspects of its organizational chart |

Source: Own elaboration

- *Overall average score*: the result of averaging the scores obtained in all the measures included in the two previous variables

The purpose, both in global terms and in terms of knowledge and contribution to it, shows an intermediate degree of compliance on average for all companies, close to 50%, although it is striking that the value obtained in the contribution aspect is higher than in the knowledge aspect. The oil and energy sector once again obtains the best results, showing even more logical results than the rest (knowledge scores better than contribution).

To reinforce these initial results, one further consideration should be added. To rate the level of engagement with stakeholders, Stocker et al. (2020) use a scale to measure the level of strategic communication. They place the optimum at a level of communication that they call collaborative or associated, which could be associated with levels 4 and 5 of the grades for the relationship with stakeholders that have been designed in this model. The previous level is called consultation, for which, among other things, surveys are used. This point may be a good contrasting element of the results obtained in relation to the scale of relationship with stakeholders, where the

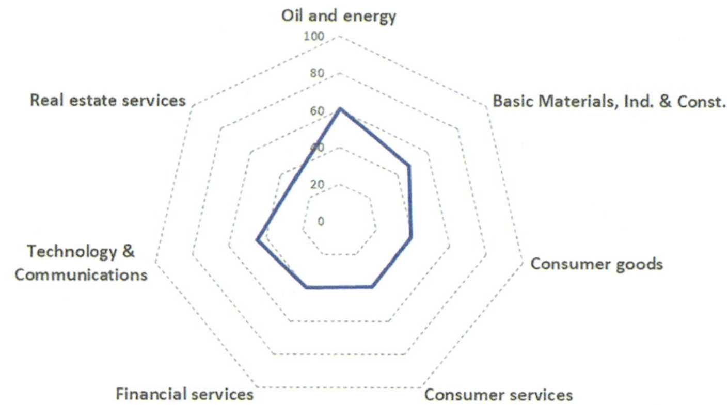


| Averages are presented in base 100 for ease of comparison | Global | Oil and energy | Basic Materials, Ind. & Const. | Consumer goods | Consumer services | Financial services | Technology & Communications | Real estate services |
|---|---|---|---|---|---|---|---|---|
| Overall average score | 43,15 | 60,83 | 47,46 | 38,91 | 39,13 | 40,13 | 44,74 | 32,11 |

**Graph 2** Category 2: Principles for Improvement. Source: Own elaboration

highest scores were achieved by defensive relationships with stakeholders (grades 1, 2, and 3). This result can be contrasted with the fact that the use of indirect means to gather feedback from stakeholders on ESG issues was evident in the process of gathering information from the companies in the sample. Thus, of the 35 IBEX35 companies, 19 use some kind of indirect mechanisms to engage with stakeholders (mechanisms that can be associated with the surveys conducted by Stocker and colleagues), which means that 54% use external instruments to justify engagement with stakeholders. From this, it can be inferred that at least this percentage does not reach the optimal level of communication and therefore the results obtained through the model for the degrees of relationship are plausible. Thus, IBEX35 companies, presented in Table 8, do not use an active dialog with their stakeholders in the majority of cases, which is consistent with Kumar et al. (2021), who argue that large companies consider engagement initiatives with stakeholders as a way to improve their reputation and competitiveness without going any further.

## 6 Conclusion and Contribution

There are few stakeholders' relationship models in the literature. This paper proposes a new stakeholder relationship assessment model to better understand how companies are dealing with stakeholders' relations. This model brings together the contributions of different authors, in a recently evolving field, considering four

**Table 6** Factors and items integrated in Category 3: Executive Management

| | Factors | Items to be assessed |
|---|---|---|
| 3.1 | Alignment between CEO and shareholders around a stakeholder relationship policy | The aim is to mitigate agency problems around ESG issues |
| 3.2 | Board of directors responsible for policies and risks of the relationship with stakeholders | Legal obligation. For assessment purposes, this item is broken down into two other items: **(3.2 a)** on policies and **(3.2 b)** on risks |
| 3.3 | Designation of the owners of the commitment with a council unit | Existence of a person responsible for executing the relationship strategy with the stakeholders |
| 3.4 | Stakeholder map (under a preestablished systematic stakeholders identification) | Systematic identification of stakeholders under and its evaluation taking into account the purpose and scope of the commitment. This is assessed by averaging **(3.4 a)** on identification and **(3.4 b)** on mapping |
| 3.5 | Analysis of possible disputes between stakeholders | Risk management of the relationship with the stakeholders |
| 3.6 | Establishment of a relationship framework that incorporates purpose | Existence of a purpose for the momentum of the relationship concretized in a plan. This item is assessed by averaging two other items: **(3.6 a)** on the purpose and **(3.6 b)** on the existence of a plan for engagement |
| 3.7 | Establishing and documenting a mechanics of the relationship | Existence of specific procedures for each stakeholder to reach agreement on ESG issues |
| 3.8 | Provision of a budget for the relationship with stakeholders | Existence of a budget allocation |
| 3.9 | Achievement of ESG objectives | Outcomes of the dialog with stakeholders |
| 3.10 | Implementation, measurement of agreements and incentives (of the relationship with the stakeholders) | The concrete plan for the relationship with each stakeholder should be evaluated, incentivized, and documented |
| 3.11 | Transparency in reporting, for the company and the stakeholders (relating to the relationship with the stakeholders) | Shareholders, and other stakeholders, should have access to information on action plans |
| 3.12 | Periodic review of the process | The stakeholder's relationship process is a process of continuous improvement |
| 3.0 | Committed shareholders | Shareholder involvement is necessary as many ESG expenditures involve expectations of long-term returns that may mean losses today |

Source: Own elaboration

dimensions: the situation of the organization's relationship with its stakeholders; its predisposition to improve it; the stakeholder management from an executive perspective; and purpose, understood as a requirement to move from ethical leadership to an open, innovative, and active relationship with all stakeholders.

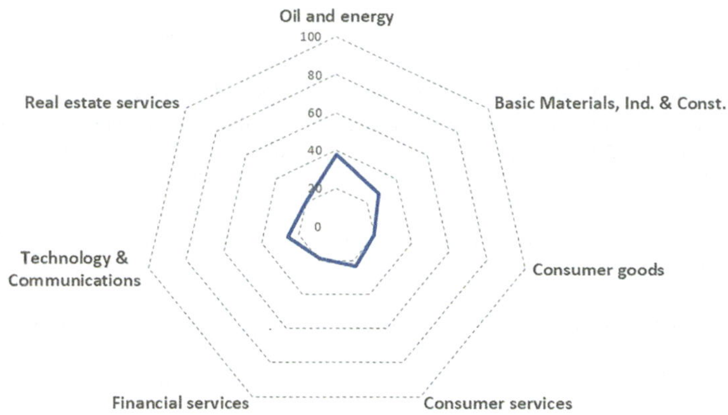


| Averages are presented in base 100 for ease of comparison | Global | Oil and energy | Basic Materials, Ind. & Const. | Consumer goods | Consumer services | Financial services | Technology & Communications | Real estate services |
|---|---|---|---|---|---|---|---|---|
| Overall average score | 23,90 | 37,50 | 27,94 | 19,71 | 23,08 | 18,96 | 26,04 | 20,19 |

**Graph 3** Category 3: Executive Management. Source: Own elaboration

In fact, one of the premises identified by the literature for an adequate management of CSR in the long term, i.e., for the maintenance of CSR as a competitive advantage that creates value, is the need to maintain an active relationship with all stakeholders. This paper identifies and includes the purpose of the company as a key issue of the model, and it is able to assess it on the basis of the three previous elements.

However, the main aim of the model is not "measuring on a scale": It tries to understand the internal dynamics of a company's relationship with its stakeholders, as well as the aspects that can enhance it. One of the issues that have emerged from the literature review is that not all companies adequately manage and give relevance to all their stakeholders, as this is not one of the priorities required by sustainability rating agencies and neither helps them to position themselves in their ratings.

There is academic literature that indicates the convenience of using primary information (extracted directly from companies). However, there is an additional problem, which is the lack of homogeneous information on the organizations' relationship with their stakeholders. In particular, this information is not standardized. The proposed model is designed to use this type of primary information, although its proper application would require further development of homogeneous reporting and corporate transparency systems. New regulatory developments in this regard, already applied in Europe (i.e., Directive 2014/95/EU, known as Nonfinancial Reporting Directive (NFRD), or the Corporate Sustainability Reporting Directive (CSRD) proposal to come into effect from 2024), may help

**Table 7** Items and factors integrated in Category 4: Purpose Management

| Knowledge of purpose | Contribution (turning purpose into action) | Degree of internalization (determines how the purpose influences the motivation of stakeholders) |
|---|---|---|
| **2.4.b** The company establishes the basis for an honest and open dialog<br>**2.7.a** The company has a purpose/mission that seeks to satisfy a need in a sustainable way<br>**2.7.b** The management of the relationship with the stakeholders is aligned with the purpose<br>**2.17** Existence of a company culture that facilitates the relationship with stakeholders<br>**3.6.a** There is a purpose, in the form of an objective, for the development of the relationship with each stakeholders, taking into account the map and controversies | **1.4** It incorporates the exercise of ethical leadership by the company with the proactive exercise of positive moral ideas<br>**2.7.c** CEOs serve this purpose<br>**2.10** Act under a purpose that satisfies commitment to the stakeholders<br>**3.6.b** There is a plan for the engagement, which specifies the purpose, managed by the engagement owner for each stakeholders | For this third block, no valid items were detected among those evaluated |

Source: Own elaboration

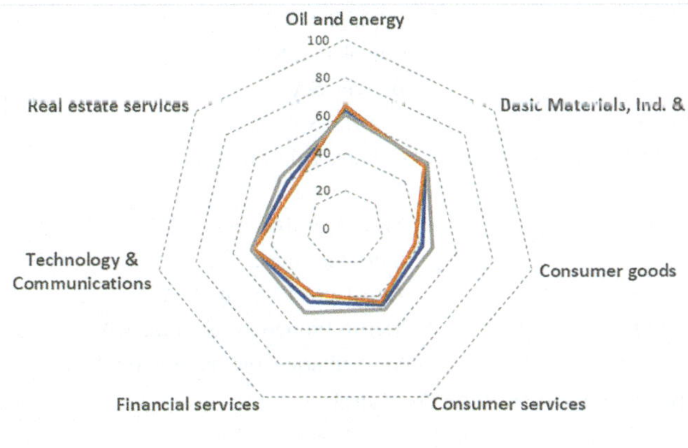


Averages are presented in base 100 for ease of comparison

| | Global | Oil and energy | Basic Materials, Ind. & Const. | Consumer goods | Consumer services | Financial services | Technology & Communications | Real estate services |
|---|---|---|---|---|---|---|---|---|
| Overall average score | 47,62 | 62,96 | 53,92 | 41,67 | 45,37 | 43,65 | 49,77 | 38,89 |
| Average knowledge ratio | 45,71 | 65,00 | 52,94 | 37,50 | 43,33 | 38,57 | 49,17 | 35,00 |
| Average contribution ratio | 50,00 | 60,42 | 55,15 | 46,88 | 47,92 | 50,00 | 50,52 | 43,75 |

**Graph 4** Category 4: Purpose Management. Source: Own elaboration

**Table 8** Details of the sample

| | | |
|---|---|---|
| 1. Oil and energy | Repsol, S.A. | 1 |
| | Enagas, S.A. | 2 |
| | Endesa, Sociedad Anonima | 3 |
| | Iberdrola, S.A. | 4 |
| | Naturgy Energy Group, S.A. | 5 |
| | Red Electrica Corporacion, S.A. | 6 |
| 2. Basic materials, industry and construction | Acerinox, S.A. | 7 |
| | Arcelormittal, S.A. | 8 |
| | Cie Automotive, S.A. | 9 |
| | Siemens Gamesa Renewable Energy, S.A. | 10 |
| | Acciona, S.A. | 11 |
| | Acs, Actividades de Const. Y Servicios, S.A. | 12 |
| | Ferrovial, S.A. | 13 |
| 3. Consumer goods | Viscofan, S.A. | 14 |
| | Industria De Diseño Textil, S.A. "Inditex" (Inditex) | 15 |
| | Ence Energia Y Celulosa, S.A. | 16 |
| | Grifols, S.A. | 17 |
| 4. Consumer services | Melia Hotels International, S.A. | 18 |
| | Mediaset España Comunicacion, S.A. | 19 |
| | Aena, S.M.E., S.A. | 20 |
| | International Consolidat. Airlines Group | 21 |
| 5. Financial services | Banco Bilbao Vizcaya Argentaria, S.A. | 22 |
| | Banco De Sabadell, S.A. | 23 |
| | Banco Santander, S.A. | 24 |
| | Bankinter, S.A. | 25 |
| | Caixabank, S.A. | 26 |
| | Bankia, SA | 27 |
| | Mapfre, S.A. | 28 |
| 6. Technology and telecommunications | Cellnex Telecom, S.A. | 29 |
| | Telefonica, S.A. | 30 |
| | Masmovil Ibercom, S.A. | 31 |
| | Indra Sistemas, S.A. | 32 |
| | Amadeus IT Group, S.A. | 33 |
| 7. Real estate services | Merlin Properties, Socimi, S.A. | 34 |
| | Inmobiliaria Colonial Socimi, S.A. | 35 |

Source: Own elaboration based on Bolsa de Madrid

the model to be applied more effectively and improve the possibilities for future research (European Commission, 2014, 2021).

The model has been usefully applied to draw conclusions for IBEX35 companies. The results show that the best scores are obtained by grouping the factors that have a purely defensive capacity in the face of reputational risk, and the poorest scores are obtained by combining the factors associated with management development for the relationship with the stakeholders. Low scores are also obtained when considering the factors that reflect the ability to create value and those factors relating to the relationship enhancer elements with the stakeholders.

All these results are in line with what has been suggested by both the academic literature and the business dynamics logic, where a defensive strategy seems to be prevailing in the face of social requirements on sustainability. Thus, the potential to create value from the relationship with stakeholders and their executive management would not be relevant elements. The reasonable result of the purpose would coincide with the idea of showing a good social image within a more defensive strategy. All this reinforces the good performance of the model, which presents coherent and logical results.

Following this analysis, we can conclude, as hypothesized, that the stakeholder engagement model of IBEX35 companies is still at an early stage of development. There is still considerable scope for the approach to stakeholder engagement to become one of shared value creation, rather than defensive. This may also have relevance in other environments, beyond Spain, given that IBEX35 companies are positioned as leaders in different international sustainability rankings and indices.

The latter opens the debate and the need for future research, namely, from the investor's perspective. In this sense, the potential of each sector to develop financial value in the future could be assessed based on its current relations with the stakeholders and its capacity to improve them.

## References

Accountability. (2015). *AA1000 Stakeholder engagement standard.* Retrieved March 21, 2017, from https://www.accountability.org/standards/

Amel-Zadeh, A., & Serafeim, G. (2018). Why and how investors use ESG information: Evidence from a global survey. *Financial Analysts Journal, 74*, 87–103.

Andreu, A. (2017). *Revisión crítica de la dimensión "Gobierno Corporativo" en los índices de sostenibilidad. Análisis del ejercicio de la responsabilidad del Consejo de Administración en materia de RSC en las compañías del Ibex 35 español* (Doctoral Thesis). Universidad Pontificia Comillas, Madrid.

Andreu, A., Fernández-Fernández, J. L., & Fernández Mateo, J. (2018). Critical review of the corporate governance dimension in sustainability index questionnaires. *Revista de Comunicación,* 17(2).

Argandoña, A. (2010). La dimensión ética de la crisis financiera. *Documento de Investigación* DI-872. IESE Business School – Universidad de Navarra.

Barton, D. (2017). Refocusing capitalism on the long term: Ownership and trust across the investment value chain. *Oxford Review of Economic Policy, 33*(2), 188–200.

Berman, S. L., Wicks, A. C., Kotha, S., & Jones, T. M. (1999). Does stakeholder orientation matter? The relationship between stakeholder management models and firm financial performance. *Academy of Management Journal, 42*(5), 488–506.

Business Roundtable. (2019). *Statement on the purpose of a corporation.* Retrieved February 25, from https://s03.s3c.es/imag/doc/2019-08-19/Business-Roundtable.pdf

Canvas (2019). *Purpose: The raison d'être of companies*. Retrieved November 1, 2020, from https://www.canvasconsultores.com/wp-content/uploads/2020/01/RADAR-CANVAS_An%C3%A1lisis-del-Entorno_Prop%C3%B3sito_diciembre-2019.pdf

Carroll, A. B., & Shabana, K. M. (2010). The business case for corporate social responsibility: A review of concepts, research and practice. *International Journal of Management Reviews, 12*(1), 85–105.

Chevreux, L., Lopez, J., & Mesnard, X. (2017). The best companies know how to balance strategy and purpose. *Harvard Business Review*. Retrieved March 3, 2020, from https://hbr.org/2017/11/the-best-companies-know-how-to-balance-strategy-and-purpose?language=es

Chinchilla, N., Lleó, A., Rey, C., Alloza, A., & Pitta, N. (2019). Purpose strength model: In search of a shared purpose. Carmina Roca and Rafael Pich-Aguilera Chair of Women and Leadership, IESE.

Choi, J., & Wang, H. L. (2009). Stakeholder relations and the persistence of corporate financial performance. *Strategic Management Journal, 30*(8), 895–907.

Clark, G., Feiner, A., & Viehs, M. (2015). *From the stockholder to the stakeholder: How sustainability can drive financial outperformance* (Working Paper, Smith School of Enterprise and the Environment, University of Oxford). Retrieved August 25, 2018, from http://ssrn.com/abstract=2508281

Dewettinck, K., & Defever, E. (2020). The case for purpose: Demystifying the field. Vlerick Business School. Retrieved November 2, 2020, from http://hdl.handle.net/20.500.12127/6525

Dow Jones Sustainability Index. (2020). Retrieved September 1, 2020, from https://www.spglobal.com/esg/csa/yearbook/ranking/

Durand, R., Paugam, L., & Stolowy, H. (2019). Do investors actually value sustainability indices? Replication, development, and new evidence on CSR visibility. *Strategic Management Journal, 40*(9), 1471–1490.

Escrig-Olmedo, E., Fernandez-Izquierdo, M. A., Ferrero-Ferrero, I., Rivera-Lirio, J. M., & Munoz-Torres, M. J. (2019). Rating the raters: Evaluating how ESG rating agencies integrate sustainability principles. *Sustainability, 11*(3), 915.

European Commission. (2014). Directive 2014/95/EU of the European Parliament and of the Council of 22 October 2014 amending Directive 2013/34/EU as regards disclosure of non-financial and diversity information by certain large undertakings and groups. Retrieved February 25, 2022, from https://eur-lex.europa.eu/legal-content/EN/TXT/?uri=CELEX%3A32014L0095

European Commission. (2021). Proposal for a Directive of the European Parliament and of the Council amending Directive 2013/34/EU, Directive 2004/109/EC, Directive 2006/43/EC and Regulation (EU) No 537/2014, as regards corporate sustainability reporting. COM/2021/189 final. Retrieved February 25, 2022, from https://eur-lex.europa.eu/legal-content/EN/TXT/?uri=CELEX:52021PC0189

Fink, L. (2018). *A sense of purpose*. BlackRock Annual Letter, (2018). Retrieved March 3, 2020, from https://corpgov.law.harvard.edu/2018/01/17/a-sense-of-purpose/; http://www.corporance.es/wp-content/uploads/2018/01/Larry-Fink-letter-to- CEOs-2018-1.pdf

Fink, L. (2019). *Purpose and profit*. BlackRock Annual Letter, (2019). Retrieved March 10, 2020, from https://www.blackrock.com/corporate/investor-relations/larry-fink-ceo-letter

Font, X., Guix, M., & Bonilla-Priego, M. J. (2016). Corporate social responsibility in cruising: Using materiality analysis to create shared value. *Tourism Management, 53*, 175–186.

Freeman, R. E. (1984). *Strategic management: A stakeholder approach*. Pitman.

Freeman, R. E., & Velamuri, S. R. (2008). *A new approach to CSR: Company stakeholder responsibility*. Retrieved May 2, 2020, from https://papers.ssrn.com/sol3/papers.cfm?abstract_id=1186223

Freudenreich, B., Luedeke-Freund, F., & Schaltegger, S. (2020). A stakeholder theory perspective on business models: Value creation for sustainability. *Journal of Business Ethics, 166*(1), 3–18.

Friede, G. (2019). Why don't we see more action? A metasynthesis of the investor impediments to integrate environmental, social, and governance factors. *Business Strategy and the Environment, 28*, 1260–1282.

GlobeScan and SC Johnson (2019). *Building trust: Why transparency must be part of the equation*. Retrieved November 1, 2020, from https://globescan.com/wp-content/uploads/2019/01/SCJ-GlobeScan-Transparency-Whitepaper-Jan2019.pdf

GRI. (2018). GRI 101: Fundamentals 2016. Retrieved February 15, 2020, from https://www.globalreporting.org/standards/media/1439/spanish-gri-101- foundation-2016.pdf

Hill, A. (2019, September 23). The limits of the pursuit of profit. *Financial Times*. Retrieve June 18, 2021, from https://amp.ft.com/content/c998cc32-d93e-11e9-8f9b-77216ebe1f17

Hsu, A., Koh, K., Liu, S., & Tong, Y. (2019). Corporate social responsibility and corporate disclosures: An investigation of investors' and analysts' perceptions. *Journal of Business Ethics, 158*, 507–534.

Ikram, A., Li, Z., & Minor, D. (2019). *CSR-contingent executive compensation contracts*. Retrieved May 4, 2021, from https://papers.ssrn.com/sol3/papers.cfm?abstract_id=3019985##

Jáuregui, R. (2019). The four contradictions of CSR. *Responsible journal.* Retrieved February 15, 2020, from https://diarioresponsable.com/noticias/27622-ramon-jauregui-las-cuatro-contradicciones-de-la-rse

Karpoff, J. M. (2021). On a stakeholder model of corporate governance. *Financial Management*, Early access 3/2021. Retrieved May 13, 2021, from https://papers.ssrn.com/sol3/papers.cfm?abstract_id=3642906

Kumar, K., Batra, R., & Boesso, G. (2021). Difference in stakeholder engagement approach of small and medium enterprises and large companies and its performance implications. *Corporate Social Responsibility and Environmental Management, 28*(3), 992–1001.

Lleo, A., Bastons, M., Rey, C., & Ruiz-Perez, F. (2021). Purpose implementation: conceptualization and measurement. *Sustainability, 13*(4), 1921.

Lo, A. (2017). *Adaptive markets: Financial evolution at the speed of thought*. Princeton University Press.

Mattingly, J., & Berman, S. (2006). Measurement of corporate social action: Discovering taxonomy in the Kinder Lydenburg Domini Ratings Data. *Business and Society, 45*(1), 20–46.

Mena, J., Hult, G., Ferrell, O., & Zhang, Y. (2019). Competing assessments of marketdriven, sustainability-centered, and stakeholder-focused approaches to the customer-brand relationships and performance. *Journal of Business Research, 95*, 531–543. ISSN 0148-2963. Retrieve May 6, 2021, from https://doi.org/10.1016/j.jbusres.2018.07.038

Mir, C. (2004). *Sistemas de Información contable de Instrumentos Para la gestión de riesgos* (Doctoral thesis). University of Alcalá.

Pfitzer, M., Bockstette, V., & Stamp, M. (2013). Innovating for shared value. Companies that deliver both social benefits and business value rely on five mutually reinforcing elements. *Harvard Business Review*, pp. 100–107.

Porter, M. E., & Kramer, M. (2006). Strategy and society: The link between competitive advantage and corporate social responsibility. *Harvard Business Review,* December 2006.

Porter, M. E., & Kramer, M. (2011). Creating shared value. *Harvard Business Review*, January–February 2011.

Schoenmaker, D., & Schramade, W. (2019). Investing for long-term value creation. *Journal of Sustainable Finance and Investment, 9*, 356–377.

Sewchurran, K., Dekker, J., & McDonogh, J. (2019). Experiences of embedding long-term thinking in an environment of short-termism and sub-par business performance: Investing in intangibles for sustainable growth. *Journal of Business Ethics, 157*, 997–1041.

Stocker, F., de Arruda, M. P., de Mascena, K. M. C., & Boaventura, J. M. G. (2020). Stakeholder engagement in sustainability reporting: A classification model. *Corporate Social Responsibility and Environmental Management, 27*(5), 2071–2080.

Sulkowski, A., Edwards, M., & Freeman, R. E. (2018). Shake your stakeholder: Firms leading engagement to cocreate sustainable value. *Organization and Environment, 31*(3), 223–241.

Torelli, R., Balluchi, F., & Furlotti, K. (2020). The materiality assessment and stakeholder engagement: A content analysis of sustainability reports. *Corporate Social Responsibility and Environmental Management, 27*(2), 470–484.

**Manuel Moreno** is a lecturer at the EAE Business School and consultant under the ARM actuaries brand. He is also a graduate of economics and business studies at the University of Barcelona and an actuary with Chartered Enterprise Risk Actuary (CERA) credential. He subsequently completed his studies with a general management programs (PDG) at IESE Business School and a master's degree at UNED on CSR. For 8 years, he has been secretary of the Council of the College of Actuaries. He was also a student of the doctorate program at the Faculty of Economics, Business and Tourism, University of Alcala, Spain. He currently combines his doctorate with his work. His areas of professional activity are business risk management and retirement plans.

**Elena Mañas** is associate professor in the Department of Economics and Business Management at the University of Alcalá (UAH). She is also researcher in the Institute for Economic and Social Analysis (IAES) and director of the Corporate Social Responsibility Chair of the UAH. Her research interest has focused in CSR, sustainable development, economics and gender, organization management with a gender perspective, services and economics, and economic policy. She holds a Ph.D. in applied economics from the University of Alcalá.

**Oscar Montes-Pineda** is assistant professor in the Department of Economics and Business Management at the University of Alcala (UAH); he is also a researcher in the Institute for Economic and Social Analysis (IAES) and the CSR chair of the UAH. His research interest has focused in service and service innovation, sustainable technological change, new management models, and entrepreneurship, collected in several articles published in scientific journals and international book chapters. He holds a Ph.D. in economic analysis from both the Universities of Alcala and Complutense de Madrid.

**Beatriz Fernández-Olit** is assistant professor in the Department of Economics and Business Management at the University of Alcala (UAH), Spain. She is also a researcher in the Institute for Economic and Social Analysis (IAES) and the CSR chair of the UAH. Her research interest has focused on sustainability and financial sector, financial inclusion, and responsible banking. She holds a Ph.D. in economics from the UNED.

# Communication Professionals as Social Change Agents in Times of Crisis: How Pandemic Situation Has Changed Initiatives in CSR and Sustainability

**Renata Matkevičienė and Lina Jakučionienė**

## 1 Introduction

The global pandemic has forced organizations to rethink sustainability strategies and focus on communication that delivers a social impact to the organization's key stakeholders. Due to the conditions of the pandemic, a hybrid mode of activities and communication in organizations has been introduced and led to the changes of everyday communication practices.

The communication tasks and roles in organizations require the professionals to keep the balance between communication with a variety of stakeholders—engagement of internal audiences through communication and harmonizing communication within the organization and involving external stakeholders to implement the organization's strategies. According to Guthey et al. (2006), the task of a communication professional is to orchestrate "productive social cooperation in [and outside] organizations through the balancing of a variety of interests" (Guthey et al., 2006, p. 55). The role of a communication professional in an organization can be explored through a model that outlines two perspectives incorporating an organization's communication strategy in engagement of stakeholders in taking decision on CSR/sustainability activities and the role of communication professionals in search for possible decisions of CSR/sustainability (see Fig. 1).

Based on the interpretation of the role of communication professionals proposed by Grunig and Hunt (1984) according to the communication process model (one-way or two-way), when considering the role of communication professionals in the implementation of CSR/sustainability initiatives and activities in organizations, we can distinguish between a number of possible models of how an

R. Matkevičienė (✉) · L. Jakučionienė
Vilnius University, Vilnius, Lithuania
e-mail: renata.matkeviciene@kf.vu.lt; lina.jakucioniene@kf.vu.lt



I. G. Arraiano et al. (eds.), *Corporate Social Responsibility in a Dynamic Global Environment*, CSR, Sustainability, Ethics & Governance,
https://doi.org/10.1007/978-3-031-24647-0_19

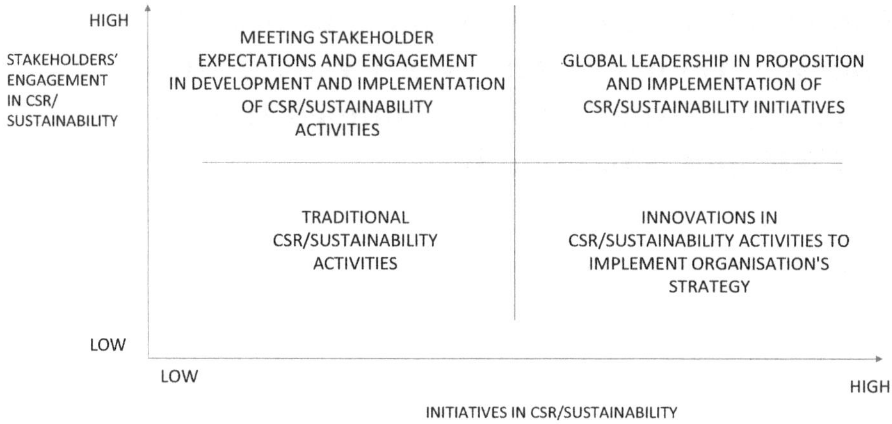


**Fig. 1** Matrix of corporate strategy and the role of communication professionals in CSR/sustainability (proposed by authors)

organization may interact with its stakeholders in the implementation of its CSR/sustainability activities (based on Morsing & Schultz, 2006) (see Fig. 1, "Stakeholders' engagement in CSR/sustainability"): (1) informing stakeholders about the CSR/sustainability activities chosen and carried out by the organization (some authors call this an "egoistic or self-centered approach" (Coombs & Holladay, 2012, p. 41), emphasizing that organizations decide on and implement CSR/sustainability activities themselves, without any consultation with stakeholders), (2) involving stakeholders in the selection of CSR/sustainability activities in the organization, taking into account stakeholder surveys, trends observed in the competitive environment, etc. (Guthey et al., 2006), and (3) the involvement of stakeholders in the development of the organization's CSR/sustainability strategy and its implementation activities, including the organization's strategic ambition based on the values of the organization to engage with stakeholders to ensure the embedding of the results of CSR/sustainability activities and the creation of societal well-being, including the requirements of a triple bottom line. Equally relevant for analysis of CSR/sustainability activities is the leadership or the role of driver of innovations assumed by the organization, which reveals the organization's approach to CSR/sustainability activities—whether this is seen as mandatory conditions and requirements to be met when operating in the field of interest in order to ensure the organization's profit, or whether this is more than the implementation of the essential requirements of CSR/sustainability for organizations operating in the field of interest, but also a voluntary assumption of the role of leadership and commitment to addressing global sustainability issues. Organizations carry out many different activities and operate in different contexts, and the task of communication and CSR/sustainability professionals is to reconcile all interests, and especially to include "the corporate bottom line" (Guthey et al., 2006, p. 56) in the implementation of corporate initiatives (see Fig. 1, "Initiatives in CSR/sustainability").

The role of communication professionals in activities of initiating, motivating, and inspiring various actions in the context of corporate CSR/sustainability is essential, as they are responsible for choosing appropriate tools for communication, inspiring and maintaining corporate CSR/sustainability-driven culture, developing the context for involving and engaging relationships with employees, and ensuring that employees can not only offer initiatives but can get involved in and benefit from them (Du et al., 2010; Duthler & Dhanesh, 2018; Gill, 2015; Jiang & Luo, 2020; Schaefer et al., 2020).

Corporate social initiatives and management fashions generally do not always clash with employees' concerns (Guthey et al., 2006, p. 53). "Employees rather than shareholders deserve to be treated as the primary concern of managers" (Heal, 2008, p. 17) simply because the organization's reputation, employee prestige, and, to a large extent, work depend on the organization's role in CSR/sustainability. In this regard, communication professional operates not only as developer of environment who supports sustainability initiatives but also as an innovator and leader who promotes CSR/sustainability issues by herself/himself (see Fig. 1).

When choosing CSR/sustainability strategies, organizations often focus their activities on external stakeholders, in formulating CSR/sustainability strategies, developing initiatives, and implementing them. Organizations seek not only to respond to market needs, but also want to involve external stakeholders in cooperation, in creating and developing of joint projects for public welfare, thus responding to the essential goal of communication: "dealing with latent, manifest and active publics through communication and dialogue, interaction and partnership" (Vallentin, 2006, p. 75).

By choosing such strategies, organizations position themselves as leaders in a certain CSR/sustainability area through initiatives and activities they perform and communicate that are designed to address the problems faced by a particular external stakeholder or global issues. The task of communication professionals in this case is not only to present the position of organizations in initiating and implementing CSR/sustainability ideas and activities but also to involve external stakeholders both in discussions on the topic raised by the organization and in communication activities in collaboration with external stakeholders.

This approach highlights the role of a communication professional as a social change agent in promotion, development, and implementation of CSR/sustainability innovations and activities in organizations and in society at large. Such a development of the profession could be seen through the statements of professional PR/communication associations underlying the competencies of the communication professional in relation to activities in the CSR/sustainability area: ethical leaders and role models in CSR/sustainability (CPRE, 2006, 2012, 2018) acting "professionally, i.e., in a socially responsible manner, within their own societies as well as within an emerging global community" (CPRE, 2018), and responsible drivers of sustainable decisions having abilities "to consider business objectives in the light of society's expectations; conduct formative and evaluation research including the collection of intelligence/insights and environmental scanning" (Global Alliance (for Public Relations and Communication Management), 2016; Fawkes et al., 2018).

The choice of such an approach to CSR/sustainability activities within the organization, which indicates the respective role of communication professionals in them, allows us to evaluate the activities of communication and sustainability of the organization, highlighting the role of communication professionals as social change agents. The social change agent responds to stakeholders' needs and demonstrates to them the benefits of social impact, whether he or she acts as a maintainer of the status quo, ensuring continuity and coherence of activities without pursuing dramatic change, on the contrary, by ensuring stability in the relationship (including communication) with stakeholders.

## 2 Methods

Based on the approach proposed in the article on analysis of CSR/sustainability activities and communication with stakeholders based on the level of engagement with stakeholders and the level of initiatives driven by a communication professional, this chapter seeks to answer the question of how CSR/sustainability activities are initiated, planned, and implemented in usual conditions and how the status of organizations and communication professionals' changes in the conditions of an uncertainty—period of crisis (pandemic).

The aim of the study is to determine the role of communication professionals as social change agents in promoting and implementing CSR/sustainability initiatives and activities in communication and engagement with different stakeholders to achieve the strategic goals of organizations.

The study raises the following questions:

- Q1: Which CSR/sustainability initiatives and activities are chosen to be implemented in the organization—those that are aligned with the organization's strategy and values and that meet key market requirements for the organization's activities, those that involve stakeholders in identifying relevant CSR/sustainability activities, or those that are undertaken by a communication professional, creating opportunities for the organization's stakeholders to be actively involved in the proposal and implementation of CSR/sustainability initiatives and activities. The answer to this question will allow us to understand and compare the CSR/sustainability activities chosen by the leading organizations in Lithuania, assessing the operational and communication tactics.
- Q2: Has the organization's CSR/sustainability initiatives and activities and communication changed during the crisis (pandemic) and how? It is assumed that during a crisis, organizations review their CSR/sustainability portfolio and focus only on those that help to achieve strategic goals of the organization and that concentrate on stakeholders and keep them motivated, engaged in the organization's activities, thus supporting and strengthening organization's reputation.

- Q3: What is the personal role and responsibility of a communication professional in initiating and implementing CSR/sustainability activities? It is assumed that CSR/sustainability activities carried out by the organizations largely depend on the personal involvement of the communication professional, of his or her interest in CSR/sustainability initiatives, activities, commitment, and leadership. The answer to this question will help to identify the role (assumed or assigned) of the communication professional as a social change agent.

The research methods used are content analysis of CSR/sustainability activities and/or yearly reports of TOP sustainable brands (business organizations) in Lithuania (2019, 2020) and qualitative in-depth expert interviews with communication professionals working in those corporations.

For the research, the TOP Lithuanian sustainable brands (business organizations) in 2020 were selected: brands acting on international level: Lidl Lietuva, Eurovaistinė, IKI, Swedbank, Rimi, and locally based brands: Camelia, Maxima, Norfa, Ignitis, and AUGA group.

Annual CSR/sustainability activity reports for 2019–2020 of the corporations and/or info on the websites on these activities selected for the study in 2019–2020,[1] with text coded by keywords such as "CSR" (in Lithuanian language ĮSA, SA, "socialinė atsakomybė"), "Sustainability" (in Lithuanian language "tvarumas," "darnus vystymasis," "tvarus verslas"), and related topics such as "Charity," "Support," and "Code of business ethics" (in Lithuanian language "labdara," "parama," "verslo etikos kodeksas").

The criteria for the selection of the informants are as follows: top-level communication managers (heads of communication/public relations/corporate communications departments of business organizations), responsible for CSR/sustainability activities. They represent the top 10 most sustainable brands of 2021 according to Lithuanian consumers (retail chains, pharmacies, banking, energy, and agriculture), in which CSR/sustainability performance is under the responsibility of top-level communication managers.

The interviews were conducted in March–April 2021.

The results of the study are presented through impersonalized data from interviews and by summarized statements taken from the reports and information on the websites of organizations selected for the study.

[1] https://imone.lidl.lt/socialine-atsakomybe?_ga=2.9631989.1939807992.1623836055-676208951.1617882259; https://camelia.lt/blog; https://www.apie-eurovaistine.lt/apie-mus/verslo-etikos-kodeksas/; https://iki.lt/esame-atsakingi/; https://www.swedbank.lt/about/swedbank/about/financialResults?language=LIT; https://maximagrupe.eu/lt/socialine-atsakomybe; https://www.rimi.lt/apie/tvarumas; https://www.norfa.lt/apie-mus/labdara-ir-parama/; https://ignitisgamyba.lt/atsakingas-verslas/soc.-atsakomybes-ataskaitos/111; https://auga.lt/tvarumas/tvaraus-verslo-ataskaita/#tabs.

## 3 Results and Discussion

As shown by the analysis of the information on CSR/sustainability activities (2019–2020) available on the websites of the surveyed business organizations (see Table 1), they have been developed consistently in line with the CSR and sustainability policies and strategies focusing on such areas as environmental protection (reduction of CO2 and other emissions, energy saving, green energy consumption, etc.); reduction of food waste; refusal of disposable plastic products; health (wide selection of health-friendly products); youth (educational projects); technology (innovation and investment for sustainable growth and productivity); locality (increasing the range of locally produced goods); animal welfare (abandonment of live fish in aquariums; supermarkets no longer buying endangered fish and seafood and other animal welfare projects; support for animal welfare organizations), diversity, equal rights, human (consumer) rights, ethical business, anti-corruption and transparency, and quality assurance.

Due to the pandemic, changes in CSR/sustainability activities are related to customer support (new activities), internal staff (employees), and more active cooperation with NGOs/communities (see Table 2).

**New Activities in the Field of Customer Service** Support and restructuring measures were offered to customers during the pandemic, depreciation exemptions were applied, prices of bank-leased terminals were reduced, traders were supported to transform their business into e-commerce solutions, charges were deferred (preferential payment terms for services were provided as well as deferred payments for consumed natural gas and electricity for the quarantine period and one additional month), free helplines have been set up, assistance has been provided to women victims of violence during quarantine, and vaccination rooms have been set up in pharmacies.

**Internal Staff (Employees)** Due to the pandemic, business organizations devoted a lot of effort and resources to the safety of employees (and customers) (organic glass walls in supermarkets and pharmacies separating purchasers and cashiers, queue locations at cash desks, as well as floors marked safe, 1 meter, distance, etc.); ensuring financial stability of employees (increased amount of incentive bonuses for employees); and providing employees with all the equipment needed for work at home, preventive measures (face masks), regular disinfection, differentiation of employee teams, and work from home. The organizations paid a lot of attention to the psychological health of employees (free psychological helpline, psychological health recommendations, online meditations, sports activities, measures to take care of psychological well-being). New benefits have been included in employees' social packages (employees are reimbursed for COVID-19 tests when they become commercially available; employees with COVID-19 are given the opportunity to undergo a comprehensive health check, which is fully funded by the employer).

**Table 1** Reported CSR/sustainability activities (2019, 2020)

| CSR/sustainability activities | Consistent CSR/sustainability activities 2019–2020 |
|---|---|
| Environmental protection | • Reduction of CO2 and other emissions<br>• Energy saving<br>• Green energy consumption<br>• Biodiversity and ecosystem protection<br>• Waste management<br>• Transport fuel economy and accounting<br>• Water conservation and accounting<br>• Biogas production from waste<br>Activities related to the implementation of environmental requirements or the leadership of the organization in this field:<br>• Advanced energy consumption monitoring, energy-efficient heating, ventilation, and cooling; modern lighting systems are being implemented<br>• Renewable energy loans, financing the purchase of solar panels for private customers, as well as helping small business customers to purchase solar panels, windmill turbines, or geothermal heat sources<br>• "Green" home loan offered to private customers—a new lending product that promotes energy efficiency in private housing; the "sustainable leasing" proposal, which aims to encourage customers to choose greener cars with lower CO2 emissions<br>• Reduction of food waste (optimization of forecasts, production of goods, shelf life of goods on shelves)<br>• Refusal of disposable plastic products<br>• Animal welfare (abandonment of live fish in aquariums; supermarkets no longer buying endangered fish and seafood and other animal welfare projects; support for animal welfare organizations) |
| Diversity, equal rights, human (consumer) rights | • Prevention of discrimination and control of infringements<br>• Ensuring and monitoring gender equality<br>• Protection and guarantees of human rights<br>• Involvement of employees in volunteering (paid as a working day)<br>• Educational projects |
| Ethical business, anti-corruption, and transparency | • Principles of sustainability in business processes<br>• Ensuring financial stability for employees (fair payment of salaries)<br>• Employee's safety and health<br>• Ensuring good working conditions<br>• Compensation for accidents and injuries at work |
| Product quality assurance | • Quality services, competitive products, and products that meet customers' needs and expectations<br>• Innovation and investment for sustainable growth and productivity<br>• The range of locally produced goods is being increased |

**Table 2** Reported CSR/sustainability activities (2019, 2020) and change in activities during the pandemic

| CSR/sustainability activities | Consistent CSR/sustainability activities 2019–2020 | Change during the pandemic |
|---|---|---|
| *Customers, traders* | *Principles of sustainability in business processes* | • Proposed support and restructuring measures, reduced prices<br>• New services are provided<br>• Deferred taxes<br>• Free helplines have been set up |
| *Internal staff (employees)* | *Ensuring the welfare of employees*<br>• Ensuring financial stability for employees (fair payment of salaries)<br>• Employee's safety and health<br>• Ensuring good working conditions<br>• Prevention of discrimination and control of violations<br>• Ensuring and control of gender equality<br>• Compensation for accidents and injuries at work<br>• Protection and guarantees of human rights<br>• Involvement of employees in volunteering (paid as a working day)<br>• Development of employee competencies | • The amount of incentive bonuses for employees has increased<br>• Greater attention to ensuring safe working conditions for employees<br>• Provision of all equipment necessary for work at home, preventive measures with face masks, regular disinfection, differentiation of teams of employees, work from home<br>• Attention to psychological well-being (free psychological helpline, psychological health recommendations, online meditations, sports activities, measures to take care of psychological well-being)<br>• New benefits included in employees' social packages (employees are reimbursed for COVID-19 tests when they become commercially available, employees with COVID-19 are given the opportunity to undergo a comprehensive health check, which is fully funded by the employer) |
| *Communities: Social partners and supporters* | *Support for NGO's and communities* | • The amount of support provided to NGOs/communities has increased<br>• New activities—support for various organizations with food parcels (i.e., staff of medical and paramedical organizations, specialists of the Public Health Bureau, staff of Ambulance Stations, General Emergency Centre 112, and mobile coronavirus checkpoints)<br>• Support for medical institutions (provision of necessary protective equipment and medical equipment)<br>• Assistance to women victims of domestic violence during quarantine |

**Cooperation with NGOs/Communities** During the pandemic, the amount of financial support provided by business organizations to NGOs/communities increased, and new activities appeared—support for employees of various organizations with food parcels (i.e., staff of the Public Health Office, Ambulance Stations, General Assistance Centre 112 and mobile coronavirus checkpoint staff) and the provision of necessary protective equipment and medical equipment to medical facilities.

The analysis of the interviews with the informants shows that for the vast majority, the personal involvement and belief of the communication specialist in the topic of CSR/sustainability is the main driver for initiating, supporting, and implementing initiatives in this field. How much a communication professional can inspire an environment depends on his or her personal characteristics. "This position requires a person who is hurt (by sustainability issues)." Informants believe that a communication professional who does not personally care much about CSR/sustainability will find it quite difficult to achieve the organization's sustainable goals. "In that case, you will simply work to do your duty. "If you are genuinely involved in these issues, you will do your job sincerely, seek to resolve issues promptly."

According to the informants, a communication professional is a generator of sustainability ideas in their organization. This is the opinion of most communication professionals surveyed. "If your job is ONLY to communicate about sustainability and not to be a generator of social change, it will not be good. In this case, the organization will look at such activities as follows: write a report, organize an event and so on." In terms of leadership, an important role is played by the extent to which leadership is related to a person's personality to be a leader, faith in his or her actions, and help from leadership. "Yes, communication is definitely the driver. (...) we are the driving force. But only when we have faith in that activity ourselves and when there is support from management and other functions."

Most of the informants state that communication in organizations is a strategic level position; communication manager is a top-level manager who is involved in the development of board strategic issues (including sustainability issues) at an early stage and their implementation. According to some informants, all possible functions (including internal and product communication, sustainability office, and event organization) are integrated into the communication department, or the whole environment is under communication's responsibility; thus, they have the freedom to make decisions related to the social agenda and to be catalysts for various processes. "An empowered leader brings much better results."

Sustainability in the communication professional agenda, according to the informants, is a rational decision of the organization not only because communication is an intermediary between the organization and its stakeholders and knows their needs and expectations, but also because it has the quality—empathy, which is most needed for CSR/sustainability. On the other hand, CSR/sustainability is and must be part of the communication professional's activities, but only if sustainability is the organization's path, direction, strategy, and DNA. Otherwise, one communication unit will be not able to achieve the sustainable goals of an organization: "I am against it (sustainability) becoming just an area of communication. Communication is not a goal in itself; it is means of achieving the company's business objectives."

Representatives of international corporations mention that CSR/sustainability guidelines are centralized at the international level, but it is usually up to the country to decide on the implementation of CSR/sustainability ideas, considering each country's specifics. The head of communication has a great deal of freedom to decide for himself or herself and to suggest to the management board which sustainability ideas are most suitable for their country. However, sustainability initiatives do not come ONLY from the communication manager/communication unit but are often promoted by the organization's internal staff (employees), organizational units, the board, shareholders, or external partners. For example, in a specific sector (energy, which is a state-owned enterprise strongly influenced by political decisions), the main impetus for implementing sustainability themes comes from the board.

According to the informants, the most valuable initiatives are those that are born within the organization, because when employees themselves are involved in the implementation of sustainable initiatives, it generates more value for the organization itself (stronger employee faith in the organization, better climate, etc.). "One of the most important areas that employees emphasize as the greatest strength of the organization is its activity in the field of sustainability. Employees appreciate that the organization is not afraid of losing profits and refuses to work with a supplier that pollutes nature." On the other hand, here the role of the communication professional is particularly important: "There is a special mailbox where employees can submit their ideas. (. . .) Behind this intermediary activity between the manager and the employees is always the communication people because they are those who collect the employee ideas and pass them on to the board."

Most communication professionals see their organizations as innovators dictating certain sustainability trends to the market (e.g., the first to choose not to sell endangered species; the first to refuse to sell fish in aquariums; the only trader in Lithuania who did not sell alcohol on the actual day of September 1 and the day when the students actually went to the school although it has lost € 100,000 from turnover as a result (in Lithuania, the law does not allow the sale of alcohol on September 1 when officially the study year starts); the first to implement the idea of sustainable agriculture despite being one of the most polluting sector in the world in terms of emissions; the first to set up a Money Museum and Finance Laboratory with the aim of increasing digital literacy in the bank). Communication professionals participate in these initiatives as their initiators or active implementers.

The analysis of CSR/sustainability activities carried out and the data of the interviews of the communication professionals allows us to state that:

- Organizations tend to choose stakeholders' involvement in the selection of CSR/ sustainability activities and seek to respond to their views on relevant issues of stakeholders by offering assistance (employee volunteering, sponsorships, etc.), but in times of crisis, organizations independently decide to change CSR/sustainability activities in response to environmental changes. "At the very beginning of the first quarantine, there was a lot of confusion in the market, we increased sponsorships, provided more food, gift cards, sponsored volunteers on duty at the airport; it was on-top unplanned sponsorship actions. But later we stuck to consistent things"; "The pandemic has affected changes in CSR/Sustainability activities, but not fundamentally because sustainable activities are our DNA. The business decision to defer loans is not economically justified. So, it was a social action in the face of a pandemic."

- Organizations tend to address local issues and implement their CSR/sustainability activities, ensuring innovative CSR/sustainability solutions in the organization's activities. Representatives of international corporations mention that CSR/sustainability guidelines are centralized at the international level, but it is usually up to the country to decide on the implementation of CSR/sustainability ideas, considering each country's specifics. The head of communication has a great deal of freedom to decide for himself or herself and to suggest to the management board which sustainability ideas are most suitable for their country.
- Organizational leadership in the field of CSR/sustainability depends to a large extent on the personal leadership of communication professionals and their involvement in discussing CSR/sustainability issues, raising problems, and finding possible solutions, and, of course, solving them. According to most of the informants, the personal involvement and belief of a communication professional in the topic of sustainability is a key driver in the organization's development and implementation of initiatives in this area. How much a communication professional can inspire an environment depends on his or her personal characteristics. According to the informants, it will be quite difficult for a communication professional who does not personally care much about CSR/sustainability to achieve the sustainable goals of the organization. "In that case, you will simply work to do your duty. If it hurts you sincerely, you will do your job sincerely, seek to resolve issues promptly." According to the informants, it is especially important for communication professionals that their organizations are innovative; they take it for granted: "if we are not the first, I immediately ask myself and other members of the organization what happened that we are not the first?" According to the informants, a great tone to be an innovator or a cautious market player is posed by the organization itself; however, innovation-promoting initiatives coming from communication professionals are accepted at the highest level and encouraged to develop further. Thus, communication professionals are involved as initiators of innovation or active members in their implementation.

Based on the survey data and qualitative in-depth interviews with communication professionals, the distribution of the role of communication professionals in the implementation of CSR/sustainability activities chosen by the organizations is presented (see Fig. 2), according to the diagram describing the role of communication professionals developed by the authors of the paper (see Fig. 1).

The analysis of the results of the research highlighted the following roles of communication professionals, as revealed through the description of their roles and the discussion of best practices during the in-depth interviews (see Fig. 3):

- Implementor—a communication professional who plans and implements CSR/sustainability activities and their communication in the context of the CSR/sustainability strategy set out by the organization's managers (owners, board members, shareholders). The communication professional sees himself or herself

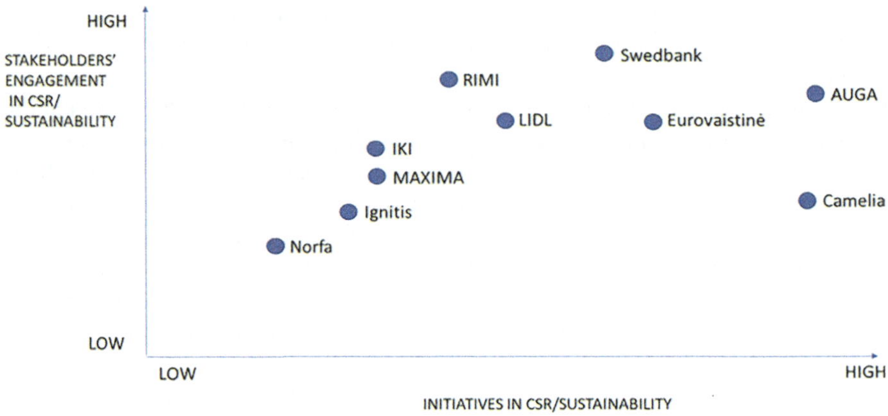


**Fig. 2** CSR/sustainability matrix based on analysis of roles of communication professionals in top 10 Lithuanian CSR/sustainability organizations

**Fig. 3** Roles of communication professionals in CSR/ sustainability practices (proposed by authors based on the interviews)

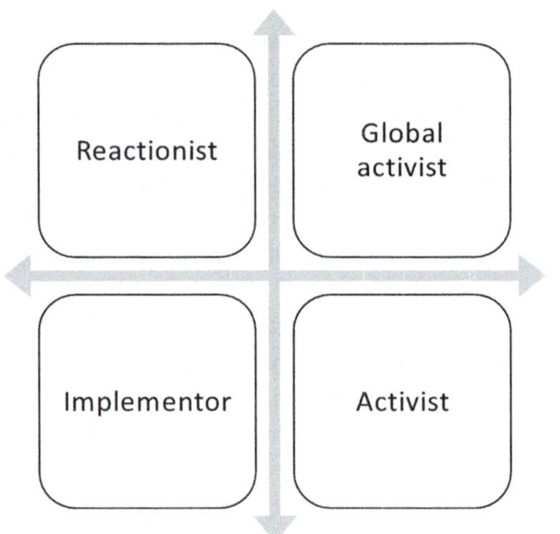


as the implementer of CSR/sustainability activities proposed by the various stakeholders of the organization, ranging from the organization's employees to the various groups of stakeholders external to the organization, who arrange the opportunity to propose them, and who not only implement the activities, but also communicate about them. ("The role of the communication professional is particularly important: (. . .) [the communication professional's role as] an intermediary between the manager and the employees").

- Reactionist—a role undertaken or assigned to a communication professional by the organization's management, with the task of reacting to external factors: challenges from outside the organization or demands for CSR/sustainability activities from relevant strategic stakeholders. In this way, the communication professional is like an observer of the organization's environment, offering the organization strategic and tactical solutions for the implementation of CSR/sustainability activities and communication, in response to changes in the environment related to the organization's field of activity ("Employees appreciate that the organization is not afraid of losing profits and refuses to work with a supplier that pollutes the environment").

- Activist—a communication professional who tries to encourage employees in the organization to engage in CSR/sustainability activities by personal example, usually linked to personal involvement, belief in CSR/sustainability ("Everything in life is up to us: if you don't personally believe in the things you do, nothing will come of it"; "if your job is ONLY to communicate about sustainability, not to be a generator of social change yourself, then nothing will work").
- Global activist—a communication professional who not only takes personal leadership in the promotion and implementation of CSR/sustainability activities within the organization but also creates an environment that ensures the implementation of the organization's CSR/sustainability strategy in line with international/global trends and demonstrates the organization's leadership in these trends. ("Sustainability activism," "communication really is a driver. (...) we are the driving force").

The research data show that communication professionals tend to take personal leadership in CSR/sustainability initiatives and activities but propose these activities in line with shareholders' position, taking into account customers' attitudes and expectations, and in collaboration with NGOs promote CSR/sustainability goals to society in general, to make it aware of issues, prospective solutions, and actions needed to take.

The pandemic highlighted the importance of the performance of communication professionals employing hybrid instruments to demonstrate organizations' CSR/sustainability initiatives and activities through involvement and engagement of various stakeholders while taking a personal lead in sustainability: if in the past communication professionals were able to communicate face-to-face through various communication tools and encourage stakeholders to engage in various organizational activities, the changed conditions not only forced to search for new forms of communication building and strengthening relationships with stakeholders in the "new reality" but also to review the organization's activities, strengthen the motivation of individual stakeholders, and encourage them to keep in touch with the organization. Nevertheless, the CSR/sustainability initiatives the organization relates itself with allocating human, financial, and other resources in most cases are cascaded by the shareholders and only then implemented with the help of communication professionals.

## 4 Conclusions

Communication professionals, driven by the organization's values, take on the role of social change and CSR/sustainability "ambassadors" in the pursuit of the organization's corporate strategy, promoting/sharing sustainability ideas and encouraging responsible actions.

The actions of communication professionals are related to integrating the relevant values into the organization's activities and communication, promoting CSR/sustainability initiatives, motivating the launching of activities, and engaging actively in the implementation of CSR/sustainability activities. The role of communication professionals in CSR/sustainability can be seen in the context of the organization's system and culture for initiating and implementing CSR/sustainability activities but is also determined by the communication professional's personal leadership and individual engagement in initiating sustainability activities, finding CSR/sustainability solutions, and implementing activities involving different stakeholders in the organization.

As the research data demonstrated, in performing CSR/sustainability strategies communication professionals employ all communication functions (including internal communication, product communication, sustainability office, and organization of events) taking the wide range of issues under the responsibility of communication; thus, communication professionals have the freedom of decisions related to CSR/sustainability. Therefore, having such a portfolio, communication professionals can influence all processes related to the social agenda and to be catalysts for various processes. As it was stated by several informants, responsibility for CSR/sustainability initiatives and strategies is mostly implied by personal drive and deep concerns on the issues.

The proposed approach allows us to analyze the role of a communication professional in the development and implementation of CSR/sustainability initiatives and activities in the organization and to analyze how these initiatives and activities are performed (on national level or globally).

The proposed model of CSR/sustainability activities and communication allows us to assess not only an organization's CSR/sustainability initiatives and activities (achieving profit for the organization, securing and creating a competitive advantage for the organization by meeting the general market and societal requirements for organizations' performance) but also value-based leadership and the promotion of local or global initiatives that create opportunities for societal transformation and sustainable well-being.

## References

Coombs, W. T., & Holladay, S. J. (2012). *Managing corporate social responsibility. A communication approach*. Wiley-Blackwell. 180p.

CPRE. (2006, November). *The professional bond. Education. Public relations. The practice*. The Report of the Comission on Public Relations Education. Virginia Commonwealth University. http://www.commissionpred.org/wp-content/uploads/2018/03/report2-full.pdf

CPRE. (2012, October). *Standards for a master's degree in public relations: Educating for complexity*. The Report of the Commission on Public Relations Education. http://www.commissionpred.org/wp-content/uploads/2018/03/report5-full.pdf

CPRE. (2018). *Fast forward. foundations + future state. educators + practitioners*. The Commission on Public Relations Education 2017 Report on Undergraduate Education. http://www.commissionpred.org/wp-content/uploads/2018/04/report6-full.pdf

Du, S., Bhattacharya, C. B., & Sen, S. (2010). Maximizing business returns to corporate social responsibility (CSR): The role of CSR communication. *International Journal of Management Reviews, 12*(1), 8–19. https://doi.org/10.1111/j.1468-2370.2009.00276.x

Duthler, G., & Dhanesh, G. S. (2018). The role of corporate social responsibility (CSR) and internal CSR communication in predicting employee engagement: Perspectives from the United Arab Emirates (UAE). *Public Relations Review, 44*(4), 453–462. https://doi.org/10.1016/j.pubrev.2018.04.001

Fawkes, J., Gregory, A., Falkheimer, J., Gutiérrez-García, E., Halff, G., Rensburg, R., Sadi, G., Sevigny, A., Sison, M. D., Thurlow, A., Tsetsura, K., & Wolf, K. (2018). *A Global Capability Framework for the public relations and communication management profession*. Global Alliance for Public Relations and Communication Management (GA) Research Report. University of Huddersfield, Huddersfield. https://prinz.org.nz/wp-content/uploads/2020/09/Global_Capability_Framework.pdf

Gill, R. (2015). Why the PR strategy of storytelling improves employee engagement and adds value to CSR: An integrated literature review. *Public Relations Review, 41*(5), 662–674. https://doi.org/10.1016/j.pubrev.2014.02.012

Global Alliance (for Public Relations and Communication Management). (2016, May). *From the global body of knowledge (GBOK) to a global capabilities framework*. https://static1.squarespace.com/static/561d0274e4b0601b7c814ca9/t/574ae46707eaa010521f0d31/1464525936413/08_GBOK+Progress+report+%26+global+capabilities+framework+May+2016.pdf

Grunig, J. E., & Hunt, T. (1984). *Managing public relations*.

Guthey, E., Longer, R., & Morsing, M. (2006). Corporate social responsibility is a management fashion. So what? In M. Morsing & S. C. Beckmann (Eds.), *Strategic CSR communication* (pp. 39–60). Djøf Forlag.

Heal, G. (2008). *When principles pay. Corporate social responsibility and the bottom line*. Columbia Business School.

Jiang, H., & Luo, Y. (2020). Driving employee engagement through CSR communication and employee perceived motives: The role of CSR-related social media engagement and job engagement. *International Journal of Business Communication*. https://doi.org/10.1177/2329488420960528.

Morsing, M., & Schultz, M. (2006). Stakeholder communication strategies. In M. Morsing & S. C. Beckmann (Eds.), *Strategic CSR communication* (301p.). Djøf Forlag.

Schaefer, S. D., Terlutter, R., & Diehl, S. (2020). Talking about CSR matters: Employees' perception of and reaction to their company's CSR communication in four different CSR domains. *International Journal of Advertising, 39*(2), 191–212. https://doi.org/10.1080/02650487.2019.1593736

Vallentin, S. (2006). Corporate social responsiveness and public opinion. In M. Morsing & S. C. Beckmann (Eds.), *Strategic CSR communication* (301p.). Djøf Forlag.

**Renata Matkeviciene** is a researcher at Vilnius University (Lithuania) since 2002. Her main area of research is corporate communication (organizational communication, reputation, trust, and CSR) and media sociology. Currently, she is a chair of the Department of Organizational Information and Communication Research at the Faculty of Communication, Vilnius University. As a researcher (after defense of doctoral dissertation in 2007) she has written several articles in scientific monographs analyzing quality of information and communication activities of organizations, trust, and sustainability. She has published articles in scientific peer-reviewed journals mostly related to the political communication issues such as political media discourse, political image, political corruption and political election campaigns, and political populism as a communication style. She has participated in a number of scientific conferences in the field of communication (PR and CSR) research. For almost 10 years she is a member of Lithuanian Communications Association. In 2018, together with colleagues from Lithuanian Communications Association, she worked on the first official communication/PR dictionary. Since 2016, she is actively involved in social entrepreneurship as consultant and mentor for young entrepreneurs: together with partners from NGOs and the public and private sector she participates in educational and research projects, and incubator or accelerator for young entrepreneurs as well.

**Lina Jakučionienė** is a Ph.D. candidate at the Faculty of Communication of Vilnius University. Lina Jakučionienė is researching the issues of the communication market, the changing requirements of a communication professional, and the value created by communication in the context of today's environment. Lina is Lecturer at Vilnius University, as well as the member of the Management Committee for the International Communication Master's Degree Study Programme at Vilnius University.

Lina Jakučionienė counts more than 20 years of experience in the communications field, both internal and external, as well as international communication. Lina has been the Director of Corporate Communication at ERGO insurance group in the Baltic States. She has worked as the Head of the Public Relations Department of the Ministry of Health of the Republic of Lithuania and as the CEO of "European Innovation Projects." Lina also has work experience in the German Parliament and the European Parliament.

Lina is the Ambassador of the European Association of Communication Directors (EACD) for Lithuania; she has served as the Chairwoman of the Lithuanian Communication Association.

# Environmental Information Disclosure and Profitability: The Environmental Certification Influence

**Albertina Paula Monteiro, Catarina Cepêda, and Francisco Barbosa**

## 1 Introduction

During the last few decades, pollution and excessive use of natural resources have caused environment destruction, putting future generations at stake. Indeed, society (people and companies) should be environmentally responsible, where focus is to meet needs of today without compromising the ability to meet future needs (Nejati et al., 2011).

However, concern for meeting future generations needs is not the only reason for companies to direct their attention toward the environment. These have already realized that they can be more competitive, improve reputation, and achieve economic growth if they are environmentally responsible and consequently contribute to sustainable development (Elshabasy, 2018).

In a constantly changing world, sustainable development challenges are increasingly important, with environmental protection and pollution prevention being current concerns for any organization. Thus, environmental problems solution, or their minimization, requires a new attitude from managers, who must start to consider environment in their decisions and adopt technological methods that contribute to the expansion of the planet's carrying capacity (Barbieri, 2017).

With the concern of obtaining a better environmental performance, companies implement Environmental Management Systems (EMS). In fact, an EMS implementation is something that companies have been betting, more and more, in recent years. Literature suggests that environmentally certified entities are interested in

A. P. Monteiro · F. Barbosa
Porto Accounting and Business School, Polytechnic of Porto, Porto, Portugal
e-mail: amonteiro@iscap.ipp.pt

C. Cepêda (✉)
University of Trás-os-Montes and Alto Douro, Vila Real, Portugal



I. G. Arraiano et al. (eds.), *Corporate Social Responsibility in a Dynamic Global Environment*, CSR, Sustainability, Ethics & Governance,
https://doi.org/10.1007/978-3-031-24647-0_20

disclosing their environmental achievements (Archel & Lizarraga, 2017), as well as companies with environmental certification tend to disclose more environmental information in their reports (Barbosa et al., 2021).

The literature suggests that there are other factors, such as industry, profitability, and corporate governance attributes, influencing environmental disclosure level on annual and sustainability reports (e.g., Ahmadi & Bouri, 2017; Barbosa et al., 2021; Monteiro & Guzman, 2010; Sumiani et al., 2007). On the other hand, attributes, such as company size, capital requirement, profitability, and capital expenditures, are positively associated with the extent or quality of environmental disclosure (Barbosa et al., 2021; Sutantoputra et al., 2012).

Regarding the relationship between environmental disclosure and financial performance, Nor et al.'s (2016) study, for 100 largest listed companies in Malaysia and for the year 2011, shows mixed results between environmental disclosure practices and financial performance. Nor et al. (2016) reinforce that the need for environmental information disclosure still exists if companies want to legitimize their position in society. Sutantoputra et al.'s (2012) study analyzes environmental disclosure level of 53 ASX200 Australian listed companies and whether better performing companies related to lower or more positive emissions disclose more information about their environmental performance. Their results indicate a generally low level of environmental disclosure, and that environmental performance was not significantly associated with the level of environmental disclosure. However, the same results indicate that larger companies, those in polluting industries and with higher levels of capital investment, disclosed more environmental information. Huang and Kung (2010) and Kathy Rao et al. (2012) verify a negative relationship between both variables in a study performed for Taiwan and Australian listed companies.

The literature review identified a gap in the scope of studies that investigate relationship between environmental certification, environmental disclosure level, and company's performance, more specifically studies that analyze, simultaneously, the direct effect of environmental certification on companies' economic performance and indirect through environmental information disclosure level.

To fill this gap, this study, besides investigating environmental information disclosure evolution in annual and sustainability reports, aims at analyzing whether companies with environmental certification are those that disclose more environmental information and whether this has an impact on the economic performance (profitability) of Portuguese companies. In addition, this research analyzes whether the Environmental Disclosure Index (IDA) mediates the relationship between environmental certification and profitability.

Results empirically prove that environmental certification does not directly contribute to profitability, but indirectly influences it through IDA.

This chapter is structured as follows. The first section refers to the introduction and the second contains a literature review where the theoretical framework of central themes of the study is carried out and research hypotheses are developed. The third and fourth sections present the research methodology and results discussion analysis. The last section highlights the main conclusions and makes final considerations.

## 2 Theoretical Framework and Research Hypotheses

According to Branco and Delgado (2011), pressures to consider a social and environmental impact in the business sector have gradually grown in recent decades. Caseirão (2000) states that public opinion and legislation exert pressure and make companies feel the need to take into account social and environmental implementation measures in a more systematic and integrated manner in overall organization management. For Bhattacharya and Sen (2004), consumers have a favorable attitude toward companies that engage in social and environmental responsibility. It is in this sense that the concept of corporate social responsibility (CSR) has started to develop over the last few decades.

Howard Bowen is considered by Carroll (1991) the first to address the issue of CSR in the 1950s. Bowen and Johnson (1953) state that entrepreneurs should have a set of social obligations toward citizens and environment.

Throughout the second half of the twentieth century, many authors took various approaches to this concept. Davis defined CSR as the consideration of and response by companies to issues that go beyond their economic, technical, and legal aspects. Years later, Carroll (1991) suggested a pyramid model to elucidate the concept of CSR. This model is based on four levels: economic responsibility, legal responsibility, ethical responsibility, and philanthropic responsibility. In this model, the main objective of the entity is no longer only financial profit and legality but also ethical and philanthropic which is displayed at its top. Ethical responsibilities incorporate standards or expectations that reflect a concern with what consumers, employees, shareholders, and society consider fair or understand to be in line with the respect and protection of moral rights. Philanthropic level refers to awareness of social context involving organization actions in the surrounding communities in order to provide an increase in quality of life.

In the twenty-first century, according to the Green Paper of the Commission of the European Communities (EC, 2001), CSR is essentially a concept whereby companies decide voluntarily to contribute to a fairer society and a cleaner environment. Also, according to the EC (2001, p. 6), "being socially responsible means not only fulfilling legal expectations, but also going beyond compliance and investing "more" into human capital, the environment and the relations with stakeholders."

With the concern of obtaining a better environmental performance, companies implement Environmental Management Systems (EMS). An EMS aims to improve overall economic development of companies through continuous improvement of their environmental performance and compliance with environmental legislation application (Monteiro, 2013). This system involves the entire organizational structure in a planned approach to identify correct procedures with a view to improving different environmental items. The main goal of adopting an EMS is continuous improvement of company's environmental performance. To achieve this, the organization must define specific objectives based on its environmental policies and adopt different practices and procedures that lead to these goals' achievement.

EMS process implementation is time-consuming and involves going through several stages (Pinto, 2005). Initial phase involves surveying the situation in which company finds itself; that is, an exhaustive diagnosis is made of all activities involving environmental matters. This is followed by the company's environmental policy definition that will serve as a basis for the whole process called "project team." This team will have as its functions EMS documentation preparation and project's execution. The final phase comprises planning, implementation, and system operation. This stage is considered most laborious because it is here that everything that was previously discussed and projected is put into practice. Note that environmental certification by an external entity ensures that system meets all requirements and all organization's stakeholders guarantee that company's activities are in line with EMS implemented.

Companies that choose to adopt an EMS have several possibilities of which standard to follow, taking into consideration factors such as characteristics, stipulated objectives, and conditions that company presents. Internationally, there are five standards available to companies: standards produced by the United States Environmental Protection Agency (US EPA), EMS British Standard 7750 (BS 7750), Community Ecolabel System, certification by International Organization for Standardization (ISO) 14001, and Eco-Management and Audit Scheme (EMAS). In Portugal, both ISO and EMAS standards are normally applied.

ISO 14001, published in 1996, was developed to be applicable to all types of organizations. This standard provides guidelines for EMS implementation in order to guarantee continuous improvement of environmental aspects and is based on four pillars: plan, execute, verify, and act. Its general purpose is to balance environmental protection and pollution prevention with socioeconomic needs.

For Zeng et al. (2005), ISO 14001 implementation provides several benefits for an organization, including:

- Opening new national and international markets
- Improvement in management as a whole
- Increased customer satisfaction
- Reduction of waste and costs
- Improvement in the company's image
- Improvement in environmental performance as a whole

EMAS is a voluntary environmental management instrument aimed at improving and assessing environmental impacts of companies in the industrial sector. These organizations intend to commit to manage their environmental performance according to existing rules, in order to guarantee to third parties the credibility of their EMS and to provide relevant information to the outside.

In addition to contributing to environmental responsibility and compliance with legislation, these systems enable the identification of opportunities to reduce materials and energy use and improve process efficiency (Chan & Wong, 2006). Given the global market environment, which brings to companies the need for cost reduction and adequacy of products and production processes for market needs, organizations, particularly industrial ones, are pressured to modernize their management systems so that they provide higher product quality, enable and support technological innovations, contribute to sustainable development, and ensure increased competitiveness and, consequently, profitability (Oliveira & Pinheiro, 2010).

Companies are pressured to develop actions within the scope of sustainability and to disclose information about their performance to their stakeholders. For Saremi and Nezhad (2014), environmental accounting is a difficult concept to define, given the variety of meanings it can encompass, such as those related to environment. According to paragraph 6 of the Portuguese Financial Reporting Accounting Standard (NCRF) 26—Environmental matters, the term environmental refers to "the natural physical environment, including air, water, land, flora, fauna and non-renewable resources such as fossil fuels and minerals."

For Pires (2014), accounting has an obligation to concern itself with environmental reality because, providing a service, it must be willing to address what is relevant and of concern to civil society. The information to be provided must reflect company's assets truth, taking into account that the latter can be strongly affected by environmental issues because their risks, in addition to causing ecological damage, can also cause economic and financial damage to the entity itself (Ferreira, 2000).

Eugénio (2004) summarized advantages of environmental accounting for companies both internally and externally, namely:

- To respond to new accounting control needs
- To make financial reporting more useful to its users and being an additional element of the performance demonstration
- To improve the planning and control of the company's environmental management activities
- To enhance cost rationalization
- To stimulate new perspectives on the productive process
- To allow the correct determination of production costs and marketing prices
- To bring the environment into the management process
- To facilitate the collection and presentation of elements necessary for the fulfillment of obligations, such as:
  - Accountability documents for the national market
  - Reporting documents for the international market
  - Information requested by statistical authorities
  - Information requested by regulatory bodies

Environmental information disclosure to outside world is mandatory, according to paragraph 47 of NCRF 26, when those matters "are materially relevant for the assessment of the performance or the financial position of the entity." This information may be disclosed through the management report, providing an explanation of the facts, a description of the valuation methods used, and a perspective of company economic situation. According to paragraph 48 of NCRF 26 (effective until 2015), management report should include the following information:

- Policies and programs adopted by the entity relating to environmental protection measures, including pollution.
- Improvements made in key areas of environmental protection.
- Information on environmental performance (consumption of raw materials, water, gas emissions, waste removal). Figures shall be expressed in physical units rather than monetary terms.
- The entity may issue a separate environmental report containing more detailed quantitative and qualitative information and the management report shall make reference to such a document.

In relation to the information disclosed in financial statements notes, it is important to highlight that the paragraph referring to these matters in NCRF 26 was removed in 2015 due to SNC amendment. However, these same indicators are currently explicit in paragraph 28 of *Portaria* 220/2015, published in the Decree-Law n.° 143/2015, being the following:

- Measurement bases adopted for environmental assets and liabilities
- Public incentives related to environmental protection
- Accounting policy adopted in long-term expenditures
- Undiscounted amount of the liability as well as the discount rate used
- Contingent environmental liabilities
- Significant expenditures incurred with fines and penalties
- Emission of gases with greenhouse effects

Portuguese listed companies apply international accounting standards adopted by the European Union. As they are not subject to any environmental reporting standard, they disclose environmental information on an optional basis in annual, integrated, or sustainability reports (which may have different designations).

Various entities have emerged to promote non-financial information reporting. *Global Report Initiative* (GRI) standards are the most widely used in sustainability reporting. According to Eugénio and Gomes, GRI is the first organization, on a global scale, that aims to reach a consensus on sustainability reporting guidelines or standards, one of its fundamental objectives being to increase reports quality to a level that is comparable, consistent, and useful on a global scale.

Environmental dimension of sustainability concerns organization's impact on natural living and non-living systems, including land, air, water, and ecosystems (GRI, 2015). There are 34 environmental indicators for the aspects described above that contain performance related to expenses such as raw materials, energy and water, production of emissions and waste, and impact on biodiversity. Environmental factors' influence on corporate decision-making is increasing day by day. This greater influence forces companies to minimize their environmental costs, to include environmental factors in strategic management decisions, and to try different ways of dealing with competition (Tanc & Gokoglan, 2015). Thus, it is essential that companies have access to organized and relevant information for management decision-making, and GRI indicators are a valuable tool in this regard.

In addition to the importance of disclosing organized and relevant environmental information to various stakeholders, literature suggests that this is influenced by several factors. EMS implementation is something that companies have increasingly invested in and several authors have indicated this management system as a determining factor in environmental information disclosure, given that companies want to disclose their benefits/performance (Archel & Lizarraga, 2017; Barbosa et al., 2021; Sumiani et al., 2007). For Archel and Lizarraga (2017), environmentally certified entities have an interest in disclosing their achievements which results in an increase in environmental information disclosure. Also, Barros states in his research that companies with environmental certification tend to disclose more environmental information in their reports. In this context, the first research hypothesis is formulated.

*H1: Companies with environmental certification present a higher EDI.*

The sub-hypotheses associated with this hypothesis are:

*H1.1: Companies with environmental certification present a higher EDI on management reports and notes to financial statements* ($EDI_{MR\&N}$).
*H1.2: Companies with environmental certification have a higher EDI on sustainability reports* ($IDA_{SR}$).

Profitability is defined as an indicator of a company's performance in managing its assets (Juhmani, 2014). For Elshabasy (2018), companies with higher profits tend to disclose more environmental information unlike companies with lower profits that opt for more secrecy. Brammer and Pavelin (2006) point out the same theory, as companies with high profits are more able to pay for the costs associated with environmental disclosure. On the other hand, Huang and Kung (2010) and Kathy Rao et al. (2012) find a negative relationship between both variables in a study applied to Taiwanese and Australian listed companies. Al-Tuwaijri et al.'s (2004) study, which interrelates environmental disclosure, environmental performance, and economic performance, evidences that better environmental performance is significantly related to better economic performance and more extension of the environmental disclosure (of specific pollution measures and occurrences).

However, the literature tends to look at the impact of economic performance on information disclosure rather than the opposite relationship. Considering that companies that disclose more information tend to be more profitable, due to the advantages that come from being environmentally sustainable, in this study we formulate the following research hypothesis:

*H2: More profitable companies have higher EDI.*

The two sub-hypotheses are:

*H2.1: More profitable companies have higher* $EDI_{MR\&N}$.
*H2.2: More profitable companies have higher* $EDI_{SR}$.

Figure 1 presents the theoretical model.

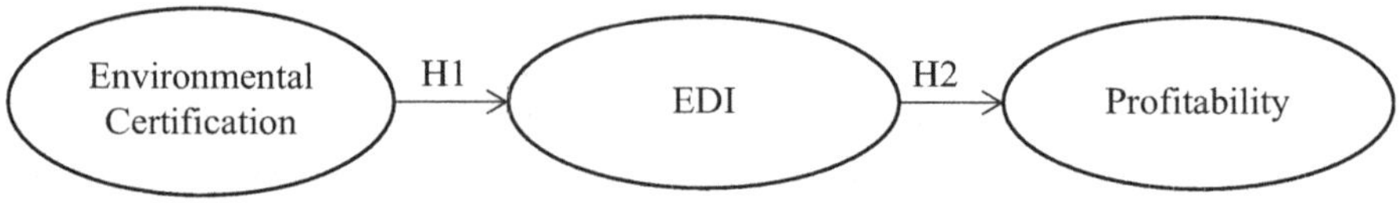


**Fig. 1** Theoretical model

**Table 1** Composition of the sample

| | |
|---|---|
| Altri | Media Capital |
| Commercial Bank | Mota Engil |
| Cofina | AT |
| Corticeira Amorim | Novabase |
| CTT Post Office | Orey Antunes |
| EDP | Pharol |
| EDP Renováveis | Ramada |
| Estoril Sol Casinos | Ren |
| Galp Energia | Semapa |
| Glintt | Sonae |
| Ibersol | Sonae Indústria |
| Impresa | Teixeira Duarte |
| Inapa Inv.Management | Toyota Caetano |
| Jerónimo Martins | The Navigator Company |
| Martifer | Vista Alegre |

## 3 Methodology

The population group under study are the companies belonging to *Euronext Lisbon*, currently composed of 43 companies. However, companies that have not exhibited activity on the stock exchange in the last 3 months and sports corporations (FC Porto SAD, SL Benfica SAD and Sporting CP SAD) have been removed from the study because they have a purely sporting scope. Therefore, the final sample is made up of the 30 companies, i.e., 70% of Portuguese companies listed on the stock exchange. Table 1 shows the composition of the sample.

We choose listed companies because these firms tend to disclose more environmental information than other firms (Monteiro & Guzman, 2010).

Content analysis technique was performed to codify qualitative information into categories and to provide a synthetic analysis using a quantitative scale (Mion & Adau, 2019; Monteiro & Guzman, 2010). The data were obtained from annual reports (management report and notes) and sustainability reports of the companies analyzed, in the period from 2015 to 2017, because they provide environmental information. It should be noted that data were previously collected on the SABI platform in order to collect data from companies under study (general and financial data) more quickly. The remaining information or that which does not exist in the SABI database, described above, was obtained.

In this survey two indexes will be calculated: Environmental Disclosure Index (EDI), called $EDI_{MR\&N}$, which will allow us to know environmental disclosure level by companies in documents of a mandatory nature, namely, management report and

**Table 2** $EDI_{MR\&N}$ indicators

| Management report | Notes to the financial statements |
|---|---|
| Environmental policy | Measurement criteria |
| Environmental Management System (EMS) | Environmental incentives |
| Environmental protection measures (EPM) | Environmental expenditures |
| Training | Extraordinary costs |
| Environmental auditing | Environmental liabilities |
| Environmental performance | Contingent liabilities |
| Environmental investments | |
| Other environmental elements | |

**Table 3** $EDI_{SR}$ indicators

| | |
|---|---|
| Materials | Compliance |
| Energy | Transport |
| Water | General |
| Biodiversity | Supplier evaluation |
| Emissions | Environmental complaints |
| Effluents and waste | |
| Products and services | |

notes to the financial statements, and designed $EDI_{SR}$, which will indicate environmental disclosure level in documents of an optional nature, in this case, sustainability report.

In $EDI_{MR\&N}$ development, 14 indicators were used (adopted from the study conducted by Monteiro & Guzman, 2010). These indicators are those described in paragraphs 48 and 49 of NCRF 26 (in force until 2016) and refer to information that must be disclosed in the management report and in the annex to financial statements (Table 2). It should be noted that information from the standard that was in force until 2016 was used because the study covers years prior to 2016 and also because it is easier to collect information according to the items in the said standard.

For $EDI_{RS}$, indicators were selected according to the environmental category of GRI guidelines, which are 12 (Table 3).

EDI results from the following formula:

$$EDI_n = \sum_{n=1}^{e} \frac{ej}{e}$$

where:

*ej*—Number of indicators included in reporting.
*e*—Total number of indicators.
When one indicator is disclosure, it is set to 1; otherwise it is 0.

*Environmental certification* is set to “1” if a firm has environmental certification implemented and “0” otherwise.

*Profitability* is measured through the earnings before interest and taxes (EBIT), according to Janošová.

In this study, a descriptive analysis of the sample will be carried out first. Statistical tests are performed in IBM Statistics SPSS software, version 21, using the Process Macro of SPSS software.

# 4 Analysis and Discussion of Results

## *4.1 Descriptive Analysis*

The sample, which is composed of 30 listed companies from the Lisbon Stock Exchange, is mostly represented by companies from the "services" and "industry" industries (56.7%). Table 4 presents the frequency of companies per sector of activity.

On average, most analyzed companies (76.7%), in the period between 2015 and 2017, disclose environmental information in their annual reports. There is also an increase of 28.6% in the number of companies that disclose environmental information from 2015 to 2017. In 2017, only three entities do not disclose any type of environmental information.

As for the structure of the sustainability report, it can be seen that 25 of the 30 reports analysed (9 in 2015, 7 in 2016, and 14 in 2017); i.e., more than 80% of the companies follow GRI guidelines. Thus, companies, in sustainability report preparation and in part that concerns environmental matters, follow the 12 indicators as previously mentioned to disclose the environmental performance.

In relation to the document type where the company discloses environmental information, it can be seen that 56% of the companies disclose said information only in management report and annex, 9% only in the sustainability report, and 35% in both documents.

Regarding environmental certification, 50 of the 90 companies (56%) are certified under the EMS (ISO 14001).

**Table 4** Number of firms by industry and their relative position

| Industry | Absolute frequency (*n*) | Relative frequency (%) |
|---|---|---|
| Services | 9 | 30.0 |
| Manufacture | 8 | 26.7 |
| Basic materials | 5 | 16.7 |
| Electricity | 3 | 10.0 |
| Technology | 2 | 6.7 |
| Business | 1 | 3.3 |
| Oil and gas | 1 | 3.3 |
| Telecommunication | 1 | 3.3 |
| Total | 30 | 100% |

## 4.2 Results Discussion

In Graph 1, it is possible to observe the results obtained from $EDI_{MR\&N}$. There is a decrease in the index in the year 2016 compared to 2015. However, there is an increase in the last year analyzed (2017). In percentage terms, there is a negative variation rate of 4.6% between 2015 and 2016, while between 2016 and 2017 the variation rate was positive in the order of 20.13%. Overall, the figures for 2015 and 2017 show a positive variation of 14.6%

Regarding $EDI_{SR}$, taking into account GRI indicators, results, according to Graph 2, show a gradual and positive evolution in 3 years analyzed, with a positive variation of 25.8% from 2015 to 2017.

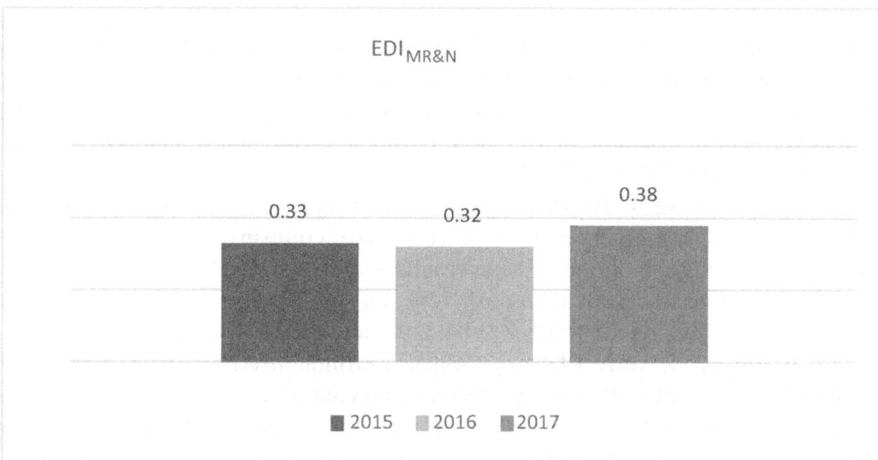


**Graph 1** $EDI_{MR\&N}$ 2015, 2016, and 2017

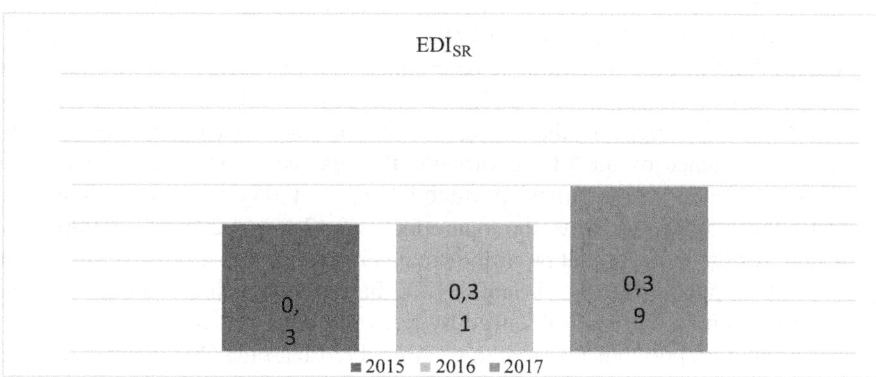


**Graph 2** $EDI_{RS}$ in 2015, 2016, and 2017

**Fig. 2** Results of model 1. * indicates significance at 0.1%

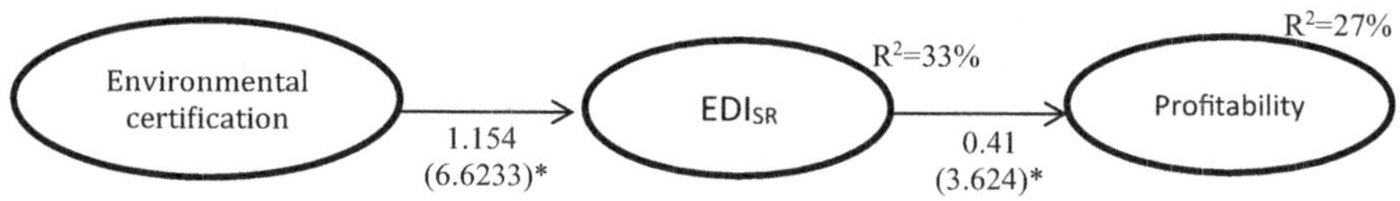


**Fig. 3** Results to model 2. * indicates significance at 0.1%

Results show a positive evolution in the disclosure of environmental matters by listed companies in annual report and accounts and sustainability report from 2015 to 2017. Results meet the conclusions drawn by Monteiro and Guzman (2010). These authors verified, in a sample of 109 large Portuguese companies, that IDA evolved positively between 2002 and 2004.

The main objective of this investigation is to analyze whether the environmental certification directly influences EDI and indirectly company performance (profitably). Figure 2 presents the regression results for $EDI_{MR\&N.}$ Results show that the type of environmental certification positively and significantly influences $EDI_{MR\&N}$ ($\beta = 1.40$; $t = 9.195$; $p = 0.0000$; $F = 84.549$) and that the first variable determines 49% of $EDI_{MR\&N}$. In turn, $EDI_{MR\&N}$ impacts profitability ($\beta = 0.71$; $t = 6.219$; $p = 0.000$; $F = 12.493$). $R^2$ suggests that a significant portion of the variance of the $EDI_{MR\&N}$ variable (42%) is explained by the $EDI_{AR}$ variable. Results show that $EDI_{MR\&N}$ does not directly influence companies' profitability ($\beta = -0.21$; $t = -0.9184$; $p = 0.3609$), but have an indirect effect through $EDI_{MR\&N}$. The total effect is 0.994 ($t = 4.205$; $p = 0.0001$).

Figure 3 presents the results of the regression for $EDI_{SR}$. Results show that the type of industry positively and significantly influences $EDI_{SR}$ ($\beta = 0.82$; $t = 2.091$; $p = 0.0457$) and that the first variable determines 14% of $EDI_{SR}$. In turn, $EDI_{RS}$ has an impact on profitability ($\beta = 1.154$; $t = 6.6233$; $p = 0.000$). The coefficient of determination ($R^2$) suggests that a significant portion of $EDI_{SR}$ variable (33%) variance is explained by the $EDI_{SR}$ variable. Results also indicate that $EDI_{SR}$ does not directly influence companies' profitability ($p > 0.05$), but does influence it indirectly through $EDI_{SR}$. Thus, the total effect is 0.47 ($t = 4.205$; $p = 0.0001$).

In sum, results indicate that environmental certification has a positive impact on EDI, which supports H1 (H1.1 and H1.2). In both the annual report and the sustainability report, IDA is influenced by environmental certification. Our results are consistent with previous research (e.g., Archel & Lizarraga, 2017; Barbosa et al., 2021). On the other hand, EDI (annual report and the sustainability report) positively influences companies' profitability. Our empirical evidence supports H2.1 and H2.2.

Results lead to the conclusion that disclosure of environmental information on a voluntary basis by companies contributes to their performance. Our study provides statistical support to the existing foundations in the literature; i.e., companies with higher EDIs become more competitive and gain advantages in terms of reputation and economic growth (Elshabasy, 2018).

Results also show that environmental certification does not directly affect corporate profitability, but nevertheless has an indirect effect through EDI, both for the annual report (management report and annex) and the sustainability report.

## 5 Conclusions

During last few decades, pollution and the excessive use of natural resources have caused environment destruction, putting future generations at stake. Indeed, in a constantly changing world, the challenges for sustainable development are increasingly important, with environmental protection and pollution prevention being current concerns for any organization. Thus, environmental problems solution, or their minimization, requires a new attitude from managers, who must start to consider the environment in their strategic decisions.

Stakeholders are focused on corporate environmental responsibility. In this sense, pressures to disclose environmental impact in the business sector have grown over the last few decades. The literature suggests environmental information disclosure tends to increase with the implementation of environmental management system implementation. On the other hand, the companies with the highest EDI are the most profitable. In this context, this study aims to develop and analyze a theoretical model which seeks to measure the relationship between environmental certification, level of environmental information disclosure, and profitability of Portuguese companies listed on stock exchange.

Results show that companies are increasingly involved in responding to stakeholder pressures. In sum, results indicate that environmental certification has a positive impact on EDI. On the other hand, EDI (annual report and the sustainability report) positively influences companies' profitability. Results lead to the conclusion that disclosure of environmental information on a voluntary basis by companies contributes to their performance.

An important conclusion emerges from this study. Environmental certification by itself does not guarantee business success, but this, together with the environmental information disclosure, has a positive impact on the profitability of Portuguese listed companies.

Our study provides statistical support to the existing foundations in the literature; i.e., companies with higher EDIs become more competitive and gain advantages in terms of reputation and economic growth (Elshabasy, 2018).

The main study limitation is the sample size and period of analysis, which is relatively limited (2015–2017). Environmental information disclosure context is analyzed to explain influences on environmental reporting practice, as well as certification environmental influence on IDA and business performance.

In future studies, we suggest a more extended analysis in order to analyze the entry of new directive 2014/95/EU that obliges large companies and public interest companies to report non-financial reporting and to verify if the crisis caused by COVID-19 had an impact on production of this type of report.

## References


Ahmadi, A., & Bouri, A. (2017). The relationship between financial attributes, environmental performance and environmental disclosure. *Management of Environmental Quality: An International Journal, 28*(45), 490–506.

Al-Tuwaijri, S. A., Christensen, T. E., & Hughes Ii, K. E. (2004). The relations among environmental disclosure, environmental performance, and economic performance: A simultaneous equations approach. *Accounting, Organizations and Society, 29*(5–6), 447–471.

Archel, P., & Lizarraga, D. (2017). Algunos determinantes de la informacion medioambiental divulgada por las empresas espanolas cotizadas. *Revista de Contabilidad: Spanish Accounting Review, 4*(7), 129–154.

Barbieri, J. C. (2017). *Gestão ambiental empresarial*. Saraiva Educação SA.

Barbosa, F., Monteiro, A. P., & Pereira, C. (2021). Determinants of environmental responsibility disclosure on mandatory and voluntary reporting of Portuguese listed firms. *Journal of Environmental Accounting and Management, 9*(3), 219–233.

Bhattacharya, C. B., & Sen, S. (2004). Doing better at doing good: When, why, and how consumers respond to corporate social initiatives. *California Management Review, 47*(1), 9–24.

Bowen, H. R., & Johnson, F. E. (1953). Social responsibility of the businessman. .

Brammer, S., & Pavelin, S. (2006). Voluntary environmental disclosures by large UK companies. *Journal of Business Finance & Accounting, 33*(7–8), 1168–1188.

Branco, M., & Delgado, C. (2011). Research on corporate social responsibility and disclosure in Portugal. *Social Responsibility Journal, 7*(2), 202–217.

Carroll, A. B. (1991). The pyramid of corporate social responsibility: Toward the moral management of organizational stakeholders. *Business Horizons, 34*(4), 39–48.

Caseirão, M. (2000). *Auditoria Ambiental: Perspetiva Contabilística-Financeira*. Áreas Editora.

Chan, E. S., & Wong, S. C. (2006). Motivations for ISO 14001 in the hotel industry. *Tourism Management, 27*(3), 481–492.

EC. (2001). *Green paper: Promoting a European Framework for corporate social responsibility*. https://eur-lex.europa.eu/LexUriServ/LexUriServ.do?uri=COM:2001:0366:FIN:en:PDF

Elshabasy, Y. N. (2018). The impact of corporate characteristics on environmental information disclosure: An empirical study on the listed firms in Egypt. *Journal of Business and Retail Management Research, 12*(2), 232–241.

Eugénio, T. (2004). *Contabilidade e gestão ambiental*. Áreas Editora.

Ferreira, C. (2000). *Da Contabilidade e do Meio Ambiente*. Vislis Editores.

GRI. (2015). *G4 - Diretrizes para Relato de Sustentabilidade: Princípios para Relato e Conteúdos Padrão*. https://edisciplinas.usp.br/pluginfile.php/3368600/mod_resource/content/1/Brazilian-Portuguese-G4-Part-One.pdf em 26/03/2019

Huang, C. L., & Kung, F. H. (2010). Drivers of environmental disclosure and stakeholder expectation: Evidence from Taiwan. *Journal of Business Ethics, 96*(3), 435–451.

Juhmani, O. (2014). Determinants of corporate social and environmental disclosure on websites: The case of Bahrain. *Universal Journal of Accounting and Finance, 2*(4), 77–87.

Kathy Rao, K., Tilt, C. A., & Lester, L. H. (2012). Corporate governance and environmental reporting: An Australian study. *Corporate Governance: The International Journal of Business in Society, 12*(2), 143–163.
Mion, G., & Adau, C. R. L. (2019). Mandatory nonfinancial disclosure and its consequences on the sustainability reporting quality of Italian and German companies. *Sustainability, 11*(17), 4612.
Monteiro, S. M., & Guzman, B. (2010). The influence of the Portuguese environmental accounting standard on the environmental disclosures in the annual reports of large companies operating in Portugal: A first view (2002-2004). *Management of Environmental Quality: An International Journal, 21*(4), 414–435.
Monteiro, T. (2013). *Planeamento de um sistema de gestão ambiental segundo a norma ISO: 140011: 2004* [Dissertação de mestrado]. Faculdade de Engenharia da Universidade do Porto.
NCRF 26. Matérias ambientais. http://www.cnc.min-financas.pt/pdf/snc/2016/normas%20com%20retifica%C3%A7%C3%A3o/NCRF_26.pdf
Nejati, M., Shafaei, A., Salamzadeh, Y., & Daraei, M. (2011). Corporate social responsibility and universities: A study of top 10 world universities' websites. *African Journal of Business Management, 5*(2), 440–447.
Nor, N. M., Bahari, N. A. S., Adnan, N. A., Kamal, S. M. Q. A. S., & Ali, I. M. (2016). The effects of environmental disclosure on financial performance in Malaysia. *Procedia Economics and Finance, 35*, 117–126.
Oliveira, O. J. D., & Pinheiro, C. R. M. S. (2010). Implantação de sistemas de gestão ambiental ISO 14001: uma contribuição da área de gestão de pessoas. *Gestão & Produção*, 51–61.
Pinto, A. (2005). *Sistemas de Gestão Ambiental*. Edições Sílabo.
Pires, Â. S. (2014). *A evolução da prática de divulgação da informação ambiental* [Dissertação de Mestrado]. Instituto Politécnico da Guarda.
Portaria 220/2015. https://dre.pt/application/conteudo/69866634
Saremi, H., & Nezhad, B. M. (2014). Role of environmental accounting in enterprises. *Ecology Environment and Conservation, 20*(3), 1257–1268.
Sumiani, Y., Haslinda, Y., & Lehman, G. (2007). Environmental reporting in a developing country: A case study on status and implementation in Malaysia. *Journal of Cleaner Production, 15*(10), 895–901.
Sutantoputra, A. W., Lindorff, M., & Johnson, E. P. (2012). The relationship between environmental performance and environmental disclosure. *Australasian Journal of Environmental Management, 19*(1), 51–65.
Tanc, A., & Gokoglan, K. (2015). The impact of environmental accounting on strategic management accounting: A research on manufacturing companies. *International Journal of Economics and Financial Issues, 5*(2), 566–573.
Zeng, S. X., Tam, C. M., Tam, V. W., & Deng, Z. M. (2005). Towards implementation of ISO 14001 environmental management systems in selected industries in China. *Journal of Cleaner Production, 13*(7), 645–656.

**Albertina Paula Monteiro** is a Ph.D. in Marketing and Strategy on 2013, School of Economy and Management, University of Minho, and Senior Lecturer of Accounting at Porto Accounting and Business School, Polytechnic of Porto, Portugal, since 2013. Her main areas of research are accounting, financial accounting, corporate social responsibility reporting, quality service, entrepreneurship, strategic management, and internationalization. She has published at international conferences and in the *Baltic Journal of Management*, *Euromed Journal of Business*, *Journal of Innovation & Knowledge*, *RBGN*, and *Journal of Business Ethics*. ORCID ID: https://orcid.org/0000-0002-2146-9807.

**Catarina Cepêda** is professor at ISCAP-IPP and UTAD, Ph.D. student at the University of Minho, researcher at the University of Trás-os-Montes and Alto Douro, Portugal, and certified accountant. Her research interests are financial accounting, corporate social responsibility, accounting information system, digital accounting, and decision-making. She has published at international conferences and in international scientific journals such as *RBGN*, *Accounting*, *Systems*, *Journal of Entrepreneurship and Sustainability Issues*, and *Universal Journal of Accounting and Finance*. ORCID ID: https://orcid.org/0000-0002-3911-0894.

**Francisco Barbosa** is a master's student in Finance Accounting at Porto Accounting and Business School, Polytechnic of Porto, Portugal. He works in the accounting area.

# Index

**A**
AA1000SES, 339, 340, 342, 343, 345, 350, 353
Absence of ventilation, 103
Academic community, 42
Academic qualification, 37, 38
Accounting, 6–8, 11, 12, 14, 15, 70, 266, 275, 309–311, 371, 385, 386
Achievement indicator, 30
Action, 3, 9, 20, 40, 65–68, 70, 72, 74–78, 82, 108, 110, 111, 120, 125–128, 142, 147, 172–174, 185, 186, 203, 206, 210, 219, 223–226, 228, 229, 269, 270, 273, 276, 306, 308, 313, 315, 346, 353, 357, 359, 367, 373–375, 377, 378, 383, 385
Active dialogue, 338, 356
Active relationship, 341, 342, 352, 357, 358
Adverse weather conditions, 90
Agency theory, 226, 268, 275, 277
Agricultural, 150, 238
Air quality, 90, 99
Almeida, 94, 98, 100–104
Analysis model, 339
Anglo-American corporate governance, 263–265, 270, 275, 276, 278
Anglo-Saxon corporate governance, 272, 278
Annual reports, 5, 42, 66–68, 72, 73, 81, 174, 175, 208, 210, 330, 331, 333, 343, 345, 347, 349, 388, 390, 392, 393
Approach, 5–14, 20, 29, 57, 66, 67, 69, 74, 80, 81, 92, 95, 110–111, 123, 129, 134, 135, 139, 145, 146, 149, 186, 199, 207, 210, 211, 218, 219, 224, 226, 227, 250, 251, 264, 266, 268, 271, 285–291, 293, 295, 299, 310, 311, 315, 317, 321, 327–329, 333, 337–361, 366–368, 378, 383
Automobile industry, 33–38

**B**
Bankruptcy, 276, 310
Banks, 8, 12, 113, 115, 119, 142, 144, 147, 236, 238, 265, 270, 274, 285–299, 374, 388
Barrier, 121, 124, 288, 291, 295
Basic Housing Law, 89
Basic needs of human beings, 89
Beiras and Serra da Estrela Region, 97, 99, 103
Belmonte, 93, 94, 98, 100–103
Biodiversity, 76, 91, 332, 371, 386, 389
Board, 4, 5, 8, 12, 78, 113, 136, 142, 263, 267–275, 277, 279, 285–299, 310, 313, 324, 329–331, 357, 373–375
Board independence, 263, 264, 267–269, 275, 277
Board members, 5, 136, 272, 330, 331, 375
Board of Directors, 21, 240, 243, 268, 272, 274, 275, 277, 286, 309, 330, 349
Budgeting, 307–309, 313
Building renovation, 93–95
Bureaucracy, 112, 144–149, 169
Business, 4, 5, 7, 22–26, 41, 49–61, 65, 67, 69–72, 76, 77, 81, 90, 111–113, 115, 124, 128, 133, 135, 138, 142, 145, 148, 149, 157, 160–164, 166, 171–173, 175, 176, 179, 185, 186, 203, 205, 206, 224, 226, 227, 230, 235, 238, 248, 264, 265, 267, 276, 286, 287, 289, 290, 293, 294, 305–317, 321, 322, 324, 328–330, 333,

 
I. G. Arraiano et al. (eds.), *Corporate Social Responsibility in a Dynamic Global Environment*, CSR, Sustainability, Ethics & Governance,
https://doi.org/10.1007/978-3-031-24647-0

338, 339, 341, 343–346, 349, 355, 361, 367, 369–373, 375, 383, 390, 393, 394
Business community, 42, 111
Business ethics, 42, 309, 321, 369
Business opportunities, 58, 60

C
Cajas rurales, 236, 238, 242
Capacity of human resources, 42
Capital chains, 186, 192, 193, 195–198
Carbon emissions, 90
Celorico da Beira, 93, 94, 98, 100–102
CEO, 77, 133, 134, 268, 275, 294, 310, 314, 357
Certification, 92–94, 96, 102, 142, 144, 147, 296, 297, 299, 384, 394
Certification energy, 103
Charisma, 6–8
Charity, 7–11, 13–15, 111, 120, 142, 185, 189, 191, 193, 198, 203, 369
Children, 4, 6, 8–15, 55, 112, 170, 183, 189, 193, 196, 226, 228, 229, 306, 307, 312, 314, 315
China securities regulatory commission (CSRC), 263–268, 271, 272, 278
Chinese capital market, 264–267, 273, 276–278
Chinese code of corporate governance for listed companies, 268
Chinese dual board system, 271
Chronic noncommunicable diseases, 223, 224
Citizen participation, 42
Classification of energy certificates by classes, 96
Climate change, 65–70, 75–78, 80–82, 90–92, 332
Cloud, network and data centre infrastructures, 30, 31, 34–36, 38
Code, 111, 161, 163, 166, 169, 173, 174, 176, 186, 188, 191–193, 195, 198, 263–265, 267, 268, 272, 276, 278, 286, 288–290, 293, 294, 298, 299, 308, 329, 349, 369
Comfortable temperatures, 89
Commissioners, 122, 123, 286, 292–299
Committees, 12, 122, 123, 145, 159, 165, 167, 175, 176, 263, 267, 279, 286, 287, 289–295, 298, 313, 330, 331, 349
Communication professional, xvii, 365–378
Communication, protocol and event organization, 34, 36–38
Community, 4, 5, 14, 29, 42, 72, 89, 93, 107–128, 165, 184, 186, 189, 191, 193, 196, 198, 203, 207, 209, 212, 217, 248, 264, 307, 308, 325, 341, 346, 349, 367, 370, 372, 383, 384
Comparability, 321, 322, 324, 328, 333
Comparative study, 176
Competitiveness, 21–23, 90, 168–169, 171, 176, 183, 205, 250, 322, 324, 341, 356, 385
Compliance, 5, 28, 117, 118, 133–149, 158, 159, 163, 165, 167, 171, 173–175, 179, 185, 276, 285, 287–289, 291–295, 298, 323, 330, 331, 338, 349, 351, 355, 383, 384, 389
Computer applications development, 30–32, 34–36
Concentrated ownership, 264, 269–271, 275, 277
Conceptual framework, 67, 69–83, 109, 113
Conclusions, 4, 14–16, 20, 24, 29, 36, 37, 41–43, 59–61, 82–83, 103–104, 110, 127–129, 148–149, 164, 171–173, 197–199, 210, 219–220, 224, 230–231, 236, 256–257, 277–279, 298–299, 317, 332, 333, 339, 350, 356–361, 378, 382, 392–394
Control, 12, 21, 94, 111, 113, 185, 186, 219, 223, 225, 228, 230, 236, 266, 269, 270, 273, 275, 277, 286, 287, 293, 294, 307, 309, 338, 341, 371, 372, 385
Controversial industries, 66, 68, 72, 81, 83
Convergence, 265, 274–276, 322, 325, 332, 333, 344
Cooking and food production, 30–32, 34–36, 38
Core capital approach, 285–299
Corporate citizenship, 207, 329
Corporate climate change communication, 66–83
Corporate communication, 68, 71, 74, 307
Corporate Governance, 8, 42, 113, 263–279, 285–287, 289, 291–294, 298, 312, 331, 349, 382
Corporate social responsibility (CSR), 3–6, 65, 66, 68–72, 75, 77, 79, 83, 107–129, 133–149, 157, 158, 171, 176, 183–199, 203, 205–210, 212, 215–218, 223–231, 248, 291, 323, 331, 337–339, 342, 343, 345, 348, 349, 352, 355, 358, 365–378, 383
Corporate websites, 66, 68, 72–74, 230, 349
Corruption, 120, 143, 309, 317
Council of the European Union, 90, 93, 94
COVID-19, ix, 30, 275, 314, 317, 370, 372, 394
Covilhã, 94, 98–104
Credit cooperatives, 235–243
Credit risk, 236

Crossvergence, 161, 172
CSR, engagement scale, 351–352
CSR practices, 70, 107–109, 112–114, 116, 120, 121, 184, 185, 188, 191, 192, 198, 230, 256, 321, 323
Cultural matrix, 42
Culture, 5, 9, 10, 12, 15, 176, 216, 287, 289, 305, 308, 314, 315, 327, 333, 338, 341, 349, 355, 359, 367, 378
Cybersecurity, 32, 34, 36

**D**
Decent work, 43, 157–172
Default, 236, 240, 242
Demographic change, 91
Developmental disability, 49, 55, 58
Dialogue, 73, 78, 115, 116, 118–122, 124, 125, 127, 158, 164, 165, 167, 169, 172, 173, 180, 338, 340, 341, 343, 345, 352, 354, 355, 357, 359, 367
Differentiation, 312, 313, 315, 317, 370, 372
Dignity, 10, 12–15
Director-manager agency issue, 275
Directors, 5, 12–14, 115, 122, 124, 133, 136, 216, 268, 269, 271, 272, 274, 275, 278, 289, 290, 292–299, 331, 349
Directors-shareholder agency issue, 275
Disabled, 4, 15, 49–52, 54–57, 112
Disclosure, 65, 68, 72, 77, 78, 270, 321–333, 339, 347, 382, 387–389, 392, 393
Discrimination, 8–10, 13, 14, 157, 164, 166, 167, 179, 371, 372
Divergence, 275
Diversity, 4, 5, 20, 28, 29, 42, 187, 188, 190, 198, 205, 218, 249, 267, 268, 279, 289, 291, 339, 370, 371
Doctrinal research, xvi

**E**
Economic and social purposes, 48, 49, 51, 56–61
Economic development, 90, 111, 114, 205, 332, 383
Economic outcome, 48, 50, 51, 55, 57, 59
Economic purpose, 48, 51, 58–60
Economic value, 58, 60
Economic value-driven, 58
Education, 9, 15, 19–25, 28, 29, 37, 38, 41, 43, 89, 103, 115, 121, 128, 137, 198, 204, 207, 216–218, 249, 252, 264
Efficiency building, xiii, 90
Electromechanical industrial maintenance, 32, 34–37, 39
Emotion, emotional, 6, 250, 306, 310–315
Employability of graduates, 19
Employee, 3, 5, 21, 37, 40, 54, 55, 58, 66, 71, 73, 123, 143, 164, 166, 185–198, 207, 239, 240, 242, 247–251, 254–257, 274, 311, 314, 315, 328, 330, 331, 339, 367, 370–372, 374, 376, 377, 383
Employment, 21, 22, 28, 30, 37, 40, 48–61, 89, 90, 175, 176, 247, 297, 307, 308, 330
Employment transition support, 49–61
Energy bills, 91, 92, 101
Energy Certificate (CE), 96, 99
Energy Certification System (ECS), 90, 94, 97, 103
Energy costs, 90, 185
Energy efficiency, 78, 89, 90, 92, 94–97, 101–103, 332, 371
Energy-efficient renovations, 92
Energy efficient systems, 90
Energy performance of buildings directive, 95
Energy Performance Regulation of Commercial and Service Buildings (RECS), 94, 97
Energy poverty, 90, 91, 95, 101
Entrepreneurship, 43, 60, 137, 138, 143, 150, 294
Entrepreneur team, 55, 57–58
Environmental, 13, 65, 66, 69–72, 74, 75, 77, 112–114, 117, 119, 120, 125, 127, 143, 159, 183, 185–190, 192, 194, 196, 198, 205, 207–209, 223, 224, 227, 229–231, 236, 239, 243, 267, 279, 323–325, 327–332, 343, 367, 370, 371, 374, 381–390, 392–394
Environmental, 367, 371, 374
Environmental certification, 382, 384, 387, 389, 390, 392, 393
Environmental communication, 68, 70, 76, 78
Environmental information, 382, 386–388, 390, 393
Environmental information disclosure, 382, 385, 387, 393
Environmental social government (ESG), 235–243, 324–326, 329, 333, 337, 338, 340, 343, 351, 355–357
Environmental stewardship, 90
Environmental sustainability, 76, 89, 208
ESG management, 337, 338
ESG reporting, 324, 338
Europe 2020, 20, 28, 29
European Centre for the Development of Vocational Training, 22

European Commission (EC), 21, 22, 30, 91–93, 96, 101, 235, 248, 324, 360, 383
European Commission's Ecological Pact, 91
European funds, 30, 104
European Jobs Monitor, 22
European Parliament, 90, 92–95
European Union, 89–91, 101, 104, 206, 386
European Vacancy Monitor, 22
Europe Union (EU), 21, 22, 90–93, 95, 163, 176, 323, 324, 330, 339, 347, 358, 394
Executive management, 343, 345, 346, 353, 361

F
Factor analysis, 252, 253
Family, 6, 7, 9, 11, 13, 89–92, 101, 165, 249, 274, 286, 287, 294, 306–307, 310–312, 314, 315
Family systems, 305, 310, 311, 314, 317
Figueira Castelo Rodrigo, 94, 97, 98, 100–102
Finance, 141, 145, 148, 158, 163, 171, 174, 235, 236, 266, 273, 296, 297, 308–312, 349, 374
Financial governance, 305, 308–310, 316, 317
Financially deprived situation, 103
Financial system, 236, 264, 277, 278
Fire safety, 89
Flexibility, 90, 161, 322, 332, 333
Food industry, 224, 229
Food systems, xv, 225
Fornos de Algodres, 93, 94, 97, 98, 100–102
For-profit corporation, 56
Framework, 3, 5–8, 15, 22, 23, 30, 55, 59, 60, 65–83, 91–93, 103, 108, 110, 111, 113, 135–149, 158, 161, 205, 208, 226, 248, 264, 265, 270, 273, 276–278, 287, 291, 293, 294, 324, 325, 327–331, 333, 340, 343, 357
Functions, 7, 11, 14, 21, 42, 73, 113, 183, 206, 208, 214, 218, 229, 237, 242, 267, 268, 279, 286, 287, 289, 291–295, 298, 307, 309, 331, 373, 378, 384
Fundão, 93, 94, 98–102

G
Gender, 4–6, 9, 24, 27, 28, 115, 158, 160, 188, 190, 268, 329, 330, 371, 372
Generations, generational, 14, 70, 76, 204, 207, 214, 225, 235, 309, 313, 381, 393
German corporate governance, 271, 272, 278
Gerontology, 31, 32, 34–36, 38
GHG emissions, 66
Global challenges, 91
Global Reporting Initiative (GRI), 77, 78, 225, 229, 322, 327–330, 333, 340, 349, 351, 386, 389–391
Global supply chains, 162
Goal, goal statement, 20, 28–30, 37, 74–77, 80, 91, 92, 111, 137, 149, 150, 157, 211, 215, 219, 223, 224, 231, 235, 251, 256, 287, 294, 311, 315, 316, 329, 341, 367, 368, 373, 375, 377, 383
Gouveia, 93, 94, 98, 100–102
Governance, 3–5, 7, 11, 12, 14, 21, 75, 162, 163, 171, 207, 208, 236, 239, 264, 267, 271, 273, 275, 277–279, 285, 286, 289–291, 294, 295, 308–310, 314–317, 323–325, 327–331, 333, 337
Governance information, 330
Government, 20, 29, 41, 56, 65, 71, 107–119, 121–128, 133–139, 141–149, 159, 163, 166, 167, 169, 172, 206, 218, 219, 266, 267, 269–271, 273, 274, 276, 321, 323, 324, 341
Green-highlighting, 80
Greenhouse gas (GHG), 67, 77, 78, 81, 90, 92
Green-washing, 66, 72, 77, 80
Guarda, 93, 94, 98–102, 104

H
Harmonization, 21, 322, 324, 332, 333
Health and well-being, 90, 224, 249
Healthy housing, 89
Higher education, 19–24, 27–29, 37, 38, 40, 41, 43, 207, 208, 219
Higher education institutions (HEIs), 20, 23, 30, 41–43, 207, 208
Home, 7, 89–91, 95, 101–103, 147, 315, 370–372
Hospital, 4, 8–15, 126, 148
Housing and health guidelines (HHGL), 90
Housing conditions, 89
HPT courses, 19–43
Human capital, 21, 23, 37, 248, 383
Human resources, 12, 57

I
IBEX35, 337–361
Illegitimate mothers, 9, 11, 13, 15
Immediate result indicator, 30
Incentives, 42, 111, 113, 115–117, 120, 127, 273, 357, 370, 372, 386, 389

Inclusion, 21, 28, 68, 81, 91, 104, 141, 263, 354
Independent director, 5, 268, 269, 274–277, 286, 290, 294
Indicators, 28, 49, 69, 91, 95, 128, 218, 229, 230, 255, 325, 327, 329–332, 340, 386, 387, 389–391
Indoor air quality, 93, 96
Infant healthcare, 6, 15
Informal economy, 309, 310, 317
Innovation, 7, 15, 20, 22, 28, 133, 134, 137, 138, 148, 225, 269, 307, 338, 340, 349, 355, 366, 367, 370, 371, 375, 380, 385
Insolvency, 236, 238, 242, 243
Institutional investors, 264, 265, 267, 273–274, 276–278, 324
Institutions, 4, 6, 8–12, 15, 20, 21, 29, 30, 42, 71, 72, 79, 107–116, 118–120, 125, 127, 128, 137, 149, 158, 161, 172, 204, 205, 207, 208, 211, 217, 228–230, 236, 238, 242, 291, 296, 297, 299, 372
Instituto Politécnico da Guarda (IPG), 29, 30, 33, 34, 40, 42
Internal audit, 292–295, 298
Internalisation degree, 347, 359
Internal social responsibility, 247–257
International Labour Organization (ILO), 157, 158, 160, 163, 166, 169, 170, 174
International trade and management, 30–38
Internship, 23, 41, 42, 215
Investors, 21, 71, 73, 117, 122, 123, 126, 169, 187, 235, 250, 264, 266–268, 271, 273, 274, 277, 279, 296, 322–325, 327, 330, 337, 338, 361

**J**
Japan, 48–61, 77, 203–220, 325, 330, 332
Japanese universities, 213, 218, 219
Job satisfaction, 248–257
Jordanian Garment Industry, 157–172

**K**
KLD Database, 339
Knowledge-based economy, 20

**L**
Labour standards, 158, 159, 161, 163, 165, 166, 169–171
Large, 11, 71, 73, 93, 128, 135, 142, 144, 145, 165, 185, 186, 204, 230, 247, 251, 264, 270, 271, 275, 289, 291–294, 298, 314, 316, 325, 347, 356, 367, 375, 392, 394
Laws, 24–26, 29, 41, 93, 97, 109, 111, 117, 118, 127, 133–149, 161, 163, 165, 168–170, 174, 179, 218, 225, 228, 263, 266, 273–275, 277–279, 285, 286, 288, 290, 293–295, 299, 308, 321–323, 332, 349, 374
Leadership, 4, 6–16, 108, 122, 124, 136, 250, 293, 311, 314–316, 329, 341, 344, 345, 352, 357, 359, 366, 369, 371, 373, 375, 377, 378
Legislation, 136, 170, 263, 322–325, 327–333, 383, 384
Legitimacy, 7, 65–69, 71, 72, 76, 78–81, 83, 110–111, 117, 125, 127, 128, 226–231, 352
Legitimation, 67, 68, 71, 80, 81
Lighting natural, 103
Listed companies, 5, 50, 266, 267, 272, 275, 278, 323, 325, 327, 331, 333, 339, 382, 386–388, 390, 392, 393
Local, 4, 9, 12, 15, 56, 65, 70, 72, 78, 91, 93, 95, 107, 109, 110, 112–115, 118–120, 122–124, 126, 145, 158, 161–167, 169–173, 184, 186, 187, 189, 191, 193, 194, 196–199, 207, 248, 269, 273, 308, 314, 341, 375, 378
Logit, 236, 237, 240, 242
Logit analysis, 236, 242
Long-run value, 337–361
Long run value creation, 337–361
Long termism, 337
Long-term result indicator, 30

**M**
Managers, 3, 21, 107, 109, 110, 114, 115, 118–121, 123, 124, 126, 127, 159, 166, 169, 170, 173, 174, 179, 183, 186, 199, 208, 248, 249, 256, 257, 268–272, 275, 277, 286, 309, 311, 313, 349, 367, 369, 373–376, 381, 393
Mandatory approach, 290, 298, 299
Mandatory reporting, 321–323
Mandatory requirements, 321
Manteigas, 93, 94, 97, 98, 100–103
Manual text analysis, 322, 325
Marginalization, 14, 16
Materiality, 230, 328, 339, 340
Meda, 94, 98, 100–102
Medium, 41, 65, 66, 71–74, 76, 95, 121, 144, 186, 230, 235, 254, 255, 276, 288, 290–295, 298, 299, 388
Meeting, 8, 21, 28, 55, 79, 111, 115, 121–125, 128, 136, 167, 269, 272, 288, 298, 314, 316, 317, 349, 378, 381

Mental health, 89, 90
Merger, 238, 242, 275, 294, 296, 316
Migratory movements, 91
Mission, 3, 4, 9, 14, 15, 42, 49, 50, 138, 188, 190, 192, 195, 203, 208, 217, 218, 266, 315, 349, 354, 359
Motherhood, 4, 8
Mothers, 4, 6, 8–14
Mountain sports, 31, 32, 34–36, 38

**N**
Narrative review, 264, 278
National System for Energy Certification, 93
New Leipzig Charter, 91, 92
New Urban Agenda, 91, 92
Nineteenth century, 7
Nomination Committee, 289, 294, 298
Non-financial, 77, 323–325, 329, 330, 337, 339, 347, 350, 394
Non-financial disclosure (NFD), 321–325, 328–330, 332, 333
Non-financial information, 275, 321, 322, 324, 325, 329, 330, 332, 339, 347, 349, 386
Non-governmental organisation (NGO), 65, 71, 72, 74, 107–116, 120–128, 135, 141, 144–149, 162, 163, 167, 248, 370, 372, 377, 380
Non-profit organizations (NPOs), 48, 52, 54, 60, 61

**O**
Obesity, 223–225, 228, 229
Oil and gas industry, 65, 68, 69, 81
Organisation for Economic Co-operation and Development (OECD), 22, 263, 269, 270
Organization, 3, 4, 8, 9, 21, 48, 51, 52, 56, 58–61, 68, 79, 83, 110, 126, 164, 166, 170, 171, 188, 191, 192, 203–206, 209, 217–220, 225, 249–251, 308, 309, 314, 316, 317, 333, 341, 358, 365–378, 381, 383, 384, 393
Organizational culture, 317
Organizational history, 315
Organizational management, 55, 59, 60
Outcome, 22, 49, 50, 60, 79, 135, 137, 149, 158, 161, 172, 248, 250, 266, 275, 276, 279, 305, 308, 309, 311, 313, 316, 328, 355, 357
Overweight, 223–225, 229
Owners, 21, 108, 159, 165, 173, 174, 186, 268, 270, 271, 278, 286, 294, 305, 309, 357, 359, 375
Ownership, 68, 113, 185, 267, 268, 270, 271, 273–275, 277, 278, 287, 293, 294, 296

**P**
Pandemic crisis, 41, 103
Pandemic situation (COVID-19), 30, 275, 314, 317, 365–378, 394
Paris Agreement, 67, 91
Park housing, 95
People's Credit Banks (PCBs), xvi, 285–299
Performance, 4, 5, 11, 21, 40, 66, 69–72, 74, 75, 77, 78, 80, 81, 83, 92–97, 103, 108, 113, 165, 203, 207, 213, 216, 218, 225, 226, 248, 256, 264, 266, 268–273, 276, 277, 291, 293, 294, 305, 307, 324, 329, 330, 342, 361, 369, 377, 378, 381–387, 390, 392–394
Pinhel, 94, 98, 100–102
Polluting industry, 65, 76, 382
Polytechnic Higher Education Institutions, 19, 20, 29, 42
Polytechnic Institute of Guarda, 19–43
Portugal, 19–43, 89–104, 238, 384
Portuguese Energy Agency (ADENE), 95, 96, 98, 100–102
Portuguese Energy Plan and Climate 2021–2030, 95
Portuguese Government Programme, xii
Position of report, 325, 327, 330, 333
Post-secondary non-higher education, 38
Post-secondary training, 42
Poverty, 8, 15, 22, 61, 90, 137, 148, 149, 189, 193, 196, 207, 224
Practical implications, 299
Principles, 11, 21, 125, 134–136, 139, 142, 149, 186, 189, 192, 194, 208, 224, 235, 263, 289, 290, 294, 308, 325, 327–330, 332, 333, 339, 340, 343–345, 352, 354, 356, 371, 372
Problem, 6, 9, 41, 43, 48, 65, 66, 68, 70, 71, 73, 91, 95, 111, 113, 123, 124, 126, 167, 169, 184–186, 189, 193, 205, 206, 218, 223–225, 229, 238, 249, 264, 266, 268, 270, 271, 275, 277, 286, 287, 293, 296–298, 308, 310–312, 314–317, 348, 357, 358, 367, 375, 381, 393
Professional education, xii
Professional training, 23, 42

Profitability, 57, 70, 93, 206, 249, 309, 337, 382, 385, 387, 389, 392, 393
Projection, 310–313
Property owners, 92
Protection from pollutants, injury hazards, mould, and pests, 90
Public health, 12, 90, 223, 225, 228, 231, 372
Purpose, 21, 41, 48, 55, 57, 60, 69, 70, 91, 99, 107, 115, 142, 172, 227, 230, 250, 264, 266, 270, 306, 310, 314, 325, 327, 328, 330, 331, 338, 341–343, 345–347, 349, 350, 352–355, 357, 359, 361, 384
Purpose contribution, 346, 347, 355
Purpose knowledge, 346, 347, 355
Purpose management, 345–347, 353, 355, 356
Purpose of legislation, 327

**Q**
Qualified Foreign Institutional Investors (QFII), 263
Qualifying human resources, 40
Qualitative research, 210
Quality of life, 90, 103, 251, 383
Quality of the air inside, 103
Questionnaire, 20, 29, 30, 34, 41, 129, 186, 252

**R**
Ratings, 328, 337–339, 343, 349–351, 353, 358
Regional, 23, 78, 95, 122, 288
Regulation, 21, 52, 74, 77, 78, 81, 92–94, 97, 99, 102, 108, 109, 111, 113, 117, 139, 143, 144, 148, 158, 161, 167, 169, 183, 185, 218, 226, 228, 238, 263–268, 277, 278, 286–288, 290–294, 296–298, 308, 309, 311, 321–334, 349
Regulation of Energy Systems for Air Conditioning in Buildings, 93
Regulatory system, 325
Rehabilitation policies, 99, 103
Remuneration Committee, 289, 292, 293
Renewable Energy and Energy Efficiency, 92
Reporting, 4, 5, 69–71, 73, 74, 77, 78, 120, 136, 142, 146, 149, 175, 176, 198, 225, 263, 286, 305, 307–309, 321, 323–325, 327–330, 332, 333, 338, 339, 345, 357, 358, 385, 386, 389, 394
Reputation, 71, 78, 125, 207, 216, 297, 337, 339, 356, 367, 368, 381, 393
Respiratory and cardiac diseases, 90
Responsibility, 3, 6, 7, 14, 69, 71, 72, 77, 81, 91, 92, 114, 116, 143, 165, 176, 187, 198, 203–208, 214, 224, 226, 249, 270, 271, 293, 305, 307, 308, 314, 315, 321, 329, 369, 373, 378, 383, 384, 393
Responsible practices, 248
Retailers, 183–187, 193, 197–199
Risk management, 287, 292–295, 298, 307, 309, 331, 357
Role of government, 108, 111–113
Roles, 3–15, 21, 24, 41, 42, 52, 65, 67, 68, 71, 72, 75–77, 80, 93, 103, 107–128, 135, 142, 149, 162, 166, 167, 173, 183, 185, 199, 204, 205, 207, 208, 224, 247, 250, 256, 267–274, 276–278, 308, 315, 316, 323, 365–369, 373–376, 378

**S**
Sabugal, 94, 98, 100–103
Sanitation and illumination, 89
Sarbanes-Oxley Act 2002, 275, 285, 290, 293
Scale, 55, 60, 77, 92, 95–97, 102, 107, 118, 126, 127, 187, 190–194, 197, 219, 250, 252–254, 256, 257, 286, 289, 331, 339, 340, 342–344, 349, 351–353, 355, 358, 386, 388
Scarcity of resources, 91
SDG 3, 91
SDG 6, 92
SDG 7, 92
SDG 8, 157
SDG 11, 91
SDG 13, 92
SDG 17, 76
Secondary education, 19, 23, 38
Seia, 94, 98–103
Seismic resilience of buildings, 89
Services and Supports for Persons with Disabilities Act, 51, 52
Severe deprivation of housing conditions, 91
Shanghai Stock Exchange, 265
Shareholders, 12, 21, 73, 206, 263, 266–275, 277, 279, 285–288, 291, 294, 305, 311, 330, 331, 338, 349, 357, 367, 374, 375, 377, 383
Share split reform, 269
Shenzhen Stock Exchange, 263, 265
Short-termism, 337, 338
Sibling, 314
Significance, 59–60, 247, 248, 256, 265, 266, 287, 291, 316, 392
Small, 6, 11, 14, 40, 66, 92, 126, 144, 147, 149, 286, 287, 290–296, 298, 299, 309, 333, 371

Smart, 20, 21, 77, 89–104
Social, 1, 19, 47, 65, 90, 107, 133, 157, 183, 203, 224, 236, 247, 267, 287, 305, 323, 337, 365, 383
Social and economic transformation, 42
Social change agents, xii, 361–378
Social contract, 71, 72, 83, 111
Social enterprise, 48, 49, 55, 137
Social entrepreneur, 48
Social entrepreneurship, 294
Social equality, 90
Social information, 328
Social injustice, 8, 9, 12, 14, 15
Social innovation, 3–15, 48
Social outcome, 48, 51, 55, 57–60
Social plague, 3, 8, 13
Social policy, 9, 48
Social purpose, 48, 50, 51, 59, 60, 345, 346
Social responsibilities, 3–15, 103, 183–186, 192, 203, 205, 207, 208, 218–220, 247–251, 257
Social value, 15, 58, 60, 203
Social value-driven, 58
Social welfare corporations, 52, 54, 60, 61
Societal, 65–83, 162, 229, 308, 341, 344, 366, 378
Society, 3, 4, 8–10, 12, 13, 15, 20, 42, 65, 66, 69, 71, 74, 75, 89, 91, 92, 103, 108, 110–112, 117, 120, 123, 135, 136, 141, 144, 146, 148, 149, 161, 176, 185–187, 189, 191, 193, 196, 203–208, 212, 214, 215, 226, 227, 229, 238, 307, 314, 321, 332, 367, 377, 381–383, 385
Software testing, 30, 32–37, 39
Solutions, 80, 149, 189, 194, 206, 224, 253, 287, 291, 317, 354, 355, 370, 375, 377, 378, 381, 393
Sound and image reporter, 30–37, 39
Spain, 236, 238, 325, 361
Stakeholder, 21, 65, 107, 134, 162, 203, 223, 247, 264, 285, 316, 321, 337, 365, 383
Stakeholder pressure, 68, 78, 81, 393
Stakeholder relationship, 128, 207, 328, 329, 356
Stakeholder theory, 72, 226, 339
Standards, 23, 35, 52, 57, 70, 72, 74, 77, 78, 81, 91, 101, 111, 113, 137, 149, 159, 164, 170, 174, 185, 187–189, 192, 194–198, 206, 207, 210, 229, 240, 263, 264, 266, 273–275, 278, 291, 294, 295, 308, 316, 321, 324, 325, 327–333, 338–340, 342, 343, 345, 349, 350, 353, 383–386, 389
State-owned enterprises (SOE), 277
State-owned enterprises (SOE) structure, 274
State ownership, 264, 269–270, 273, 277
Strategy, 3–5, 8, 20–22, 28, 29, 49, 55, 58, 60, 65–68, 71, 72, 74, 75, 78, 80, 81, 92, 95, 107, 110, 111, 113, 157, 169, 172, 183, 184, 187, 188, 190, 198, 199, 205, 206, 208, 217, 218, 224, 225, 228, 230, 248, 249, 257, 317, 329–332, 339, 340, 342, 343, 345, 357, 361, 365–368, 373, 375, 377, 378
Stress, 6, 69, 91, 112, 206
Structural tension, 56, 57
Structure of report, 325, 327, 329
Student, 19–21, 23, 24, 27, 29–32, 34–38, 40–43, 119, 207, 211, 215, 217, 374
Student worker, 37, 40
Substantive legitimacy, 66
Substantive legitimation, 67
Supervisory board (SB), 264, 267, 271–272, 275, 277, 278, 289
Sustainability, 3, 21, 66, 68–79, 83, 92, 102, 103, 145, 174, 204, 205, 207, 208, 225, 230, 236, 305, 321, 323–325, 327–330, 332, 333, 337, 338, 343, 344, 346, 348, 349, 354, 358, 361, 365–378, 385, 386
Sustainability reports, 67, 68, 71, 72, 74, 78, 81, 225, 382, 386–390, 392, 393
Sustainable development, 70, 75, 76, 78, 92, 137, 150, 186, 204–208, 215, 219, 224, 235, 381, 385, 393
Sustainable living conditions, 95
Sustainable society, 3–15
Symbolic legitimacy, 83
Systems, 16, 22, 55, 65, 79, 90, 93, 94, 96, 97, 99, 103, 106, 110, 111, 114, 120, 127, 136, 161, 168, 170, 171, 177, 178, 204, 208, 211, 219, 223, 225, 227, 228, 236, 263, 264, 268, 271–275, 277, 278, 290, 305–317, 323, 325, 347, 358, 371, 378, 381, 383–387, 393
Systems Integrated of Management, 103

**T**

Tanzania, 107–128
Technical and professional skills, 43
Technical training component, 23, 42
Tension, 48, 51, 59
Tertiary institutions, 19, 22, 23
Theoretical framework, 212, 287, 291, 293, 298, 382
Thermal behaviour, 92, 93, 99, 102
Training component in the real-work context, 42

Trancoso, 94, 98–102
Transfer knowledge to society, 21
Treaty on the Functioning of the European Union, 90
Triangles, 310, 311
Triple bottom line (TBL), 70, 225, 366
Turnover, 133, 294, 311, 374
2030 Agenda for Sustainable Development, 91
Type of information, 99, 325, 327, 329, 330, 332

**U**
Unemployed, 37, 40, 41, 112
UNESCO, 207
Unification, 293, 294
Units, 23, 29, 57, 172, 210–212, 286, 287, 289, 291–295, 298, 357, 373, 374, 386
University Social Responsibility (USR), 203–220
Unworthy conditions for the citizen, 95
Users, 50, 52, 54, 55, 57–59, 325, 327, 330, 331, 385

**V**
Value creation, 338–343, 349, 355, 361
Ventilation, 90, 99, 102, 371
Vocational and Technical Higher Education, 41
Voluntary approach, 289, 299

**W**
Weatherization, 90
Welfare services, 51, 52
Well-being of the citizen, 104
Women, 3–16, 112, 115, 165, 240, 242, 252, 330, 370, 372
Work integrated social enterprise (WISE), 47–49, 55, 59, 60
World Health Organisation (WHO), 89, 90, 225, 228

The manufacturer's authorised representative in the EU is Springer Nature Customer Service Centre GmbH, Europaplatz 3, 69115 Heidelberg, Germany. If you have any concerns regarding our products, please contact ProductSafety@springernature.com

Printed and bound by CPI Group (UK) Ltd, Croydon, CR0 4YY
28/09/2026
02224056-0007